THEOLOGY AND THE SCIENTIFIC

IMAGINATION FROM THE MIDDLE AGES TO

THE SEVENTEENTH CENTURY

AMOS FUNKENSTEIN

THEOLOGY
AND THE SCIENTIFIC
IMAGINATION

from the
Middle Ages to the
Seventeenth
Century

PRINCETON, NEW JERSEY

PRINCETON UNIVERSITY PRESS

1986

COPYRIGHT © 1986 BY PRINCETON UNIVERSITY PRESS
PUBLISHED BY PRINCETON UNIVERSITY PRESS, 41 WILLIAM STREET
PRINCETON, NEW JERSEY 08540
IN THE UNITED KINGDOM
PRINCETON UNIVERSITY PRESS, CHICHESTER, WEST SUSSEX

LIBRARY OF CONGRESS CATALOGING IN PUBLICATION
DATA WILL BE FOUND ON THE LAST PRINTED PAGE OF THIS BOOK

ISBN 0–691–08408–4

ISBN 0–691–02425–1 (PBK.)

PUBLICATION OF THIS BOOK HAS BEEN AIDED BY A GRANT
FROM THE WHITNEY DARROW FUND OF
PRINCETON UNIVERSITY PRESS

THIS BOOK HAS BEEN COMPOSED IN LINOTRON BEMBO

PRINCETON UNIVERSITY PRESS BOOKS ARE PRINTED ON ACID-FREE PAPER
AND MEET THE GUIDELINES FOR PERMANENCE AND DURABILITY OF THE
COMMITTEE ON PRODUCTION GUIDELINES FOR BOOK LONGEVITY OF THE
COUNCIL ON LIBRARY RESOURCES

★

5 7 9 10 8 6 4

PRINTED IN THE UNITED STATES OF AMERICA

To my father
and to the memory of
my mother

CONTENTS

PREFACE

For many years, I have been occupied in trying to find a way of defining, as precisely as possible, the different points of transition from medieval to early modern modes of reasoning in different fields of knowledge. At first I pursued the themes of this book independently of one another. In the course of study, I became aware not only of the ties between theology and science—these have been recognized and studied before—but also of the peculiar circumstance that, to many seventeenth-century thinkers, theology and science merged into one idiom, part of a veritable secular theology such as never existed before or after. The best way to capture both my original aims and the added insight was, I thought, to trace the change in connotations of three divine attributes from the Middle Ages to the seventeenth century. As an interpretative essay only, this book is not based on new texts or other materials. At times I had to venture into fields remote from my expertise, where I tried to follow reliable guides, and I hope that I found them. The fifth chapter is the most speculative; I hope to elaborate on the themes it touches upon in the future. Chapters two through four, the main part of the book, originate, in their present form, in three Gauss Seminars given at Princeton University in 1984.

Friends, students, and colleagues have encouraged me throughout the years. I owe special thanks to Yehuda Elkana, Richard Popkin, and Robert Westman: discussions with them throughout the various stages were invaluable, and even more so their emotional support. They also read the manuscript with a friendly yet critical eye. I thank Susannah Heschel for her constructive and critical support: without it, the God spoken of in this book would have remained a contented male, and "man" would have stood for both genders. Marilyn and Robert Adams, Jürgen Miethke, Katherine Tachau, Mary Terrall, and Norton Wise also read the manuscript and helped me to remove many ambiguities and embarrassing mistakes.

Many of my present and former students will find how much I have learned from our discussions and their works: from Susan Anderson's analysis of the *De mirabilibus*, Stephen Benin's study of the principle of accommodation, David Biale's portrait of G. Scholem's counter-history, Mati Cohen's studies in seventeenth-century historiography, André Goddu's interpretation of Ockham's physics as a reification of modal categories, Joshua Lipton's account of the role of astrology, Steven Livesey's detailed history of the injunction against *metabasis*, Josef Mali's reinterpretation of the concept of myth in Vico, Jehudith Naphtali's study of labor theories in Scholastic thought, Michael Nutkiewicz's work on the impact of natural science on political theory, Joel Rembaum's studies on religious polemics, Lisa Sarasohn's study of Gassendi's ethical-social theories, Dorit Tanay's insights into the relation of mathematics to musical theories in the Middle Ages. For the privilege and enjoyment of having been their teacher I thank them all.

Edward Tenner of the Princeton University Press was interested in the book long before its completion; without his encouragement I would not have completed it. Marilyn Campbell edited the manuscript, helped to erase many ambiguities and inaccuracies, and prepared it for print. In the typing, editing, and checking of the manuscript, I was aided by Neil Hathaway and Randy Johannessen. In the last and hardest year of work toward the publication of the book I was supported by a fellowship from the Guggenheim Foundation.

Thanks are also due to the publishers and editors of some of my previously published articles for the permission to incorporate passages and sections from them: the University of California Press; the Johns Hopkins University Press; *Studies in the History and Philosophy of Science; Viator; Medievalia et Humanistica; Miscellanea Medievalia;* Société Internationale des Études de Philosophie Médiévale; *The Israel Colloquium for the History and Philosophy of Science;* the Herzog August Bibliothek Wolfenbüttel.

My children, Daniela and Jakob, have borne with patience their father's absence and absent-mindedness, and made my life always happier and sometimes easier.

Los Angeles
JULY 1985

ABBREVIATIONS

AHDL	*Archives d'histoire doctrinale et littéraire du moyen âge*
AT	Descartes, *Oeuvres* (see bibliography)
BGPhM	Beiträge zur Geschichte der Philosophie im Mittelalter, Texte und Untersuchungen, edited by C. Baeumker
CAG	Commentaria in Aristotelem Graeca
CCSL	Corpus Christianorum series latina
CHM	*The Cambridge History of Later Medieval Philosophy*, edited by Norman Kretzmann, Anthony Kenny, and Jan Pinborg (Cambridge, 1982)
CSEL	Corpus Scriptorum Ecclesiasticorum Latinorum (Vienna, 1866ff.)
DA	*Deutsches Archiv für Erforschung des Mittelalters*
DcD	Augustine, *De civitate Dei* (see bibliography)
EW	Thomas Hobbes, *The English Works* (see bibliography)
GCS	Die Griechischen christlichen Schriftsteller des ersten Jahrhunderte (Berlin and Leipzig, 1897ff.)
GM	Leibniz, *Mathematische Schriften*, ed. Gerhardt (see bibliography)
GP	Leibniz, *Philosophische Schriften*, ed. Gerhardt (see bibliography)
HE	Eusebius, *Historia ecclesiastica* (see bibliography)
IT	Maimonides, *Iggeret teman*, in *Iggrot harambam* (see bibliography)
KdRV	Kant, *Kritik der reinen Vernunft*, ed. Weichschedel (see bibliography)
MG	Monumenta Germaniac Historica (Hannover and Berlin, 1826ff.) AA = Auctores Antiquissimi Capit. = Capitularia LdL = Libelli de lite Script. = Scriptores

	Script. in usu schol. = Scriptores rerum Germanicarum in usu scholarum
MGWJ	*Monatsschrift für Geschichte und Wissenschaft des Judentums*
MN	Maimonides, *More hanebuchim* (see bibliography)
Migne, *PG*	J. P. Migne, *Patrologia cursus completus, series Graeca* (Paris, 1857–1912)
Migne, *PL*	J. P. Migne, *Patrologia cursus completus, series Latina* (Paris, 1844–1890)
MT	Maimonides, *Mishne Tora* (see bibliography)
NE	Leibniz, *Nouveaux Essais* (GP, 5)
OC	Malebranche, *Oeuvres complètes* (see bibliography)
OF	Leibniz, *Opuscules et fragments inédits de Leibniz* (see bibliography)
OT	Ockham, *Opera philosophica et theologica* (see bibliography)
Pauly-Wissowa, *RE*	*Real-Enzyclopädie der classischen Altertumswissenschaft.* Edited by A. Pauly et al. (Stuttgart, 1894ff.)
PoC	Principle of (Non)Contradiction
PoSR	Principle of Sufficient Reason
SB	Leibniz, *Sämtliche Schriften und Briefe* (see bibliography)
SN	Vico, *Scienzia Nuova*, in *Opere* (see bibliography)
SVF	*Stoicorum Veterum Fragmenta*, edited by Johannes von Arnim, 4 vols. (Leipzig, 1903–1924)
Van Vloten-Land	Spinoza, *Opera* (see bibliography)

THEOLOGY AND THE SCIENTIFIC
IMAGINATION FROM THE MIDDLE AGES TO
THE SEVENTEENTH CENTURY

I

INTRODUCTION

A. A SECULAR THEOLOGY

A new and unique approach to matters divine, a secular theology of sorts, emerged in the sixteenth and seventeenth centuries to a short career. It was *secular* in that it was conceived by laymen for laymen. Galileo and Descartes, Leibniz and Newton, Hobbes and Vico were either not clergymen at all or did not acquire an advanced degree in divinity. They were not professional theologians, and yet they treated theological issues at length. Their theology was secular also in the sense that it was oriented toward the world, *ad seculum*. The new sciences and scholarship, they believed, made the traditional modes of theologizing obsolete; a good many professional theologians agreed with them about that. Never before or after were science, philosophy, and theology seen as almost one and the same occupation. True, secular theologians seldom composed systematic theological treatises for the use of theological faculties; some of them, mainly the Catholic, pretended to abstain from issues of sacred doctrine; but they dealt with most classical theological issues— God, the Trinity, spirits, demons, salvation, the Eucharist. Their discussions constituted *theology* inasmuch as they were not confined to the few truths that the "natural light" of reason can establish unaided by revelation—God's existence perhaps, or the immortality of the soul. Secular theology was much more than just a *theologia naturalis*.[1] Leibniz, the secular theologian par excellence, planned a comprehensive and sympathetic study on "Catholic demonstrations" of dogmas. Not only was he a layman, but also a Protestant.[2]

[1] On the classical origins of the term see Jaeger, *Die Theologie der frühen griechischen Denker* pp. 10–16. The usage of the term "theology" in the Middle Ages to characterize some pagan philosophers is rare; but see Thomas Aquinas, *Summa theologiae* I q. I a. I (to Aristotle, *Metaphysics* E.1026a19). See also Curtius, *Europäische Literatur und lateinisches Mittelalter* pp. 224–25. The name "natural theology" was revived during the Renaissance: below nn. 10, 14. In the seventeenth century it was reclaimed by deists as the sole religion.

[2] Leibniz, *Sämtliche Schriften und Briefe* (henceforth *SB*), 1.6, pp. 489–559. Cf. below II.H.6. Though this exercise was a part of Leibniz's plan to restore the unity of the Church, it was more than merely a search for the greatest common denominator or a minimum of dogma.

The secularization of theology—even in the simplest, first sense: that theological discussions were carried on by laymen—is a fact of fundamental social and cultural importance. It can be accounted for only by a variety of complementary explanations.

During the thirteenth century theology became both a distinct discipline *and* a protected profession; neither was the case earlier. Prior to the twelfth and thirteenth centuries, the term "theology" was ambiguous; it stood both for the word *of* God (the Scriptures) and for words *about* God, that is, any kind of discourse on matters divine.[3] Soon after the beginnings of its systematization, theology was established as a protected profession in the nascent universities. It was, in fact, doubly protected from the incursions of laymen. By and large, every science except medicine and sometimes law was taught by clergymen, regular and secular. But ordination and even the right to teach the arts (philosophy) did not suffice to teach theology, that is, commence with lectures on Lombard's Book of Sentences, without acquiring the proper degree.

Even though medieval philosophers could not avoid discussing matters divine, they were careful not to call by the name of theology those truths about God and the heavens accessible to mere reason. It is significant that, unlike the classical tradition, they avoided the term *theologia naturalis* and were careful not to call the ancient pagan philosophers "theologians," even while admiring their monotheism as *praeparatio evangelica*. Theology became a term reserved for supernatural knowledge. When, in the fourteenth century, Buridan suggested elimination of separate intelligences from the explanation of the motion of heavenly bodies—he favored an initial impetus instead, which keeps the heavenly bodies moving in perpetuity—he hastened to add: "But this I do not say assertively, but rather so that I might seek from the theological masters what they may teach me in these matters."[4] Buridan was only an artist (that is, a teacher of philosophy).

The first protective belt around theology eroded slowly, almost

[3] Ghellinck, *Le mouvement théologique du 12ᵉ siècle* pp. 91–92.

[4] "Sed hoc non dico assertive, sed ut a dominis theologiis petam quod in illis doceant me, quomodo possunt haec fieri": Buridan, *Questiones super octo physicorum libros Aristotelis*, ed. A. Maier, *Zwei Grundprobleme der scholastischen Naturphilosophie* p. 212; trans. Clagett, *The Science of Mechanics in the Middle Ages*, p. 536. To have opined something *disputandi more, non asserendi more* was often the defense of schoolmen when tried for false teaching.

imperceptibly, in the sixteenth century, when ever more disciplines in the universities ceased to be taught by clergymen. Nor did the university remain the only center of research and scientific communication: courts, academies, and printers became places of meeting and sources of sustenance. The rising number of educated laymen, as a reading public, as authors, and as teachers, was bound to increase instances of trespassing into the domain of theology; the case of Galileo was not unique, only the most scandalous.

The second protective belt around theology as a profession eroded with the spread of religious movements in the later Middle Ages,[5] and collapsed with the spread of Protestantism. Of the authority of the Holy Church, Augustine once said that, unless moved by it, he would not even believe the Sacred Scriptures.[6] The counterclaim of the Reformation—*sola scriptura, sola gratia, sola fide*—secured knowledge of God and access to him without the mediation of a priestly hierarchy. Protestants were encouraged, in various degrees, to read the Scriptures for themselves and to be for themselves ministers of grace. Theology became "secularized" in many parts of Europe in the original sense of the word: appropriated by laymen.

Again under the impact of Protestantism, theology became secularized in yet a deeper sense. To various degrees, it encouraged the sacralization of the world, even of "everyday life." Human labor *in hoc seculo* was not perceived anymore as a mere preparation for the future life; it acquired its own religious value in that, if well done, it increases God's honor.[7] So also does the study of this world, by ex-

[5] The late medieval proliferation of theological literature for laymen was mostly the work of theologians, as, e.g., the *Dives et pauper* (now available in the excellent edition of E. Heath Barnum [London 1976]), though not always: Dante and Marsilius of Padua were not theologians. Medievalists sometimes distinguish "secular" from "regular" theologians, i.e., theologians belonging to an order; obviously, this is not the sense in which I use the term "secular theology." Cf. n. 7.

[6] "Ego vero evangelio non crederem nisi me catholicae ecclesiae commoveret auctoritas": Augustine, *Contra epistulam Manichaei* 5, p. 197.22. For a strongly psychological interpretation of *commovere* see Oberman, *The Harvest of Medieval Theology*, p. 370 (refers also to Gregory of Rimini).

[7] Weber, "Die Protestantische Ethik und der Geist des Kapitalismus," *Gesammelte Aufsätze zur Religionssoziologie*, 1:17–206, esp. pp. 63ff., 84ff. This, of course, is true independent of the merit of Weber's central thesis. I hope that my use of the terms "secular" and "secularization" are clearly defined without being anachronistic. Originally, "secularization" was a legal term; since the Carolingian age it stood for the expropriation of Church property by worldly powers. One should be careful not to confuse "secular"

posing the ingenuity of its creator. The world, too, was not perceived as a transitory stage. It became in and of itself, as indeed attested to by the Scriptures, "very good" (Gen. 1:31), if not outright sacred. The world turned into God's temple, and the layman into its priests.

Finally, the barriers separating various scientific disciplines were fundamental to the peripatetic program of systematic knowledge. Within the Aristotelian and Scholastic tradition, it was forbidden to transplant methods and models from one area of knowledge to another, because it would lead to a category-mistake.[8] This injunction suited the social reality of medieval universities well, separating theology from philosophy to the benefit of both; but it eroded considerably from the fourteenth century, when mathematical consideration started to be heavily introduced into physics, and even into ethics and theology. What was a methodological sin to Aristotle became a recommended virtue in the seventeenth century. Since then we have been urged to transport models from mathematics to physics and from physics to psychology or social theory. The ideal of a system of our entire knowledge founded on one method was born. Aristotle never entertained it; neither did Scholasticism. Indeed, the very word "system" stood, until the seventeenth century, not for a set of interdependent propositions but for a set of *things*—for example, *systema mundi* or *systema corporis*.[9] The ideal of one, unified system of knowledge could hardly exclude theological matters, down to Spinoza's treatment of God *more geometrico*. These are some of the reasons why God ceased to be the monopoly of theologians even in Catholic quarters.

---

with "anti-religious" tendencies. Recent usages of the term among sociologists and historians are often vague. See Gusdorf, *Dieu, la nature, l'homme au siècle des lumières*, pp. 19–38; D. Martin, *A General Theory of Secularization*.

[8] Cf. below II.B.1–2; V.B.2.

[9] So already in Antiquity: see, e.g., (Pseudo)Aristotle, *On the Cosmos* 319b.9–12, p. 346. In part, "system" continued to be employed in this sense also in the seventeenth century (e.g., in the third part of Newton's *Principia*). It came to mean, among other things, an edifice of integrated propositions. Leibniz is particularly fond of speaking of "his system" (below II.H.n.3). Bayle tried, systematically, to elaborate the "system" behind all points of view—even the most abstruse or appalling. It is, I believe, a necessary component of his "critical" attitude. In the definition of Condillac, *Traité des systèmes*, in *Oeuvres complètes*, 2:1, "une système n'est autre chose que la disposition des différents parties d'un art ou d'une science dans un ordre où elles se soustiennent toutes mutuellement, et où les derniers s'expliquent par les premiers."

The Catholic response to the secularization of the divine seldom restored the fine medieval balance between philosophy and theology. To the contrary, whenever skeptical or fideistic arguments were invoked to undermine the faith in unaided reason, the medieval understanding of theology as a *rational* endeavor (albeit proceeding from premises inaccessible to the *lumen naturale*) was also undermined. Montaigne's "Apology for Raymond Sebund" is an excellent example of these opposing trends—the defense of the theologian's reserve as well as (against the theologian's wishes) the secularization of theological issues. Sebund's extreme claims for the evidence of natural theology (this name was given to the book later)[10] were censured by the Church. Montaigne believed he would be even better able to defend the Church if he were to destroy (as did Hume later) the notion that there exists an innate, self-evident core of theological truths. Man, by no means superior to brutes either emotionally or intellectually, needs a supernatural source of guidance even in daily, mundane affairs. The value of Sebund's natural theology can at best be relative: sometimes it may serve polemics. The only plausible proof for the veracity of Christianity that Montaigne elaborates at length is taken from the irrational rather than rational domain, and may be called an ethnographic proof: "I have often marveled to see, at a very great distance in time and place, the coincidence between a great number of fabulous popular opinions and savage customs and beliefs, which do not yet seem from any angle to be connected with our natural reason"—such as circumcision, the cross as sacred symbol, stories of primordial mankind, of an original sin, of a flood. "These empty shadows of our religion that are seen in some of these examples testify to its dignity and divinity,"[11] and they do so precisely because they are *not*

---

[10] Friedrich, *Montaigne*, pp. 94–96, 316 n. 58. Sebund's original title was *Liber creaturarum seu naturae seu liber de homine propter quem sunt creaturae aliae* (Lyon, 1484). The double truth theory, which Friedrich attributes to the Latin Averroists, was rather an invention of their adversaries from Étienne Tempier to Sebund.

[11] Montaigne, *Essais* 2.12: "Je me suis souvent esmerveillé de voir, en une très grande distance de lieux et de temps, les rencontres d'un grand nombre d'opinions populaires monstrueuses et des moeurs et creances sauvages, et qui, par aucun biais, ne semblent tenir à nostre naturel discours. C'est un grand ouvrier de miracles que l'esprit humain; mais cette relation a je ne sçay quoy encore de plus hétéroclite; elle se trouve aussi en noms, en accidens et en mille autres choses. . . . Ces vains ombrages de nôstre religion qui se voyent en aucuns examples, en tesmoignent la dignité et la divinité" (ed. Rat, pp. 644–45; trans. Frame, pp. 432–33).

accountable by reason. Montaigne turned natural theology on its
head while using some if its own ancient arguments. Previously,
some of these "coincidences" were invoked to show that polytheism and fetishism were just historical perversions of man's original,
natural monotheism. Montaigne denies it, denies that *anima natu-
raliter Christiana*. He believes that the "light of reason" only leads to
confusion, to a Babel of creeds.

Yet Montaigne himself was a layman. Moreover, he unwillingly
shared with Sebund the urge to abolish the demarcation line between natural and supernatural knowledge—although with opposite intents. Throughout the following century, the zeal for the defense of the doctrinal authority of the Church created critical
arguments more dangerous than their target. Richard Simon promoted biblical criticism to refute the claim that the Bible can be
understood by itself alone, *sine glossa*. Jean Astruc, wishing to defend (against Spinoza) Moses' authorship of the Pentateuch, invented the most destructive tool of biblical criticism yet: the philological distinction among the various original documents from
which the Masoretic text was composed, by Moses, as he believed,
or by others later (as we do).[12] How much more deadly to theology
were such helpers than its enemies! Yet, without being exposed to
these and other dangers, theology would never have contributed as
much as it did to the sciences and letters in the seventeenth century.

Finally, the secular theology of the seventeenth century was also
a distinct phenomenon inasmuch as it was not so universally accepted as to be beyond challenge and identification. Not all who
had a share or interest in the advancement of the new sciences ap-
proved of it. Fellows of the Royal Society, said Sprat, "meddle no
otherwise with *divine Things*, than only as the *Power*, and *Wisdom*,
and *Goodness* of the creator is displayed in the admirable Order and
Workmanship of the creatures. It cannot be deny'd, but it lies in the
natural Philosopher's Hands, best to advance that part of *Divinity*;
which, though it fills not the Mind with such *tender* and *powerfull*
Contemplations, as that which shews us Man's *Redemption* by a *Me-
diator*; yet it is by no means to be pass'd by unregarded, but is an ex-

---

[12] Lods, *Jean Astruck et la critique biblique au xviiie siècle*, esp. pp. 56–62; Eissfeld, *The
Old Testament, an Introduction*, trans. P. R. Ackroyd, pp. 160–61; on the earlier history of
biblical criticism see below, IV.B.3.

cellent Ground to establish the other. . . . These two subjects, *God* and the *Soul*, being the only forborn, in all the rest they wander at their Pleasure."[13] The *separation* of science from religion may have been as often demanded as it was violated; yet even those who demanded it with sincerity (rather than as a matter of prudent tactics) did not do so on medieval grounds. If previous generations distinguished between "natural" and "sacred" theology,[14] Sprat and others distinguished between science (or philosophy) and religion: religious contemplation, albeit more "powerfull," was placed outside the boundaries of scientific discourse. Deists were soon to recognize in "natural religion" the only true religion.

My aim in the present study is not to describe the secular theology of the seventeenth century in its breadth and in its manifold manifestations; I rather chose from it a few significant themes. When Christian Oetinger, the Pietist theologian, came to deal with God's attributes, the traditions he discussed were not those of Scholastic theology—Catholic or Protestant—but those of secular theology. "The attributes of God are ordered in one way by Leibnizians, in another by Newtonians; it is not irrelevant to compare their methods."[15] Some divine attributes and their relevance to natural science, political theory, and historical reasoning form the topic of my study. The secular theology in which these and other themes were embedded still awaits a detailed and comprehensive description as a new cultural phenomenon. My treatment of these themes is not even construed to prove the existence of a secular theology (if proof is needed) but to call to the attention of the reader the changes of connotation that some divine attributes underwent in a new intellectual climate.

[13] Thomas Sprat, *History of the Royal Society* (London, 1667), pp. 82–83. See Cragg, *From Puritanism to the Age of Reason*, pp. 96–97.

[14] "Partiemur igitur scientiam in *Theologiam* et *Philosophiam*. Theologiam hic intelligimus Inspiratam sive Sacram, non Naturalem": Bacon, *De dignitate et augmentis scientiarum* 3.1, *The Works of Francis Bacon*, 2:252.

[15] "Aliter attributa Dei ordinant Leibniziani, aliter Newtoniani. Non abs re erit, instituere comparationem inter methodum illorum et horum. Leibniziani incipiunt a contingentia ad ens absolute necessarium . . . et haec dicitur aseitas. . . . Methodus ordinandi Newtoniana incipit a libertate Dei, qua usus est in formando universo. . . .": Oetinger, *Theologia ex idea vitae deducta*, p. 50. In recent literature, Kolakowski, *Chrétiens sans église: La conscience religieuse et le lieu confessionel au xviiᵉ siècle*, comes closest to the study of secular theology as a cultural phenomenon.

B. THE THEMES

Whether or not God is immutable, our perceptions of God are not. In the following three chapters I wish to examine the changes in the meaning and usage of three divine attributes between the Middle Ages and the seventeenth century. It will serve as a convenient way to describe the changes in the nature of theological speculations vis-à-vis other disciplines—physics, history, political thought. It is also a convenient way to gauge changes in these disciplines themselves.

The divine predicates to be discussed are the omnipresence, the omnipotence, and the providence of God. They were not chosen at random. Divine predicates pose general as well as particular problems. Common to all is the problem of legitimacy of every positive mode of locution about God, or conversely, the efficiency of merely negative predicates. Of the particular problems, some are more time-bound than others. God's goodness and justice are hard to defend at all times from the vantage point of our painful world, which is the only vantage point we have. Such are not the problems I shall deal with here. I am rather concerned with those predicates that posed time-specific difficulties in the seventeenth century, and along with the difficulties opened up new opportunities of thought.

Because the seventeenth century[1] wished language to become precise and thoroughly transparent, God's *omnipresence* became a problem. If it could no longer be given a symbolic or metaphorical meaning, how else could the ubiquity of God be understood, God's being "everywhere"? The problem was compounded by the new commitment of the seventeenth century to a view of nature as thoroughly homogeneous and therefore nonhierarchical. God's omnipresence became an almost physical problem for some. Never before nor after were theological and physical arguments so intimately fused together as in that century. Why this was so and how it came about is the subject of the second chapter.

Medieval theologians engaged in a new and unique genre of hypothetical reasoning. In order to expand the logical horizon of God's *omnipotence* as far as could be, they distinguished between

[1] Throughout this book, I use "seventeenth century" as a shorthand term. The characteristics and developments I ascribe to it became more pronounced toward the second half of that century. Whether or not there was a *crise de la conscience européenne*, Hazard's periodization is reasonable and valid on many counts—including the decline in belief in witchcraft in the second half of the seventeenth century.

that which is possible or impossible *de potentia Dei absoluta* as against that which is so *de potentia Dei ordinata*. This distinction was fleshed out with an incessant search for orders of nature different from ours which are nonetheless logically possible. Leibniz's contraposition of the *nécessité logique* (founded on the law of noncontradiction) and the *nécessité physique* (founded on the principle of sufficient reason) has its roots in these Scholastic discussions, and with it the questions about the status of laws of nature in modern philosophies of science. But medieval hypothetical reasoning did not serve future meta-theoretical discussions alone. The considerations of counterfactual orders of nature in the Middle Ages actually paved the way for the formulation of laws of nature since Galileo in the following sense: seventeenth-century science articulated some basic laws of nature as counterfactual conditionals that do not descirbe any natural state but function as heuristic limiting cases to a series of phenomena, for example, the principle of inertia. Medieval schoolmen never did so; their counterfactual yet possible orders of nature were conceived as incommensurable with the actual structure of the universe, incommensurable either in principle or because none of their entities can be given a concrete measure. But in considering them vigorously, the theological imagination prepared for the scientific. This is the theme of my third chapter.

New in the seventeenth century was the critical-contextual understanding of history. Historical facts were no longer seen as self-evident, *simplex narratio gestarum*. Instead, they obtain significance only from the context in which they are embedded—a context to be reconstructed by the historian. And the meaning of historical periods or of their succession was likewise, since the revolution in historical thought, to be derived from internal connections within history rather than from a transcendental premise or promise. Indeed, the "fitting together" of events and institutions in any given period and the evolution of periods from each other constituted the new sense of divine providence. Yet, some modes of interpretation that were essential to the new historical-contextual reasoning were already present in medieval Jewish and Christian discussions about the working of providence in history. Exegetical as well as historical speculations since Antiquity were guided by the principle of accommodation, by the assumption that revelation and other divine institutions were adjusted to the capacity of men at different times to receive and perceive them. Once the principle of accommodation was secularized, as indeed it was since the seventeenth century, the "cunning of God" could become "the secret plan of history," the "invisible hand," the "cunning of reason." This transition is the subject of chapter four.

These different themes converge in a new assessment of the difference between divine and human knowledge with which my study will conclude. *Verum et factum convertuntur*—the identity of truth with doing, or of knowledge with construction—had been seen, in the Middle Ages, at best as the character of divine knowledge. In the seventeenth century it became also the mark of human knowledge, epitomized in the mathematical physics that showed not only how things are structured, but also how they are made. The identity of truth and fact was also claimed by a new brand of political theoreticians for whom the body politic seemed through and through a man-made artifact: human society is a spontaneous human construction. A new ideal of knowledge was born—the ideal of knowledge-by-doing, or knowledge by construction.

Some general remarks concerning methodological presuppositions are in order. My study is concerned with the shift from medieval to early modern modes of thought. In what sense was continuity preserved? What was revolutionary? Do the categories of continuity and change still have a heuristic value? The study is concerned with the scientific imagination, with ideals of science no less than with science itself. This distinction ought to be justified.

C. A DIFFERENTIAL HISTORY

Various parts of this study deal with questions of continuity and change; so much so that I may be accused of chasing indiscriminately after medieval precursors and anticipations. How much did seventeenth-century science and scholarship owe to the Middle Ages? How revolutionary was it? Never before or after did so many works praise themselves as "new observations," "new discoveries," "new method," "new science." If revolutions are a conscious and "resolute attempt . . . to break with the past,"[1] then the revolution-

[1] Tocqueville, *L'Ancien régime et la révolution*, trans. S. Gilbert, p. vii. The connotation of purposeful assent distinguishes the modern political usage of the term "revolution" from the more passive connotation of complete change attached to this astronomical

ary consciousness was certainly present: from the vantage point of the seventeenth century, the Aristotelian-Scholastic science was a barren enterprise from its outset, a dead-end. Its main concern had been with definition rather than with the precise relation between phenomena; *philosophia philologia facta est*[2]—it had lost itself in the search for essences and obscure qualities. Yet ever since the creative energies of medieval thought were rediscovered by recent historians, this seventeenth-century assessment has been called into question. Was Galileo's law of free fall the first concrete proof since Archimedes that "nature is written in a mathematical language"?[3] Was not its mathematical apparatus anticipated by the *calculatores* in the fourteenth century? Was not the principle of inertia underlying Galileo's law anticipated by some medieval impetus theoreticians? Was the radical search for self-evident truths really a new beginning in philosophy, or did Descartes borrow his seemingly new broom from the despised arsenal of later medieval Scholasticism? Was the discovery of historical anachronisms so new, or did it derive from

metaphor earlier. See Griewank, *Der neuzeitliche Revolutionsbegriff, Entstehung und Entwicklung*, pp. 143–58; Rosenstock-Huessy, *Die europäischen Revolutionen*, pp. 7ff.; the literature on the origins of the notion has focused solely on the astronomical connotation. Another, *medical* connotation is worth mentioning: turning of the sick from one side to another. See Otto of Freising, *Chronicon sive historia de duabus civitatibus* 5.36, ed. A. Hofmeister, p. 260 ("febricitantis, mundialis dignitas volvi et revolvi; crebris revolutionibus se iactant huc et illuc"). On him, see below IV.D.4.

[2] "Postremo, ut coarguantur verba magis curare, quam sensus, efficere ut cum Seneca exclamare merito liceat 'nostra, quae erat, Philosophia, facta Philologia est, ex quo disputare docemus, non vivere,' non memoro quot quaestiones apud Aristoteleos de nomine fere sunt": Gassendi, *Exercitationes paradoxicae adversus Aristoteleos* 1.14, ed. Rochot, pp. 45–46. While there is no denying that Aristotle often makes the meaning of words the beginning of philosophical inquiry, this does not mean that he remains on the level of words only. E.g., his physics changes the common meanings of "downwards" and "upwards" into "towards and away from the center of the universe." Cf. Wieland, *Die aristotelische Physik: Untersuchungen über die Grundlegung der Naturwissenschaften und die sprachlichen Bedingungen der Prinzipienforschung bei Aristoteles*, pp. 1–10; and also below III.C.1.

[3] Galileo Galilei, *Il Saggiatore*, in *Opere*, EN, 6:197–372, esp. p. 232. The schools, by contrast, view nature not as a *grandissimo libro scritto in lingua mathematica*; they think that truth can be found not in the world or nature, but in the confrontation of texts (*confrontatione textuum*): letter to Kepler, 19 August 1610, *Opere*, 10:421–23. Cf. Favaro, *Galileo Galilei, Pensieri, motti e sentenze*, pp. 27ff. The *topos* of the "two great books" (nature and Scripture) has attracted two recent philosophers of culture: Blumenberg, *Die Lesbarkeit der Welt* (Galileo, pp. 71–80); Derrida, *Of Grammatology*, trans. G. Ch. Spivak, pp. 14–18 (the identity of writing and the written text). Derrida relies, for his historical information, on Curtius, *Europäische Literatur*, pp. 323–29.

the medieval attention to the *qualitas temporum*? But these and similar questions are, I believe, misleading. They assume continuity and innovation to be disjunctive, mutually exclusive predicates. The "new" often consists not in the invention of new categories or new figures of thought, but rather in a surprising employment of existing ones.

Of the variety of ways in which a new theory can be said to have been prepared by an older one, two ideal modes are particularly pertinent to our discussion: the *dialectical* anticipation of a new theory by an older, even adverse, one; and the *transplantation* of existing categories to a new domain—employing them under a new perspective. Whatever remains vague in the preliminary explanations will, I hope, win more precision in the following chapters.

(i) A good number of examples can be gathered to illustrate the following circumstance. Well reasoned, elaborated theories may, or may not, specify possible instances of falsification; the demand of an *experimentum crucis* is, after all, relatively modern. Since the beginning of consistent theoretical reasoning, however, sound theories have often specified explicitly that which, in their own terms, must be regarded as a wrong, if not impossible or absurd, position. A conceptual *revolution* consists more often than not in the deliberate adaptation of such well-defined "absurdities" (or, better yet, the absurd consequences of contradictory assumptions) as the cornerstone of a new theory. Such were the beginnings of the atomistic theory. Parmenides had proved that to ascribe any degree of reality to negation amounts to attributing being to nonbeing. Being suffers no differentiation or change. That which "is not" cannot be "thought of." The atomists committed themselves consciously and deliberately to this absurdity in order to save movement and variety. Their atoms were Parmenidean "beings" embedded within the void, i.e., within a nonbeing endowed with "a kind of being."[4] Similarly, Aristotle's theory of motion may be said to have paved the way toward the principle of inertia more than any of its alleged forerunners, in-

---

[4] Diels and Kranz, *Fragmente der Vorsokratiker*, 2:2, 67(54)A6 = Aristotle, *Metaphysics* A4.985b4): "Λεύκιππος δὲ καὶ ὁ ἑταῖρος αὐτοῦ Δημόκριτος στοιχεῖα μὲν τὸ πλῆρες καὶ τὸ κενὸν εἶναί φασι, [λέγοντες τὸ μὲν ὂν τὸ δὲ μὴ ὄν] τούτων δὲ τὸ μὲν πλῆρες καὶστερεὸν τὸ ὄν, τὸ δὲ κενὸν τὸ μὴ ὄν." Cf. Simplicius on Aristotle's *Physics* 28.4ff. (ibid. 54 A8). ". . . was in Wahrheit bei den Eleaten vorhanden war, spricht Leukipp als seiend aus": Hegel, *Vorlesungen über die Geschichte der Philosophie*, in *Werke*, ed. Moldenhauer and Michel, 18:355; this remained the accepted interpretation.

cluding the impetus theory. For he anticipated some of its conceptual implications as the absurd (or impossible) consequence of a misleading assumption, the (atomistic!) assumption of movement in the void.[5]

The audacity to think the unthinkable is well known to historians of mathematics. All expansions of the realm of numbers beyond the rational numbers were once considered to be such impossibilities of thought and, at the time of their conception, "amphibians between being and nonbeing" (Leibniz), tolerated only by virtue of their performance.[6] The history of mathematics may be read as a running commentary on the incompleteness theorem. Time and again the inability to solve problems within one field led to the construction of new fields, since "no antecedent limits can be placed on the inventiveness of mathematicians in devising new rules of proof."[7] New mathematical disciplines have often accompanied scientific revolutions. Some grew out of a conceptual revision in science (the calculus), some made revisions within a science possible (non-Euclidean geometry). Nevertheless, conceptual revolutions in the sciences or in philosophy are different from those in mathematics even where they, too, involve the assertion of the absurd. The inherited body of mathematical theorems is not proven to be wrong, or only approximately true, but rather richer, by the legitimation of a mathematical entity or operation that was previously taken intuitively to be a non-number or a nonprocedure. Yet, physical theories are concerned not only with consistency and richness, but with truth and meaning. Where such theories introduce an absurdity in terms of a previous explanatory endeavor, the latter is destroyed, or at least proven inaccurate.

Nor should the dialectical preparation for scientific revolutions be confused with the readiness, already manifested in Greek astronomy, to entertain several explanatory models and to operate with those explanations best capable of "saving the phenomena" (σώζειν τὰ φαινόμενα), disregarding the question of their physical reality. I do not underestimate the emancipatory value of the recognition of a plurality of models in spite of, or if you wish, because of, the ep-

[5] Cf. below III.C.1.

[6] Leibniz, *Die Mathematische Schriften*, ed. Gerhardt, 5:357.

[7] Nagel and Newman, *Gödel's Proof*, p. 99; less optimistic Weyl, *Philosophy of Mathematics and Natural Science*, p. 235.

istemic resignation involved in it.[8] I agree with Feyerabend that the pursuance of a plurality of alternative explanations is, at least today, imperative. At any rate, the history of astronomy is a paradigmatic case of the benefits of theoretical "anarchy."[9] But the history of astronomy in antiquity and the Middle Ages shows also that it is one thing to look for many alternative explanations within given assumptions, and another to become conscious of such assumptions and revise them. In spite of its liberality, astronomy had as many difficulties as mechanics in becoming aware of its most deeply rooted preconception, the assignment of circular, "perfect" motion to the planetary orbits.[10]

We ought, then, to pay close attention to the terms in which a theory defines "improbabilities" and, still more important, "impossibilities." The more precise the argument, the likelier it is to be a candidate for future revisions. Once the impermissible assumption is spelled out with some of its consequences, it is but a matter of time and circumstances (a different climate of opinion, tensions within the old theory, developments in other fields, new factual evidence) until the truly radical alternative is reconsidered. The starting point of scientific reasoning, the Socratic curiosity ($\vartheta\alpha\nu\mu\acute{\alpha}\zeta\epsilon\iota\nu$), consists not only in asking why and how within a given theory or as if no theory existed. It consists rather at certain critical junctures in asking why not? despite a definite, enduring, argued consensus to the contrary.

It is tempting to describe the rise of early modern mechanics or astronomy as dialectically prepared by earlier theories. But the historical perspective is bound to blur such a schematic exposition considerably. The historian, as so often, finds himself entangled in the web of nuances after embarking from a clear-cut thesis. We shall find that, although Copernicus feared that traditional astronomers would condemn his model for its absurdity, and although Descartes regarded the inertial principle to be inconceivable in the terms of

---

[8] Simplicius, *In Aristotelis quatuor libros de caelo commentaria* 1.2, ed. Heiberg, p. 32: οὐδὲν οὖν θαυμαστόν, εἰ ἄλλοι ἐξ ἄλλων ὑποθέσεως ἐπειράθησαν διασῶσαι τὰ φαινόμενα. P. Duhem, *To Save the Phenomena: An Essay on the Idea of Physical Theory from Plato to Galileo*, trans. Donald and Maschler, p. 23.

[9] Feyerabend, "Consolations for the Specialist," in *Criticism and the Growth of Knowledge*, ed. Lakatos and Musgrave, pp. 197–229. On the whole, the guiding principle, "anything goes," fits best the history of mathematics, and least of all, say, zoology.

[10] Cf. below II.B.1. n. 1 (the Atomists as exception).

Aristotelian physics, nevertheless the main tenets of neither the Copernican theory nor Galilean mechanics were mainly formulated out of such a contraposition with past theories, and certainly not out of literary reminiscences of passages where "absurdities" were defined. In effect, many of the "absurdities" in the terms of Aristotelian or Ptolemaic science had already become, through the medieval exercises in hypothetical reasoning, mere improbabilities. In one sense, however, the "assertion of the impossible" will, I believe, stand the test of modification by nuances. Even where schoolmen in the Middle Ages traded Aristotle's "impossibilities" for possibilities *de potentia Dei absoluta*, they regarded them only as incompossible with our universe. With the usage of ideal experiments in the seventeenth century many such incompossibles became limiting cases of our universe; even if they do not describe our universe, they are necessary to explain it.

(ii) In other cases—by far more numerous—the mediation between new theories and the theories they replaced consists in the persistence of dominant figures of thought that are given a new perspective, placed in an unexpected new context. Early modern physics inherited many of the medieval techniques of hypothetical reasoning that involved, in questions mechanical, the beginning of a new mathematical technique. But it gave them a concrete, new interpretation. Early modern historical thought inherited some of its key categories from the medieval theological reading of history. But it applied them to secular history in a radically new way. Finally, the view of the state as a human artifact through and through rather than as a natural product of a built-in *inclinatio ad societatem*, though it had never before been defended so radically and systematically, replaced pure natural law traditions. "Sciendum est quod civitas sit aliquo modo quid naturale eo quo naturalem impetum ad civitatem constituendam: non tamen efficitur nec perficitur civitas nisi ex opera et industria hominum":[11] Aegidius Colonna, like many before and after him, believed that states have both a natural and an artificial aspect to them. By eliminating the human *natural* social impetus altogether, Hobbes stressed more than anyone before him that "man maketh his commonwealth himself," just as

---

[11] Quoted by Gierke, *Johannes Althusius und die Entwicklung der naturrechtlichen Staatstheorien*, p. 95 n. 52.

Marx would later do by eliminating the natural desire to barter by reducing all economic relations to human, historical conditions. The transition from the old to the new theory was a case of radicalization of already present possibilities of interpretation.

By these and other means, I shall try to discern and differentiate such points of transition as precisely as I can. By differentiating points of transition, questions of continuity or change lose much of their edge. As to the question why such transitions came about at the time they did, more often than not I do not know. Perhaps it is the sign of revolutionary periods that radical departures, "paradigm shifts," take inspiration and take courage from each other.

D. IDEAS AND IDEALS OF SCIENCE

The transition from medieval to early modern science and thought was not only a transition of ideas, methods, and arguments; the very ideals of science changed. Ideals of science differ in many ways from ideas in science. They indicate how a scientific community imagines science as it ought to be if ever completed; they express the ultimate criteria of rationality of their time. The very same body of measurements and procedures, assumptions and explanations—in short, ideas—accepted by an entire generation of scientists may be judged as wanting by some of them in the name of an ideal. In the name of an ideal of monocausality, or the elimination of all but mechanical causes from the consideration of nature, the seventeenth century felt uncomfortable with its most successful physical theory, the general law of gravity, even while accepting it. Newton himself was certain that gravity is not an "obscure quality" even though it seems to act *in distans*; but he hoped that it would one day be explained mechanically. In the name of the ideal of a complete description of reality, a reality independent of the observer, Einstein led a fierce controversy against the ultimate value of the uncertainty principle in quantum physics, a principle that, he admitted, explains and predicts certain phenomena most successfully.

Kant, who may have been the first to distinguish between ideals and ideas of science, also recognized that ideals may be at odds with each other without jeopardizing the actual scientific enterprise they guide. Ideas constitute science; ideals—Kant called them regulative ideas or principles—chart its goals. If we mistake constitutive for regulative ideas, he argues in a concluding chapter of his first critique, science may seem to us contradictory even where it is not. "The interest of reason,"[1] while demanding consistency from any concrete scientific explanation (or set of constitutive ideas), is often compelled to sustain regulative ideas which would be incompatible and would lead to contradicting results if employed with equal rigor in every scientific explanation. His examples are the principle of parsimony against the principle of plenitude.

Kant could not admit that ideals of science, let alone the categories by which we secure the intelligibility of the universe, can change with time. But they do: even the partial abandonment of final causes as a legitimate criterion of rationality proves it, or, more recently, the introduction of statistical causality. Who still speaks, except in a speculative vein, of the harmony of the universe? At times, the ideal may seem unchanged, but its meaning has been radically transformed. In the name of the ideal of consistency, theories of light until recently forced one to choose between either a corpuscular or a wave model of explanation. Both analogies have now ceased to be subsumed under the principle of the excluded middle and have become complementary. Neither the wave nor the corpuscular analogy can explain the nature of light exhaustively; light shares the characteristics of both.

In one important sense, the distinction between ideals of science and ideas in science must again be blurred and relativized. All criteria of rationality are, in a way, ideal, and for the following reasons. The nineteenth century believed in a steady, organic growth of the scientific enterprise under the same canon of rational principles. Very little of this confidence remains today. Historians and philosophers of science have argued for the relativization of science, even of rationality itself. The decision between competing theories, we have learned, depends in fact not on their intrinsic merits only, or perhaps not at all, but on external factors: ideology, faith, social conditions, economic forces, generational changes. Nor perhaps

[1] Kant, *Kritik der reinen Vernunft*, in *Werke*, ed. Weichscnedel 4:B670–696. It is noteworthy that, while the transcendental analytics is oriented toward the explication of Newtonian physics, this discussion takes biology as its paradigm. For a modern argument that calls for balancing the principle of parsimony with a counter-principle, see Menger, "A Counterpart of Ockham's Razor in Pure and Applied Mathematics: Ontological Uses," p. 415. One could say that Kant pronounced a principle of complementarity—but, unlike Bohr, on the metatheoretical level only.

can any two theories really be compared to each other, strictly speaking: competing theories, some argue, are always incommensurable. If even truly analytical propositions cannot be construed,[2] then no translation is possible from one language to another or from the terms of one theory into the terms of another.

But, you may ask, does not any extreme relativistic position defeat itself on purely logical grounds? Does it not lead us straight into paradoxes such as the liar's paradox? The ideological critique of Marx and Mannheim, the pragmatic epistemology of James and Vaihinger, the relativistic theories of science of Cassirer and Feyerabend, so it seems, must abrogate themselves because they ultimately refer also to themselves.[3] Now, this is not at all the case. Relativization should not be confused with the accusation of error. The proposition, "all propositions are only conditionally true (valid)" or "all propositions are relative," avoids the pitfalls of the proposition "all propositions are false." The latter, by including itself, becomes self-contradictory; the former does not. The assertion that all truths are relative may itself be relative, that is, only conditionally true, and yet be universally valid for as long as we cannot name the conditions under which it is false. The relativization of all criteria of rationality—of psychological, historical, sociological, epistemological—is not paradoxical. At the worst, it is doomed to be infinitely regressive or eternally provisional.

Yet, something positive should also be learned from these accusations of self-reference. Those who argue that the choice between theories depends on cultural, economic, or other extrinsic exigencies must still recognize that the arguments actually advanced for and against theories *claim* to be founded on reason alone. And more than that: even their very arguments for relativization appeal to reason. Not Vaihinger, not Kuhn, not even Feyerabend urge us to accept *their* point of view because it is beautiful or because it serves the

---

[2] Quine, "Two Dogmas of Empiricism," in *From a Logical Point of View*, ch. 2; "Reference and Modality," ibid., pp. 20–46. Cf. Bennett, "Analytic-Synthetic," pp. 163–88. The contention that analytic propositions cannot be construed may well serve as an ultimate argument against the possibility of any kind of translation between theories.

[3] E.g., Merton, "Karl Mannheim and the Sociology of Knowledge," in *Social Theory and Social Structure*, p. 503: "This leads at once, it would seem, to radical relativism with its familiar vicious circle in which the very propositions asserting such relativism are *ipso facto* invalid." They are not. I must, though, concede that an absolutely relativistic point of view will make it impossible to identify a common meaning or referent of a proposition, and thus defeat all judgment.

economy best or because it is good for faith or health. By appealing to rational criteria, I repeat, they are not inconsistent, as long as they also recognize the difference between judgments of origin (or causes) and judgments of validity.[4] The latter, whatever in fact the origin or causes for its acceptance may be, can only be defended on its own grounds. "*P* is valid because *q* is valid and *q* implies *p*" is a valid scheme of presenting an argument, even if it is clear that *p* was in fact accepted not by virtue of *q*. "*P* is valid because a scientific community accepted it once under such-and-such historical conditions or accepts it today" is perhaps a good historical account, but not an *argument* for *p*. Validity can only be examined in terms of itself. Whatever the origins of criteria for rationality may be, ideally they stand apart from the actual forces that shaped them. Science is a rational endeavor because both are true: it is true that it should examine its criteria and find that in fact they are not absolute or self-sufficient; and nonetheless it is also true that science must invoke only such criteria to mediate the business of persuasion. Rational procedures and proofs, albeit an ideal construct, distinguish science from other performances of persuasion such as rainmaking or play-acting.

And so, in a certain sense, all science, every scientific argument or procedure, has an ideal—and, if you wish, fictional—aspect to it. It is the ultimate justification why the historian of science ought to distinguish between ideals and actual arguments, and then detect the former even in the latter.[5] But ideals of science are hard to iden-

---

[4] James, *The Varieties of Religious Experience*, pp. 4–18. James, whose pragmatic stance could hardly afford it, turned this distinction against the "neurological" account of the genesis of religious states and ideas.

[5] My (quasi-Kantian) use of "ideals" differs form Holton's "themes" in that it can be articulated precisely as a demand: Holton, *Thematic Origins of Scientific Thought: Kepler to Einstein*, pp. 47–68. It also differs somewhat from Elkana's "images" in that it refers not only to science, but also to that which science refers, and in that it is, again, capable of precise formulation: Elkana, "Science as a Cultural System: An Anthropological Approach," in *Scientific Culture in the Contemporary World*, ed. Mathieu and Rossi, pp. 269–89. Nor, of course, are they coextensive with Kuhn's paradigms. If we agree that the *episteme* of the French structuralists is not a monolithic entity, that distinct, independent elements of "discourse" can be detected in it, then, I would suggest, regulative ideals form its backbone. They fulfill the demand of being regulations but not formalizable rules, unstructured structures. Foucault, in his *Archéologie de savoir*, has mitigated considerably the monolithic interpretation of the shift from one *episteme* to another described in his *Les Mots et les choses*. In his later conception, the unity of discourse is "vertical" only; continuities from one *episteme* to another are thinkable. Cf. also M. Frank, *Was ist Neu-strukturalismus?* pp. 135–237.

tify. They are often vague. Ideals, like virtues, are most spoken of when in doubt or danger. The wish to articulate them goes hand in hand with the need to defend them; yet polemics can also distort them. It is my aim to identify some of the leading ideals of science in the seventeenth century, to trace their origin and examine the connections and tensions between them. It seems advisable, at least to the historian, not to seek immediate definitions for something which seventeenth-century thinkers did not define clearly either. Instead of defining some early modern ideals of rationality, I shall try to describe their genesis and interaction.

II

GOD'S OMNIPRESENCE,
GOD'S BODY, AND FOUR IDEALS
OF SCIENCE

A. THE BODY OF GOD

1. A Family of Ideas

Thomas Hobbes, trying as always to be blunt, insisted that to call God—or spirits—immaterial entities is like calling them bodiless bodies, a contradiction in terms. We may assert their existence not by an act of reason, but only by decree of a sovereign.[1] Spinoza's *Deus sive natura*, the one and only substance, must possess infinitely many attributes. Of these, only two are known, namely thought (*cogitatio*) and extension. Now Spinoza, following Descartes, regarded extension as the constitutive attribute of bodies, the only "clear and distinct" idea we have of a body if regarded by itself.[2] According to Henry More, God and other spirits, albeit incorporeal, must nevertheless be extended things. Extension is as necessary a predicate to the divine as are perfection or sui-sufficiency.[3] More developed his doctrine against Descartes's and Hobbes's mechanical outlook: only if spirits are extended could real forces be introduced into the universe. But he knew (if not from other sources, then from Descartes's response) that he could easily be misconstrued as claim-

---

[1] Hobbes, *Leviathan* 1.4, 3.33, 4.45, ed. Macpherson, pp. 108, 426, 428–30, 661; cf. Watkins, *Hobbes' System of Ideas*, pp. 68, 157, 164. In order to enhance the role of the sovereign, Hobbes emphasizes the discrepancies in Scriptures: see Strauss, *Spinoza's Criticism of Religion*, pp. 98–100 and nn. 130–31, 101–104.

[2] Spinoza, *Ethica more geometrico demonstrata* 2, def. 1, in *Opera quotquod reperta sunt*, ed. Van Vloten-Land, 1:73; Descartes, *Principia philosophiae* 2.4, AT, 8:42. Cf. below, II.E.1; F.1.

[3] More, *The Easie, True, and Genuine Notion . . . of a Spirit* 22, in *Philosophical Writings of Henry More*, ed. McKinnon, p. 213; further elaborated below, II.E.2. Leibniz alludes to Hobbes, More (and Newton) in the opening sentence of his correspondence with Clarke: "Il semble que la religion naturelle même s'affoiblit extremement. Plusieurs font les ames corporelles d'autres font Dieu luy même corporel." Leibniz, *Die philosophischen Schriften*, ed. Gerhardt, 7:352. But it was not only an English disease.

ing that God has a body.[4] And Newton called space—the empty, infinite, homogeneous, absolute Euclidian space—quasi a sense organ of God (*sensorium Dei*), without which there could be no real forces in nature, without which therefore we could not say of God that he acts upon his creatures, nor that he intuits them, let alone that he is "everywhere" (*ubique*).[5]

These and similar positions are not occasional odd formulations, metaphors chosen at random, but rather systematic pronouncements. They cross the dividing lines between rationalists and empiricists, pantheists and deists, theosophic speculators and discursive reasoners. They form a genuine family of ideas that I shall call, for the sake of brevity, "the body of God." You may object that the term distorts the intentions of Spinoza and More, Newton and Raphson. "The body of God" smacks of anthropomorphic images, which all of them rejected emphatically. Spinoza did not say of God that he *has* a body; his God *is* body. Every idea corresponds to, or is the extension of, a material constellation: *ordo rerum idem est ordo et connexio idearum*.[6] The soul is but an ideational expression for a material constellation. The sum total of all bodies and their constellations (*facies totius universi*) is an infinite mode of God.[7] God, the na-

---

[4] AT, 5:269–70: "I am not in the habit of disputing about words, and therefore if somebody wants to say that God is, in some sense, extended because He is everywhere, I shall not object. But I deny that there is in God, in an Angel, in our soul, and in any substance that is not a body, a true extension."

[5] Newton, *Opticks; or, a Treatise of the Reflections, Refractions, Inflections, and Colours of Light*, queries 28 and 31; below pp. 90–97. If extension be attributed to God, we hear from a critic, then it either means mere space, in which case "it might as well be said . . . that God is an infinite *inane* or *vacuum*, that is, in plain English, an infinite Nothing imbued with Wisdom, Goodness and Power. . . . On the other side if by Extension is understood a *thing that* in the Idea and first Conception of it is Extensive, . . . in this sense, I cannot see how it differs for *Matter*. . . . And if at any time we do endeavor to apply Extension or Space unto Mind . . . there always arises repugnance in us, upon but the thought of it; an Inch, a Foot, a Yard of Understanding, or Goodness, is a *Bull*": Burthogge, *An Essay upon Reason and the Nature of Spirits*, pp. 120–21.

[6] Spinoza, *Ethica* 2, prop. 7, Van Vloten-Land, 1:89. It should be noted that this correspondence theory accounts also for confused and indistinct ideas, inasmuch as they correspond to indistinct boundaries between bodily constellations. All single bodies are but a relative unity, inasmuch as their proportion of motion ($m \cdot v$) remains constant. Rivaud, "La Physique de Spinoza," pp. 24–27. See below II.F.1; V.C.5.

[7] Spinoza, *Epistulae* 64, Van Vloten-Land, 3:120. The *facies totius universi* is the universe "seen as" one body; Spinoza refers to *Ethics* 2,lemma7, schol.(before prop. 14), Van Vloten-Land 1:88: "facile concipiemus, totam naturam unum esse individuum, cuius partes, hoc est omnia corpora, infinitis modis variant." See below II.F.1.n.13 (*facies* = person).

tura naturans, is the idea of unity and coherence of all laws of nature.[8]
More and Newton distinguished between corporeality, which God
lacks if he is to penetrate all things, and dimensionality, which every
entity qua entity must have. Spirits, according to More, are unlike
bodies in that they are not impenetrable. Yet they are blessed with
a fourth dimension—spissitude—that bodies lack.[9]

Yet these very distinctions prove that the absolute medieval commitment to an idea of God radically purged from all material connotations, however abstract and remote, was broken in the seventeenth century. Medieval theology in most of its varieties viewed
with intense suspicion any doctrine that took God's presence in the
world too literally. So much was this true that not only physical
predicates, but also general-abstract predicates such as goodness,
truth, power, and even existence were at times considered an illicit
mode of speech when predicated of God and his creation univocally.
In the latter sense, nearly all important philosophical discussions on
the nature of God sinned against the classical, medieval-Thomistic
tradition. Not only More or Spinoza were guilty; to all of them, including Descartes, Malebranche, and Leibniz, God shared with his
creation some genuine predicates literally and unequivocally. The
body of God is only a special case of what may be called the transparency of God in the seventeenth century. I do not necessarily
mean that seventeenth-century thinkers always claimed to know
more about God than medieval theologians. To some of them God
remained a *deus absconditus* about whom little can be known. What I
mean to say is that they claimed what they knew about God, be it
much or little, to be precise, "clear and distinct" ideas.

2. Univocation and the Ontological Argument

Why was Anselm's ontological proof neglected in many quarters
during the Middle Ages, and why was it so widely acclaimed in the
seventeenth century?[10] If successful, it proves God's existence by

---

[8] Curley, *Spinoza's Metaphysics: An Essay in Interpretation*, pp. 45–81, 119–58.

[9] Below, II.E.2.

[10] On the marginal position of the ontological proof in the Middle Ages and the causes
for its revival in the seventeenth century, see Henrich, *Der ontologische Gottesbeweis: Sein
Problem und seine Geschichte in der Neuzeit*, pp. 1–22. For important medieval exceptions
in the thirteenth century cf. Daniels, *Quellenbeiträge und Untersuchungen zur Geschichte des
Gottesbeweises im Mittelalter*, esp. p. 125 (there reference to the texts); most of those who
saw merit in the argument also endorsed—as Anselm did—an illumination theory of

demonstrating that an adequate notion of God excludes, of necessity, non-existence. Yet many medieval theologians denied that we possess a notion of God adequate to sustain the ontological argument without watering it down. God, according to Thomas, is indeed a *notum per se ipsum*, but only to himself, not to us.[11] Descartes, More, Leibniz, and Wolff revived the argument because they believed in our capacity to form an adequate and precise, if incomplete, idea of God. Inasmuch as our ideas are clear and distinct, they are, we are told, the same as God's. Descartes chose the term "idea" over others because "it was the term commonly used by philosophers for the form of perception of the divine mind."[12]

In part, this change originated in the later Middle Ages. Already Duns Scotus insisted that some divine attributes, notably existence and willing, have the very same meaning whether applied to God or to man.[13] Duns Scotus reacted against the Thomistic doctrine of *analogia entis* according to which divine attributes are never univocal; he and his followers could—and did—accept the validity of Anselm's proof at least partially or hypothetically. A number of fourteenth-century theologians accepted the ontological argument as a valid proof of God's infinity.[14] But with or without the ontological argument, a renewed commitment to an unequivocal language of science—every science, including theology—was the mark of the fourteenth century as it was again that of the seventeenth. We may

---

knowledge; and most of them were Franciscans. After Scotus, the argument was usually used only partially (e.g., to prove God's infinity). Henrich's distinction between the forms of the ontological proof has been shown to be implicit in Anselm's *Proslogion*: Malcolm, "Anselm's Ontological Arguments," pp. 41–62.

[11] Thomas Aquinas, *Summa theol.* 1 q.2 a.1, in *Opera omnia* (Parma, 1855); *De veritate* q. 10 a. 12; *Summa contra gentiles* 1.10–11, in *Opera*, 5:6–8. Thomas distinguishes between *per se notum simpliciter*, which God is, and *per se notum quoad nos*, which God is not.

[12] Descartes, *Response to Objections* 3.5, AT, 7:181: "Ususque sum hoc nomine, quia iam tritum erat a Philosophis ad formas perceptionum mentis divinae significanda, quamvis nullam in Deo phantasiam agnoscimus; et nullum aptius habebam"; trans. Haldane and Ross, in *Philosophical Works of Descartes*, 2:68. Cf. below p. 291.

[13] All *transcendentalia* are predicated univocally of God and his creation, whereby *unum, verum, bonum* are *passiones entis unicae*, others *disiunctae*; Scotus, *Expositio in metaphysicam* 4, summ. ii c.2 n.9 (Wadding 4:112), and in many other places. Cf. Gottfried Martin (n. 15 below).

[14] Scotus, *Ordinatio* 1 d.2 q.2 n.8 in *Opera omnia*, ed. Balic et al., 2:120. cf. Daniels, *Quellenbeiträge*, pp. 105–107; Gilson, *Jean Duns Scot, Introduction à ses positions fondamentales*, pp. 175–79; Bonansea, *Man and His Approach to God in John Duns Scotus*, pp. 173–86.

call it the nominalistic revolution. Ockham's emphasis on unequivocal terminology was even stronger than that of Duns Scotus. Only
discrete entities, singulars, exist and they do not need the mediation
of universals either for their existence or for their immediate, "intuitive" cognition. Connotative terms, such as relations, are only
valid to the extent that they are coextensive with a set of singular
entities.[15] Ockham rejected both the *analogia entis* and the ontological argument, relativizing all proofs for God's existence. They may
prove, a posteriori, that *a* God exists, but not necessarily one God,
nor indeed the God of revelation.[16]

In other words, once the Thomistic theology of equivocation was
rejected, knowledge of God *sola ratione* had to be either very little,
or include several aspects common to God and the world besides existence. The so-called Nominalists of the later Middle Ages chose
the first position; many seventeenth-century thinkers (in which
nearly all original philosophical minds were Nominalist) chose the
second. God wills, according to Descartes, the same way we do—
namely, without bounds. Note that, for Descartes, infinity is a
"clear and distinct" idea from which the notion of finitude is derived, rather than vice versa.[17] God is extended in the same sense
that other spirits or bodies are, according to More, except that
God's extension is infinite and unchangeable. He, too, embellished
the ontological proof. Leibniz and his eighteenth-century followers
went even further: simple, univocal attributes never contradict each
other, by definition.[18] Every one of them is a perfection, and adds
reality to the entity of which it is predicated. Every real entity possesses some perfections, and God possesses them all. God is the
summa realitatum, the sum total of all realities and therefore the embodiment of a well-defined entity (*omnimodo determinatum*). He

---

[15] Moody, *The Logic of William of Ockham*, pp. 53ff.; G. Martin, *Wilhelm von Ockham,
Untersuchungen zur Ontologie der Ordnungen*, pp. 221–27; Leff, *William of Ockham: The
Metamorphosis of Scholastic Discourse*, pp. 139ff.; K. Tachau, "Vision and Certitude in the
Age of Ockham," pp. 103–105.

[16] Ockham, *Quodlibeta* I q. I, in *Philosophical Writings*, ed. Boehner, p. 126: neither the
proposition "the unicity of God" nor its negation can be proven demonstratively.

[17] AT, 6:426: "Infinitum non a nobis intelligi per limitationis negationem." For other,
similar as well as contradicting, references, see Gilson, *Index Scholastico-Cartésien*, p. 143.
Cf. below II.D.4.

[18] See below II.H.4; and also, A. Maier, *Kants Qualitätskategorien*, pp. 10–23, 34–38,
who traces the history of the problem down to the Kantian separation of the category of
"reality" (*Realität*) from that of existence (*Wirklichkeit*).

must, therefore (this was their version of the ontological proof, and their answer to Scotus's objections), exist. It was this version of the ontological argument that Kant sought to demolish.[19] He tried to prove that the *ens realissimum* is only a hypostatization, and even personification, of a necessary ideal of reason: namely the idea of unequivocal, complete determination of every thing. But this *durchgängige Bestimmung* is only a regulative ideal of reason, never a patterning precondition of all possible experience.

3. Univocation, Homogeneity, and Other Ideals

The quest for a precise and univocal language of science led to a reexamination of some further divine predicates, notably God's omnipresence, in physical terms. Since Duns Scotus they were indeed so reexamined, and "God's existence in things" was given, as we shall see, startling new interpretations. But physics itself also changed, and with it the possible world-oriented meaning of divine predicates. Two forceful impulses determined the outlook of nature in early modern science; I will call them the drive for unequivocation and the drive for homogeneity. The former seeks simplicity and coherence *a parte nominis*, the latter seeks them *a parte rei*. Scientists since the seventeenth century wanted their scientific language to be as unambiguous as possible; therefore, they emptied nature of *intrinsic* meanings. In the words of Foucault, science in the seventeenth century exchanged "similitudes" in nature for the precise comparison of "sameness and difference."[20] No longer were natural phenomena to symbolize and reflect each other and that which is beyond them; the symbolic-allegorical perception of nature as a network of mutual references was discarded as a source for protracted equivocation. The image, say, of man as a microcosm that reflects and embodies the macrocosm lost much of its immediate heuristic force. Things ceased to refer to each other intrinsically, by virtue of their "participation in" and "imitation" of each other. Only language was henceforth to refer to things and to constellations of things in a system of artificial, univocal *signs*, such as mathematics. The ultimate prospect of science was a *mathesis universalis*—an unequivocal, universal, coherent, yet artificial language

---

[19] Kant, *Kritik der reinen Vernunft* A572–584, in *Gesammelte Schriften*, ed. Weichschedel 4:515–23. See below VI.A.2.

[20] Foucault, *Les Mots*, p. 67.

to capture our "clear and distinct" ideas and their unique combinations.

On the other hand, scientists since the seventeenth century also demanded of nature itself that it be homogeneous, uniform, symmetrical. The same laws of nature should apply to heaven and earth alike—as they do "in Europe and in America."[21] No more should separate regions of the universe obey, as is the case in Aristotle's physics, different mathematical models, such as the "natural" motion in straight lines, that is, upwards or downwards, within the sublunar realm of the universe, as against the circular, eternal motion that is natural only within the celestial region. The same kind of matter ought to build all parts of the universe, and it ought to be governed by the same causes or forces. How else could we reason, as Newton expects us to, from the "analogy of Nature"?[22]

The revival of God's body, or the physical meanings attached to God's omnipresence, can best be explained in view of the tensions between these two ideals. They were the most comprehensive ideals of rationality in the seventeenth century, but not the only ones. In their proximity, perhaps subordinate yet not identical, other ideals are clearly recognizable, notably the ideals of *mathematization* and *mechanization*. The demand to see nature as "written in mathematical letters" coincides, in part, with the quest for an unequivocal and coherent language of science; only those properties and relations that are quantifiable are really unambiguous (even if, like Newton's universal gravitation, mysterious). In part, however, mathematics was more than just a source and paradigm of discourse. Nature itself was expected to reveal mathematical order and harmony—elusive properties indeed, the meaning of which could be either that

---

[21] Newton, *Philosophiae naturalis principia mathematica*, p. 402: "Hypoth. II: Ideoque effectuum naturalium ejusdem generis eadem sunt causae. Uti respirationis in Homine et in Bestia; descensus lapidum in *Europa* et in *America*; Lucis in Igne culinari et in sole; reflexionis lucis in Terra et in Planetis." The word *ideoque* refers to hypothesis I (ibid.), the law of parsimony.

[22] The "analogy of Nature" permits the extrapolation from constant qualities of bodies that are "found to be all bodies within the reach of our experiments" to "all bodies whatsoever": "We are certainly not to relinquish the evidence of experiment . . . ; nor are we to recede from the analogy of Nature, which is wont to be simple and always consonant to itself." "Rules of Reasoning in Philosophy," in *Sir Isaac Newton's Mathematical Principles of Natural Philosophy and His System of the World*, trans. Motte, rev. Cajori, 2:398. Cf. also his *Opticks*, ed. Roller, p. 376: "For nature is very consonant and comfortable to herself."

nature is homogeneous and symmetrical, or, again, that nature tends to reify those configurations and formulae that the mathematical tradition labelled "perfect," "regular," or "simple." In a later chapter I shall try to show how the ideal of mathematization came to be increasingly detached from the latter demand, why it became more and more an ideal of language only.

The ideal of mechanization, inasmuch as it calls for a monocausal explanation of nature, stands close to the ideal of uniformity although they are not altogether identical. It consists first and foremost of the demand to abolish final causes from the study of natural phenomena, to reduce all causes to mechanical causes: "I wish we could derive the rest of the phenomena of nature by the same kind of reasoning from mechanical reasoning, for I am induced by many reasons to suspect that they may all depend upon certain forces by which the particles of bodies, by some causes hitherto unknown, are either mutually impelled towards one another, and cohere in regular figures, or are repelled and recede from one another."[23]

All four ideals of science may seem, in retrospect, to be so many complementary aspects of one and the same ideal of economy and coherence: the economy of language and of the structure of things. In the seventeenth century, all of them were indeed fused into various attempts to mathematize physics. Yet they are separable, both logically and historically. Unequivocal language does not imply the homogeneity of nature: should the universe be found to be asymmetrical or in any other way nonhomogeneous to its core, it may still be capable of an unequivocal, even of a mathematical, description. A nonhomogeneous universe (in any reasonable sense of the term) is not a contradiction in terms. Furthermore, the mathematical expression of relations between variables in nature need not always be "simple." The simpler (or more general) theory may require a more complex mathematical apparatus. When Kepler opted for elliptical planetary orbits, he opted with a heavy heart for what seemed at his time a less "simple" or "perfect" geometrical representation; he overcame two millennia of obsession with circularity. None of the other ideals implies, of necessity, the abandonment of final causation. Leibniz's *vis viva*, Maupertuis's principle of least

---

[23] "Praefatio ad lectorem," in *Sir Isaac Newton's Mathematical Principles*, I:xviii.

action[24] are as mathematical as they are intended to prove teleology in nature.

Logic aside, these ideals of science also have separate origins and separate careers. They were originally embedded in systems of thought hostile to each other. The ideal of arithmetization—or geometrization—of the universe propelled the physical speculations of the Pythagoreans and Plato. Unequivocation—but not homogeneity—was the guiding postulate of Aristotelian science. The view of the universe as a dynamic-homogeneous, material continuum was common to all Stoic philosophies of nature. The physics of the Atomists eliminated all other causes from nature except for the push and pull of indivisible particles in the void. These ideals continued their independent careers, sometimes within, more often outside, their original contexts prior to their fusion in the seventeenth century. While the fortunes of two of these ideals—the ideals of mathematization and mechanization—are of lesser concern to the argument of this chapter, I shall, in view of later discussions, sketch the original setting of all four.

B. THE ORIGINAL SETTING OF THE IDEALS

1. Mathematization: Plato

Why did the the ideal of mathematization fade after Plato? Why did it continue to live only in hermetic, poetic, eccentric, or mystical traditions while being discarded by the Peripatos as well as by the Stoa and the Atomists? And with what changes was it revived in the Renaissance? The Pythagoreans wished to arithmetize the universe, but their vision was discredited by the discovery of the irrational, or precise magnitudes that nonetheless have no "*ratio*" (*logos*). Of the speculative dream in Plato's *Timaeus* all that remained unchallenged was the astronomical "obsession with circularity" (Koyré), the demand that celestial orbits be described, at all costs, as perfect circles. Being the perfect bodies that they are, immutable and describing regular paths, only the most regular, simple, perfect geometrical figures should represent their shapes and movements. So powerful was this demand that (unlike the assumption of the centrality of the earth) it was hardly ever questioned, either in Antiquity or in the

[24] Maupertuis, *Essai de Cosmologie*, pp. 109, 225ff., and *passim*.

Middle Ages, not even by Copernicus.[1] But Plato's *Timaeus* contained much deeper arguments for circularity, and more: it presented a theory of the primary, stereometrical properties of the four elements in order to account for their secondary qualities. Each of the elements corresponds to one of the four regular, perfect, solid figures (the cube, the pyramid, the octahedron, and the icosahedron). These again are composed of half-square and half-equilateral right-angled triangles, which allow water, air, fire, but not earth (because its cubic figure is composed solely of half-squares) to be transformed into each other in an eternal circle through rearrangements of the elementary triangles.[2] The cosmos at large, represented as a dodecahedron, is a translation of a perfect, ideal model into an imperfect and resilient medium—space or chaotic matter, endowed only with "necessity" (ἀνάγκη). There were very few attempts to revive, let alone expand, this elaborate theory, either in Antiquity or in the Middle Ages. One certainly cannot argue that it was abandoned because its heuristic possibilities were exhausted. In the Renaissance, during the flowering of speculative systems of nature, the *Timaeus* was admired rather than acquired. The same is the case with Pythagorean and neo-Pythagorean speculations. But why?

Plato himself alerted us to the speculative nature of his story of creation: it is only a "probable myth." Even if accepted as true, the universe it describes is but an imperfect copy of the idea of a "living body."[3] The "soul" and "body" of this universe are constructed according to mathematical principles as far as possible. Mathematical

---

[1] The closest to an objection against the apotheosis of circularity, at least as an aesthetic ideal, which I could find, came from Atomistic quarters. "Admirabor eorum tarditatem qui animantem immortalem et eundem beatum rotundum esse velint, quod ea forma neget ullam esse pulchriorem Plato: at mihi vel cylindri vel quadrati vel coni vel pyramidis videtur esse formosior." With these words, Cicero lets his Epicurean dismiss the *anima mundi*. Cicero, *De natura deorum* 1.10.24 (Vellius), ed. Plassberg, p. 10; cf. 2.18.47, p. 67: "conum tibi ais et cylindrum et pyramidem pulchriorem quam sphaera videri. Novum etiam occulorum iudicium [!] habetis." Astronomical arguments follow. The only other available source of vague objections against circularity was the contention of Nicolaus Cusanus that no actual body can have the perfection of a geometrical figure, but at best approximates it.

[2] Cornford, *Plato's Cosmology: The Timaeus of Plato*, pp. 210–19. On the Pythagoreans see Sambursky, *Das physikalische Weltbild der Antike*, pp. 44–73; for Plato's physics, ibid., pp. 411–15.

[3] Plato, *Timaeus* 30c–31a, in Cornford, *Plato's Cosmology*, pp. 39–40.

entities—measures and numbers—stand between the realms of εἶδος and of αἴσθησις, mediating between both because they partake
in both. So also the world soul: it mediates between immutable
ideas and changeable matter or chance by connecting the pure ideas
of "sameness," "difference," and "existence"—ideas that, in his
later theory, Plato believed to be expressed by "eidetic numbers"
that go "across" all ideas and generate their order (τάξις) and con-
nectability (transcending the counted, "sensible" numbers and
counting, "mathematical" numbers).[4] The world soul demands the
sphericity of the world's body and the circularity of the planetary,
besouled motions. The world *body* demands four material elements—two to secure tangibility (earth) and appearance or lumi-
nosity (fire), two to mediate between those in the only harmonic
proportion of cubes (a^3, a^2b, ab^2, b^3).[5] The stereometric properties of
the elements secure the closest approximation of the body of the
universe (the dodecahedron) to a sphere.

In short, unlike the Pythagoreans, Plato did not claim that the
universe *is* numbers, figures, or ideas, but rather that it reifies them
as best it can. It is as if a picture were translated into an alien medium; nature is, in the original sense of the poetic term, only a *metaphor* for reality, a "carrying over" of meaning from one medium to
an imperfect one through the mediation of mathematics. The perfect medium is the immutable world of ideas; the imperfect medium is matter (chaos), the realm of change and becoming (γένεσις), which must be "persuaded" by the demiurge to assume a
rational, mathematical structure contrary to its nature, contrary to
"necessity." Nature resists a thoroughgoing mathematization.[6]

---

[4] On the Platonic distinction between eidetic, mathematical, and aesthetic numbers see
Klein, *Greek Mathematical Thought and the Origins of Algebra*, pp. 79–99, esp. p. 91. Klein
follows, by and large, the assessment of Stenzel and Becker (Stenzel, *Zahl und Gestalt bei
Plato und Aristoteles*) concerning the importance of the ideal numbers in Plato's later dialectics, but corrects their interpretation of the meaning of "number." Both the ideal
numbers in Plato's later theory, and the world soul in the *Timaeus* are constituted by the
principles of sameness, difference, and being (*Timaeus* 35a). Cornford (*Plato's Cosmology*,
p. 64 n. 2) warns us, however, that a connection between the intermediate status of both
the soul and the "mathematical numbers" is speculative, precisely because the theory of
ideal numbers, known to us from Aristotle's polemics only (*Metaphysics* Λ6.987b14ff), is
a later theory.

[5] Cornford, *Plato's Cosmology*, pp. 46–47, following Heath, *A History of Greek Mathematics*, 1:305–306 (and n. 2).

[6] *Timaeus* 48a (reason overruling necessity). Cornford, who identifies the demiurge

All this was cause enough for Aristotle, as well as the Stoics and
Atomists, to discard the program of thoroughly mathematizing the
cosmos. Aristotle objected not only to Plato's later theory, absent
in *Timaeus*, of eidetic numbers that go across *genera*; he saw no merit
in a metaphoric usage of mathematics either. From Aristotle onwards, mathematical considerations were seen at best as useful for
describing perfectly regular motions or equilibrium in astronomy
or statics. Change, the heart of physics, chemistry, and biology,
could only be mathematized at the price of equivocation and inex-
actitude. The relationship of mathematics to the science of *change*
was, for the most part, a symbolic one in the eyes of both those who
sought to discover this relationship and those who rejected it. Statical mechanics and astronomy remained, until the end of the Middle
Ages, the only fully mathematized sciences; the latter, even while
known to be incapable of a causal-physical explanation, was at least
capable of "saving the appearances." Some Atomists, it seems,
went even further and challenged the validity of the extant body of
mathematical reasoning itself.[7]

Mathematical considerations, then, applied much better to the
structure of the world than to its processes, let alone the process of its
construction. In the views of most ancient and medieval natural philosophers, mathematics was least precise when it came to the discussion of change—all change, including motion. Aristotle and the
Aristotelian tradition viewed a mathematical science of change not
only as imprecise and equivocal, but as a downright category-mistake. Mathematical objects are objects abstracted from all physical
properties, and physics is, first and foremost, the knowledge of
causes of change.[8] Only within some circles in the fourteenth century, and again in the sixteenth and seventeenth centuries, did a
mathematical science of motion cease to appear as a contradiction in
terms. The transition happened not only in physics: mathematics itself was to change, in substance and ideal, from an inventory of

---

with "reason," identifies also necessity (or chance) with the world soul: the latter is the
one to be "persuaded" (Cornford, *Plato's Cosmology*, pp. 160–76). In our context it
makes little difference whether "necessity" is predicated of matter and the chaos or also
the (irrational) world soul.

[7] Notably Zenon the Epicurean. On him see Fritz, Pauly-Wissowa, *RE*, ser. 2, vol. 19,
coll. 122–27 (Zenon von Sidon), esp. 125–27.

[8] Aristotle, *Physics* B8.193b22–194a6; Γ1.200b12 (φύσις the principle of motion and
change). On mathematical reasoning in physics cf. below V.B.1–2.

ideal entities and their properties into a *language*. Kepler is a case in point. Attending to the *structure* of the universe, he envisioned a cosmic harmony well within the Pythagorean-Platonic tradition, which continued to live in mystical-theosophical speculations, in the poetic cosmologies of the twelfth-century school of Chartres, and in the hermetic tradition of the Renaissance. In the *Timaeus*, Plato discussed (and discarded) the ascription of each of the "five worlds" to each of the regular solids, a theory taken by later commentators in Antiquity to mean that there existed five cosmic zones.[9] Kepler, in turn, made the five regular solids account for the distances between the planets and made the harmonic proportions between them testify to the music of the spheres. But Kepler used and enhanced another mathematical tradition when it came to integrating the areas swept by the radii of ellipses. I shall discuss the emergence of this new employment of mathematics—which enabled a new, viable ideal of mathematization—at a later point.

2. Unequivocation without Homogeneity: Aristotle

The quest for an unequivocal description of nature lies at the heart of Aristotle's natural philosophy, and was taken over by Scholasticism with the so-called "reception" of Aristotle in the thirteenth century. The main flaw in the first part of Foucault's attempt to construct an archaeology of human knowledge is the confusion between the ideals of unequivocation and homogeneity and their separate historical careers. The seventeenth century did not initiate the demand to exchange "similitudes" for exact comparisons. Aristotle's philosophy of nature—which became, with due changes, the physics and biology of the Middle Ages—was as committed to an unequivocal language of science as any of the seventeenth or eighteenth century biologists quoted by Foucault. Comparison and definition, not similitude, described nature exhaustively and unambiguously. An exhaustive definition of an entity required, in Aristotle's terms, the identification of the closest genus and the specific difference (*definitio fit per genus proximum et differentiam specificam*). In order to make possible the exhaustive and unequivocal classification of entities in the universe through definitions, Aristotle committed him-

---

[9] Plato, *Timaeus* 55c–d; Cornford, *Plato's Cosmology*, pp. 219–21. Whether Kepler remembered this distinction, or knew of later interpretive traditions, I do not know.

self to a portentous metaphysical assumption: namely, that no specific difference can ever appear in more than one genus. In other
words, he knows a priori that there can never be a piece of thinking
metal (say, a computer) or a stone with social inclinations.[10] True,
some concepts, notably being, are intrinsically equivocal; then the
task of the scientist is to transform equivocation either into univocation (by giving more names) or into systematic equivocation
(analogy) by a precise exposition of the many senses of being, senses
that are "simultaneously different and similar."[11] In contrast to this
assumption that nature could be classified according to an unequivocal order of concepts, Aristotle by no means assumed that nature
was homogeneous. On the contrary: the universe is thought of as a
hierarchy of forms, of different qualities which characterize different regions of the universe. Aristotle's *nature* is a ladder of *natures*.

The phenomena of nature are governed by different kinds of
"causes" or principles. They are many and different for each segment of nature, even though their number "should not be increased
without necessity."[12] Science, too, cannot be any more uniform
than its subject matter; the translation of methods from one science
to another leads only to category-mistakes (μετάβασις εἰς ἄλλο
γένος).[13] In the name of this injunction Aristotle repudiated, as we

---

[10] Aristotle, *Topics* z6.144b13ff.; *Metaphysics* z12.1038a5–35. On the development of
Aristotle's theory see Nacht-Eladi, "Aristotle's Doctrine of the *Differentia Specifica* and
Maimon's Laws of Determinability," pp. 222–48. Aristotle permits, however, the *analogy*
between specific differences of different (nonsubaltern) genera; scales are to fishes what
feathers are to birds: *De parte animalium* A.644a16ff. It is closely linked to the permissibil-
ity of "analogy" (proportionality) in mathematics (and may have led Aristotle, in his
later philosophy, to the doctrine of "focal meanings"). Cf. n. 11. and below v.B.2.

[11] Aristotle, *Metaphysics* A7.1017a8–1017b6; E2.1026a33–1026b27; z4.1030a29–
1030b13; Owens, *The Doctrine of Being in Aristotelian Metaphysics*, pp. 49–63; and below
v.B.2. About the possibility of a (later) Aristotelian doctrine of "focal meanings" (of,
e.g., "one," "good," and "being") see Owen, "Logic and Metaphysics in Some Earlier
Works of Aristotle," in *Aristotle and Plato in the Mid-Fourth Century*, ed. Dühring and
Owen, pp. 164–70; Patzig, "Theologie und Ontologie in der 'Metaphysik' des Aristoteles," pp. 185–205. Against it cf. Leszl, *Logic and Metaphysics in Aristotle*, pp. 135ff.,
482ff., 530–39.

[12] Parsimony: Aristotle, *Physics* A4.188a15–18; A6.189a16 (against Anaxagoras). Simplicity: *Metaphysics* K1.1059b34–35. Cf. below p. 142.

[13] Aristotle, *Analytica poster.* A7.75a38–75b6. It is directed against the Platonic view of
mathematics as a universal method. On various aspects of the Aristotelian injunction, as
well as about its later career in the Middle Ages, see the thorough study of Livesey,
"Metabasis: The Interrelationship of Sciences in Antiquity and the Middle Ages," esp.
pp. 1–50. See also Scholz, *Mathesis universalis: Abhandlungen zur Philosophie als strenge*

saw, Plato's belief in an overarching science (dialectics), as well as
Plato's "eidetic numbers" that guarantee the order and connection
of ideas, and also Plato's geometrization of the universe. The injunction against *metabasis* stands in sharp contrast to the seventeenth-century ideal of a uniform science and its practice of trying
the principles of mechanics on every subject for size. Aristotle conceived of logic as an instrument (*organon*) of science only; since he
conceived of nature as a hierarchy of qualities (natures), his logic
was first and foremost a logic of predicates. A well-formulated scientific proposition had to obey the scheme SɛP, and the laws of inference were dependent on the range of predicates and their relation
to the subject (substrate) of a proposition; "matter," "form," and
"privation" were logical as well as ontological patterns.[14] Such were
the various implications of the insistence on unequivocation without homogeneity—or better, with an equal insistence on the heterogeneity of nature.

3. Homogeneity and Forces: The Stoics

By contrast, the Stoic universe, and the universe of the Atomists,
was homogeneous; though the physics of the former sought forces,
and of the latter only efficient causes. Where Aristotle identified
fixed qualities, "forms" definable in a hierarchy, the Stoics looked
for active *forces* to account for the variety within matter,[15] and the
Atomists looked for motion in a reduction of all causes to atoms-in-
motion. To the Stoics, each discrete portion of matter, each identifiable body, was held together by the "tension" (τόνος) of its interdependent parts;[16] and all forces were particular instances of one and

---

Wissenschaft, p. 37 and n. 25 (postulate of homogeneity). On the doctrine of proportionality, which permits analogies across genera, cf. above n. 10 and below V.B.2.

[14] Aristotle, *Metaphysics* Λ2.1069b32–34: τρία δὴ τὰ αἴτια καὶ τρεῖς αἱ ἀρχαί, δύο
μὲν ἡ ἐναντίωσις, ἧς τὸ μὲν λόγος καὶ εἶδος τὸ δὲ στέρησις, τὸ δὲ τρίτον ἡ ὕλη. On
privation see below VI.A.4.

[15] Sambursky, *Das physikalische Weltbild*, pp. 217–19; Lapidge, "Stoic Cosmology," in
The Stoics, ed. Rist, pp. 161–85, esp. pp. 163–65 (ποιόν and ἄποιος); Todd, "Monism and
Immanence: The Foundation of Stoic Physics," in ibid., pp. 140–41; Pohlenz, *Die Stoa,
Geschichte einer geistigen Bewegung*, 1:67–69, 2:38 (Zeno's two ἀρχαὶ as substitutes
for the Aristotelian λόγος and ὕλη), 39 (the Pneuma as substitute for the demiurge; Cic-
ero, *De natura deorum* 1.35).

[16] Sambursky, *Das physikalische Weltbild*, pp. 187–89; Lapidge, "Stoic Cosmology,"
pp. 173–77. The advancement in our understanding of Stoic physics, due in large part to
Sambursky, can be gauged by the fact that Pohlenz did not realize the central role of the
τόνος, and did not mention it at all.

the same force that permeates and unites all—the Pneuma. The Pneuma is both divine and mundane, spiritual and material. It is God, because it governs the world by plan and holds it together by invisible bonds of sympathy. It is also matter, because everything that exists can be said to exist only inasmuch as it is material.[17] A principle of integration must be inseparable from that which it integrates. The Pneuma permeates matter entirely and must therefore itself be matter, active matter, subtle enough to penetrate and enliven all things. Indeed two bodies—the Pneuma qua matter and the body it inspires—can, or rather must, occupy the same place in order to fuse into one continuum:[18] a violation of the Aristotelian (as well as the Atomistic) sacred principle that two bodies cannot occupy the same place. The universe, embedded in the void,[19] became a quasi-animated, organic, purposeful whole: a divine body in which each part reflects and signifies others and the whole is the purpose, the final cause, of the parts. The Stoic universe is thus full of hidden, superimposed meanings, similitudes of things. Never, for example, have magic and astrology been given a better theoretical foundation than in the Stoic doctrine of universal sympathy.

A scientific language can still describe this universe with some precision, but only if it focuses on statements of fact and their connections, not on subjects and their predicates. Stoic logic was mainly a propositional logic, concerned with the coherence of language itself. In fact, just as terms assume their precise meaning from the context of a proposition—the whole proposition is the sign of a unit of meaning (λεκτόν)—so also the propositions become clearer

---

[17] Pohlenz, *Die Stoa*, 1:65–66, 2:37–38; Rist, *Stoic Philosophy*, pp. 153ff.; Todd, "Monism and Immanence," pp. 140–41; D. E. Hahn, *The Origins of Stoic Cosmology*, p. 10. The last discusses the Stoic and Epicurean definition of "body" as that which extends in three dimensions with resistance.

[18] *SVF* 2:467 = Simplicius, *In Aristotelis phys. . . . comm.*, ed. Diels, p. 530: Τὸ δὲ σῶμα διὰ σώματος χωρεῖν οἱ μὲν ἀρχαῖοι ὡς ἐναργὲς ἄτοπον ἐλάμβανον, οἱ δὲ ἀπὸ τῆς Στοᾶς ὕστερον προσήχαντο ὡς ἀκολουθοῦν ταῖς σφῶν αὐτῶν ὑποθέσεσιν, ἃς ἐνόμιζον παντὶ τρόπῳ δεῖν χυροῦν etc. The interpenetration of bodies turned later, in Neoplatonic circles, into an explanation how light—a kind of body and perhaps the first of all bodies—can coincide with all bodies and be, like space, their receptacle: Proclus ap. Simplicius, *In Aristotelis phys. comm.*, p. 612.32. Light, of course, is also homogeneous of sorts. See below II.D.n. 22.

[19] But there can be no void within the cosmos: *SVF* 2:546 (Diogenes Laertius). Cf. Sambursky, *Das physikalische Weltbild*, pp. 337–49; Lapidge, "Stoic Cosmology," pp. 173–78; Rist, *Stoic Philosophy*, pp. 173–78; E. Grant, *Much Ado about Nothing: Theories of Space and Vacuum from the Middle Ages to the Scientific Revolution*, pp. 106–108. Bloos, *Probleme der stoischen Physik*, pp. 45–50.

only in the context of discourse, so that the various λεκτά are actually part of one grand concatenation of propositions—the *logos* itself.[20]

Dilthey, who first recognized the influence of the Stoic philosophy of nature on seventeenth-century thought, once remarked that the Epicureans and the Stoics exchanged the Platonic-Aristotelian concern with the relation of the general to the particular for a concern with the relation of the whole to its parts.[21] The Stoics always took the whole to be more than the sum of its parts. In all branches of their thought, be it logics, ethics, their theory of perception, physics, they were obsessed with the search for *contexts*.[22] In the view of the Atomists, by contrast, the whole was nothing but the sum of its parts.

4. The Elimination of Final Causality: The Atomists

The elimination of final causes, perhaps the least common denominator in the various conceptions of the mechanization of the universe, was most clearly advocated by the ancient Atomists. This is not to say that it was altogether absent from earlier physics: a fair interpretation of Aristotle cannot fail to recognize that his maxim, "nature always operates with a goal," is not to be understood as though each body in nature acts or is acted upon intentionally, or as though a master design governs the conduct of all bodies. In respect to the realm beneath the sphere of the moon—the realm of the four elements of earth, air, fire, and water—Aristotle recognizes only motion and mixture as causes; his sublunar physics is dominated by efficient causes. There is no more animism in his definition of natural motion as that motion which aims in a straight line toward the proper place of the dominant element in a body (i.e., toward the center or away from it) than there is in Newton's contention that bodies aspire to remain in their state of either rest or uniform rectilinear motion. In both cases one type of motion is exempted from the search for causes, except that, for Aristotle, all sublunar motions, natural or coerced, have a point of termination. Only in the

---

[20] Mates, *Stoic Logics*, pp. 15–19; Pinborg, *Logik und Semantik im Mittelalter: Ein Überblick*, p. 32.

[21] Dilthey, *Weltanschauung und Analyse des Menschen seit Renaissance und Reformation*, in *Gesammelte Schriften*, 2:316.

[22] On the Stoic οἰκείωσις see Pohlenz, *Die Stoa*, 1:57f., 84, 113, 190, 253f., 345, 358, 397; Schwarz, *Ethik der Griechen*, pp. 202–208 and n. 17. On the theory of perception Sambursky, *Das physikalische Weltbild*, pp. 206–13.

celestial realm does he recognize an intellective moment that actualizes the natural propensity of the fifth element to rotate eternally:
the spheres are driven by a tendency to imitate the absolute self-sufficiency of the prime mover. Even here, "purpose" is not an entirely
appropriate term, since the motion of the celestial bodies could not
have been other than it is.[23]

But nowhere in Antiquity was the elimination of final causation,
or purpose, as radically pronounced as in the physics and cosmology of the Atomists. Atomism, we remember, was a universal heuristic principle that permits no exceptions. Not only is the world
around us, as seen or otherwise perceived, composed of atoms and
clusters of atoms, no entity except atoms and space can be imagined. The earliest Atomists developed their doctrine in a dialectical
reference to the one and only "Being" of Parmenides. Their atoms
were miniature versions of his undifferentiated, indestructible, unchangeable, singular being; except that, contrary to Parmenides,
the atoms were many and of different shapes in order to account for
differentiation and motion and had, therefore, to be embedded in
space, in a "nonbeing endowed with a kind of being."[24] The soul,
too, cannot be thought of except as atoms of a more rarified, refined
matter, which enables them to move very quickly everywhere in the
body. As for the gods, Democritus at times tends to identify them
with their image in visions or dreams. Epicurus postulates real, existing gods who emit, or radiate, a continuous effluence that im-
presses itself upon our (likewise material) spirit because our spirit is
materially akin to their picture (εἴδωλον, simulacrum).[25] It follows, at
least according to Epicurus, that the gods are exactly what they appear to be in visions and dreams: human-shaped.

Purpose, in part and in the whole, governed the Stoic universe,
which was indeed thoroughly teleological. In opposition to this

---

[23] "Necessary" is that which cannot be otherwise; and though Aristotle does not distinguish between "logical" and "physical" necessities, he does distinguish between absolute (or simple) necessity and hypothetical necessity (below III.B.n.62). The prime
mover, "eternal and immovable," is absolutely necessary; that which is "capable of more
than one state" (ἐνδέχεται πλεοναχῶς ἔχειν), like the rotating spheres, can at best be
hypothetically necessary. If sublunar movements are to be, the movement of the spheres
must be (e.g., De generatione B10.336b38–337b15). That the spheres are moved "by desire" is not a teleological structure (in which the effect precedes the cause) either, but a
simultaneous cause–effect relation.

[24] Above I.C.n.4.

[25] Below II.C.n.7.

stood the Atomistic universe, dominated entirely by chance. But
the Atomists were not the only thinkers in Antiquity to tame teleological considerations of nature nor was their later impact, even in
the seventeenth century, profound. Their cosmology was utterly
unacceptable to any theological framework; and their physics did
not admit the existence of real forces. I shall discuss at a later point
the reason why relatively few of Atomistic tenets could be maintained.

But the other two models of nature and of science, the Aristotelian and the Stoic, determined, in turn, the images of science and of
nature in the Middle Ages and in the Renaissance. The Scholastic
universe was Aristotelian; the cosmos of Renaissance philosophies
of nature was, instead, Stoic, with interspersed Atomistic corrections. The former sought forms, the latter sought forces. The former considered their foremost task to be the purification of all ambiguities from scientific language—all the more so since the
Nominalistic revolution. Renaissance philosophies of nature, on the
other hand, abandoned the obsession with language but advanced
the ideal of the homogeneity of nature in all its parts, a nature constructed of one matter and of one set of forces. Only in the seventeenth century were both ideals fused into one ideal: a science that
has an unequivocal language with which it speaks and uniform objects of which it speaks. The infinite Euclidian space embodied and
symbolized both aspects.

Yet, even while these and other ideals merged into one ideal of
science, they created tensions (not unlike those tensions that Kant
had in mind in the passage quoted above). A part of the scientific
disputes of the seventeenth century—say, the controversy between
Descartes and More over bodies and space, or the controversy between Leibniz and Newton over the nature of space and forces—can
be better explained against the background of the tension between
the ideals of science involved in them, particularly the tension between the ideals of unequivocation and those of homogeneity. A
historical account of the different careers of these ideals since Antiquity may illuminate the ways in which they interacted and counter-
acted in early modern visions of science. It will also show why and
how theology and physics approached each other more closely in
the seventeenth century than they ever did before or after. We shall
trace the further fortunes of both impulses—toward univocation

and toward homogeneity—in terms of the example we started with, the ubiquity and body of God.

C. A SHORT HISTORY OF GOD'S CORPOREALITY AND PRESENCE

1. The Rejection of God's Body

Why did medieval Christian theology abhor all corporeal predicates of God? The answers that come immediately to mind are imprecise and insufficient. It is true that the Church inherited from Judaism the fear of idolatry and of anthropomorphic images. Paganism was not a remote historical reminiscence in the Middle Ages; it lived subterraneously in the countryside. It is also true that, from Jewish and Hellenistic soteriological religions, the Church inherited the contraposition of sinful flesh to the pure spirit. Christianity defined itself as "Israel in the spirit." Yet neither of these impulses necessarily had to be lost if God were to have a body. Patristic theology, much as seventeenth-century philosophy, was aware of a variety of Greek philosophical interpretations of God's body that stripped it of any definite shape ($\mu o \rho \phi \dot{\eta}$) but left a material substance ($\ddot{v} \lambda \eta$), a substance more refined than any ordinary body and therefore befitting spirits. Why were all of them eventually rejected?

Lest we be accused of musing about abstract possibilities, consider, for example, Tertullian's defense of God's body: "Nothing is, unless it is a body. Whatever is, is a body of sorts. Nothing is incorporeal, unless that which is not."[1] The language reminds us of Hobbes or, more to the point, of the Stoics and Atomists. "Who can deny that God is body, even though he is spirit? Spirit is also a body of its own kind."[2] Tertullian evidently opposed the extreme pneumatization of Christology among the Alexandrian exegetes.

---

[1] Tertullian, De anima 7: "Nihil enim, si non corpus. Omne quod est, corpus est sui generis: nihil est incorporale, nisi quod non est." This is a Stoic maxim; see Zeller, Die Philosophie der Griechen 3:1, pp. 119f.; Rist, Stoic Philosophy, pp. 153ff.; above II.B.3.

[2] Tertullian, Adversus Praxean 7: "Quis enim negaverit, deum corpus esse, etsi deus spiritus est? spiritus enim corpus sui generis in sua effigie." On the other hand, Tertullian insisted that matter was created by God, which would mean, presumably, less subtle matter. The soul, too, is material (De anima 6). It caught the attention of various authors in the sixteenth and seventeenth centuries. "Cette fierté de vouloir descouvrir Dieu pour nos yeux a facit qu'un grand personnage des notres a donné à la divinité une forme corporelle": Montaigne, Essais 2.12, p. 589; Bayle, Dictionnaire historique et critique, ed. Beuchot, s.v. Simonides (quoting Daillé, Du vrai usage des pères 2.4); Pierre Bayle Historical and Critical Dictionary, Selections, ed. Popkin, pp. 277–78.

The abrogation of a body to God may lead to the denial of his affections—anger, compassion, justice—without which God would not be able to communicate with man. "I state: God could not enter the walks of man, unless he assumed human senses and affections."[3] In order to accommodate himself to the *mediocritas humana*, he must appear human; God lowers himself to the human level so as to raise man to his level (*adequatio*).[4] But, in order that incarnation be possible, God must possess, it seems, some physical properties to begin with: a most subtle body. Why did Tertullian's defense of the corporeality of the spirit—so close, in fact, to the etymology and image of the word "spirit" both in the Bible (*ruach, neshama*) and in Greek (πνεῦμα)—not strike deeper roots in Christian theology?

Let us review the alternatives—the main Greek models for defending the divine corporeality after the devastating critique of anthropomorphic images by Xenophanes. Both models, the Stoic and the Epicurean, influenced Tertullian and influenced again the seventeenth-century defenders of God's body. In their own way, the Stoics systematized one of the oldest impulses of Greek philosophy since the Ionian φυσιολόγοι—to show that the world is "full of gods,"[5] to deify nature by depersonalizing the gods. The course of early Greek religion may well have been the opposite.[6] Against the deification of nature, the Atomists, especially since Epicurus, were even willing to defend anthropomorphic images such as the human shape and habits of the gods. The later Atomists defended anthropomorphism in order to empty the world of the presence or indeed the governance of the gods, and in order to emancipate man from his fear of them.[7]

[3] Tertullian, *Adversus Marcionem* 2.27.1, p. 505: ". . . proponam: deum non potuisse humanos congressus inire, nisi humanos et sensus et adfectus suscepisset, per quos vim maiestatis suae, intolerabilem utique humanae mediocritati, humilitate temperaret, sibi quidem indigna, homini autem necessaria, et ita iam deo digna, quia nihil tam dignum deo quam salus hominis." Cf. also *De paenitentia* 3.9, p. 325 ("mediocritas humana"); ibid. 6.1, p. 329 ("mediocritas nostra").

[4] Tertullian, *Adv. Marc.* 2.27.7, p. 507; Funkenstein, *Heilsplan und natürliche Entwicklung: Formen der Gegenwartsbestimmung im Geschichtsdenken des Mittelalters*, pp. 25–27 and below IV.D.2. (accommodation).

[5] Thales, *Aetius* 1.7.11; Kirk and Raven, *The Presocratic Philosophers*, pp. 93–97; Cornford, *From Religion to Philosophy: A Study in the Origins of Western Speculation*, pp. 127–29 (Thales), 134–36 (*Physis* as the Divine), 144–59. Stoics: Todd, *Alexander of Aphrodisias on Stoic Physics*, p. 140.

[6] Murray, *Five Stages of Greek Religion*, pp. 8–37, esp. 25ff.; Cornford, *From Religion to Philosophy*, pp. 101–102.

[7] For a general discussion of the emancipatory aims of Epicurean theology see Bailey,

Atomism, we remember, placed itself squarely against a long tradition of Greek thought that sought, since Xenophanes, to de-anthropomorphize the divine. Epicurus's gods were as human-shaped as their images in us. Yet their bodies are composed of atoms so fine that they can never interact with grosser bodies or ordinary worlds. The abode of the gods, physically as well as methodologically, is therefore the *intermundia*—the spaces between the infinitely many more solid universes like ours. There the gods lead a sui-sufficient, happy, and eternal life, a paradigm for us to emulate and adore. Not only do they have a shape and a body, it is a human-shaped body, for no shape is nobler or more beautiful; and they must even feed it, although their kitchen is better than the best of ours.[8] The gods must, then, be many, sui-sufficient, inactive, and withdrawn totally from our world if the universe is to be conceived as devoid of any τέλος or plan, so that human freedom and dignity are preserved.

It is evident why Epicurus's theology could never be modified to suit Christian needs. "Epicurean" is a synonym for those who deny God's providence. That they do so by ascribing a human shape to God adds insult to injury. But why was the theology of the Stoa rejected by Christian theologians? It had much to recommend itself to Christianity: the Stoic universe is much more teleological, guided by providence, than Aristotle's. In fact, through Philo's *logos* and other channels, the Stoa did exert considerable influence on Christian thought. Henry More undoubtedly revived many elements of Stoic theology. But why were the Stoics, unlike Plato, seldom called "ours" by Christians, never seen—as Plato was—as a "preparation of Christ"? The first answer is because their gods and God expressed a naked, immediate deification of nature. The deification of nature was seen as the real essence of paganism by both Christians and Jews. They viewed the difference between crude anthropomorphism and the identification of the gods with natural forces— or even one force—as a difference of degree only. Indeed, Aristotle

---

The Greek Atomists and Epicurus, p. 83; Strauss, *Spinoza's Criticism*, pp. 2–46; W. F. Otto, "Epikur," in *Die Wirklichkeit der Götter*, pp. 10–43.

[8] That Epicurus included in his attack against the traditional gods also the astral gods and the world soul of the philosophers (Plato, Aristotle, the Stoics) has been shown by Festugière, *Epicurus and His Gods*, trans. Chilton, esp. pp. 73ff. In this context we ought also to see the Atomist argument against the sphere as the most perfect figure (above II.B.n.1).

admits that much,[9] and so did the philosophical allegoresis of myth in which the Stoics excelled. But there is a deeper reason yet. There is one instance in which Christian religion assumed something very close to a divine body, the body of Christ. It is a heavenly body, yet a body not unlike the body of Epicurus's gods. The divine body of Christ already caused pagan polemicists to accuse Christianity of anthropomorphism.[10] The meaning attached to his body added another danger. The Middle Ages came, after painful controversy, to the decision that Christ's body is really present in every piece of the Eucharist. By eating Christ's body the believer becomes part of Christ's mystical body, the Church.[11] The doctrines of real presence and transubstantiation sometimes used the Averroistic distinction between determinate and indeterminate extension: the latter is the subject in which the accidents of bread and wine inhere after transubstantiation. This very distinction was resumed by More and Newton to explain the distinction between dimensions, which God lacks, and extension, which he must possess.[12]

Admitting the physical, equal, homogeneous presence of God everywhere—with or without a material substrate—could amount to a relativization of Christology (God would be in each of us equally) and make the sacraments and the hierarchical Church superfluous. Again, I do not speculate upon merely theoretical possibilities. The nascent University of Paris was intoxicated with a sense of intellectual freedom just gained. It seemed to the new European intelligentsia as if all ideas might be tried on for size—at least if investigated without dogmatic claims, *disputandi more, non asse-*

---

[9] Aristotle, *Met.* Λ8.1074a39–b15; Cornford, *From Religion to Philosophy*, p. 135. On worship of the cosmos (*elementa*) as the essence of paganism in Christian and Jewish Polemics Cumont, *Die Orientalischen Religionen*, pp. 186–88.

[10] Arnobius, *Adversus Nationes* 3.12–13, ed. Marchesi: "Neque quisquam Iudaicas in hoc loco nobis opponat et Sadducei generis fabulas, tamquam formas tribuamus et nos deo. . . . At vero vos deos parum est formarum quod amplectimini mensione, filo et adterminatis humano, et quod indignius multo est, terrenorum corporum circumcaesura finitis."

[11] Pelikan, *The Christian Tradition: A History of the Development of Doctrine*, III: *The Growth of Medieval Theology (600–1300)*, pp. 184–204; R. Seeberg, *Lehrbuch der Dogmengeschichte*, 3:208–18 (whose Protestant bias becomes manifest at p. 210: "Nichts war für die Geschichte der Abendmahlslehre so verhängnisvoll als die römische Formel vom Jahre 1059").

[12] A. Maier, *Studien zur Naturphilosophie der Spätscholastik*, 1: *Die Vorläufer Galileis im 14. Jahrhundert*, pp. 26–52 ("Das Problem der quantitas materiae"). Further literature below II.D.n.11.

rendi more. Amalric of Bena, following, perhaps, the pantheistic cues of Johannes Scotus Erigena, maintained that God is the *forma mundi,* the true essence of all things. This could still be given an innocuous, even Aristotelian interpretation, but look at the consequences. Amalric and the Amalricans denied the merit of any positive religion. God is in all of us, and those who know it are better off. Only true philosophy saves, and it saves Jews, Christians, and Moslems alike. "What is more absurd," says their opponent, "than that God is stone in a stone, Godinus in Godinus, that Godinus should be worshipped—not only adored—because he is God. Until now we believed in the incarnated Son of God: now they preach the ingodinated Christ."[13] (Godinus was a popular Amalrican preacher.) Amalric's contemporary, David of Dinant, even maintained that God is the *materia mundi*—the very stuff underlying all substances, spiritual as well as material.[14] The details of these doctrines are hard to discern, and their exact driving forces remain obscure because of their persecution and intended obliteration. But both Amalric and David evidently were among those who were led by their pantheism to believe "quod sermones theologi fundati sunt in fabulis."[15]

How close pantheism, antihierarchical leanings, and pagan leanings really were even in the countryside we learn from a few rare outbursts. A case in point is the vulgar pantheism of a half-literate village miller in the sixteenth century, recently discovered in Venetian inquisitorial acts.[16] The man went around preaching, with a missionary zeal, that the Church, its hierarchy, and doctrines are a self-serving fraud. We all are God in the same measure; the universe in one huge wheel of cheese, spirits and angels the worms in it. Salvation can come only through this knowledge. Against such latent

[13] *Contra Amaurianos* 24.5–6 (most probably by Garnerius of Rochefort), ed. Baeumker, BGPhM (Münster, 1926), p. 24.

[14] Überweg and Geyer, *Grundriss der Geschichte der Philosophie,* 3:251; M. Kurdzialek, "David von Dinant und die Anfänge der aristotelischen Naturphilosophie," in *La Filosofia della natura nel medioevo,* pp. 407–16.

[15] Denifle and Chatelain, *Chartularium Universitatis Parisiensis,* 1:552 (a.152).

[16] Ginzburg, *The Cheese and the Worms: The Cosmos of a Sixteenth-Century Miller,* pp. 4–5, 52–71, 102–108. The history of popular pantheism is still, to a wide extent, uncharted land.

paganism throughout medieval Europe, it was prudent to abstain from any creative interpretation of God's shape or body.

2. The Roots of Omnipresence

True, the deification of nature was seen as the quintessence of paganism, and Christological concerns were a powerful barrier against pantheistic or even panentheistic trends. Yet there is one sense in which Christianity retained, after all, reminiscences of God's body. God's presence in all things was a much more fundamental part of Christian theology than ever in classical Judaism—with the (later) exception of the Kabbala. I do not overlook the image of God's glory that fills the world in biblical passages, or the various references to God as "place" (makom) or even "the place of the world" (mekomo shel olam), since the early rabbinical literature. Nor do I claim that they necessarily manifest Hellenistic influences.[17] But it seems to me that their meaning has been too readily adjusted, in retrospect, to Hellenistic doctrines of immanence or Christian notions of ubiquity.[18] The various biblical passages attest that there is no place to hide from God (e.g., Ps. 139:7–10). Makom may have been used in instances of meeting or intimacy between man and God; but the original connotation of the word is neither space (κενόν) nor place but "abode,"[19] much as ma'on (cf. Ps. 90:1). The often misinterpreted question "Is God the abode (mekomo, me'ono) of the world, or the world the abode of God?" was phrased differently: "Is God an annex (tafel) to the world, or the world an annex to God?"[20]

The influence of the Hellenistic world was decisive in but one sense: it raised the dialectics of the divine immanence and transcend-

[17] A thorough discussion in Urbach, Ḥazal: 'Emunot vede'ot [The Sages: Doctrines and Beliefs], pp. 29–52 (shechina), 53–68 (makom); see also Landau, Die dem Raume entnommenen Synonyma für Gott in der hebräischen Literatur, pp. 90–91; Marmorstein, The Old Rabbinic Doctrine of God, p. 143.

[18] Excessively so by Abelson, The Immanence of God in Rabbinical Literature, pp. 90–92; and Jammer, Concepts of Space, pp. 27ff. Jammer mixes indiscriminately shechina and makom, rabbinical, post-rabbinical, and Kabbalistic connotations.

[19] Urbach, Ḥazal, pp. 59–61 (against Baer and others). Urbach is careful not to impose any philosophical connotation, but speaks occasionally of makom as immanence. Note that the words in Ez. 3:12, "blessed the glory of God from his place [mimekomo]," entered the daily prayer. Cf. Pirke de Rabbi Eliezer 4.

[20] Bereshit Rabba 68.8, ed. Theodor and Albeck, p. 777.

ence to a conceptual level. Hellenistic-Jewish philosophy formulated the tension between inherited, competing images of God as a
problem. The earliest systematic attempt of a philosophical mediation was Philo's *Logos*-metaphysics—the archetypal pattern to all
doctrines of mediation. Thought (or word or number) is both identical with the thinking subject and different from it; while God utterly transcends the world, his *logos* created it and permeates it as
its very essence.[21] As that which is beyond every comprehension,
God is nowhere; as *logos*, he is everywhere. He *is* the "space" that
holds himself and everything else, "containing and not contained."[22] In the contemporary rabbinical literature of Palestine the
immanent aspects of God came to be identified with God's glory
(*kavod*) or condescension (*shechina*). Of the latter it was sometimes
said that "no place on earth is empty of the *shechina*,"[23] but sometimes it was also said to the contrary that "never did the *shechina*
descend downwards, just as Moses and Elias did not ascend to the
emporaeum, for it is written: 'The heavens are heavens to God, and
the earth he gave to the children of man.' "[24] Medieval Jewish religious philosophy seldom attended to God's ubiquity,[25] and never
listed it among God's primary attributes (being, will, wisdom). A
strong construction of the divine essence in things was, of course,
always part of the Neoplatonic traditions. Only in the pre-Kabba-
listic and Kabbalistic esoteric literature (*sod*) did ubiquity play a cen-

---

[21] Wolfson, *Philo: Foundations of Religious Philosophy in Judaism and Christianity*, 1:230–
38. This identity-with-difference of thought and the thinking subject was the reason why
Plotinus refused to attribute "thought" to "the one" (τὸ ἕν), and relegated it to the νοῦς
(below III.B.1).

[22] Philo of Alexandria, *De confusione linguarum* 136; *De somniis* 1.63–64, ed. Whitaker
and Marcus, pp. 328–29; Wolfson, *Philo*, 1:247; Urbach, *Hazal* pp. 60–61; Grant, *Much
Ado*, pp. 112–13. It should be remembered that Aristotle was accused—it is hard to know
how early—of having made God the place of the universe: Sextus Empiricus, *Adversus
mathematicos* 10, ed. Mittschmann, 2:33.

[23] *Ein makom ba'arets panuy min hashechina: Numeri Rabba* 13; cf. below n. 26.

[24] "Tana R. Jose Omer: me'olam lo yarda shechina le mata, velo 'alu moshe ve'eliyahu
lamarom, shene'emar: 'hashamayim shamayim la'adonay veha'arets natan libne'adam' "
(Ps. 115–17). *Babylonian Talmud*, Tr. *Sukka* 5a. Cf. Urbach, *Hazal* p. 38 and above n. 19.

[25] Sa'adia Gaon is an important exception. As, e.g., for Maimonides, he interprets
God's "dwelling" (*shechina*) to mean the constancy of his effects, i.e., providence. God's
"place" means the "degree and intensity of participation in existence": *Guide to the Perplexed* 1.8; 25.

tral role—the notion that "no place is empty of" God (*Let 'atar panuy mine*).[26]

Not so the Christian tradition, for which God's omnipresence became a central theologumenon. Its precise history has yet to be written. Whether or not my guess that its origins were more Hellenistic than Judaic is correct, the question of how God exists "in things" seems to me to encapsulate, more than any other theological issue, the dialectics of divine immanence and utter transcendence. The attribute of omnipresence had to be differentiated from Christ's presence and the presence of the Holy Ghost—but not too much. It had to be guarded from pantheistic interpretations, but also from elimination by excessive emphasis on God's being nowhere. It had to be safeguarded against both too literal and too allegorical readings.

The most natural way to perceive God's presence in the world was symbolical. Patristic and medieval theology were inevitably led toward an interpretation of the universe as a sign, symbol, picture of God. A true symbol, to use a phrase of Durkheim, manifests a *participation mystique* with that of which it is a symbol. It is both one with, and different from, its symbol; it is much more than an image or linguistic metaphor. Nature reveals God's symbolic presence, and was seen as a system of symbols, of signatures of God; so also was man's soul; and so was history. Events, persons, and institutions of the Old Testament prefigured those of the New Testament; Adam was the prefiguration of Christ, the second Adam; the six days of creation prefigured the *aetates mundi*; the Trinity was symbolized by, and its persons acted differently in, the periods *ante legem, sub lege, sub gratia*. Deciphering the *prophetia in rebus* over and beyond the *sensus historicus vel spiritualis* was the duty of the *spiritualis intelligentia*.[27] Nature and history were a mirror of the divine[28]—

---

[26] *Zohar, Tikkune hazohar* 57, ed. Margaliot.

[27] Below, n. 51;IV.D.1. and nn. 18–21.

[28] In the famous verse of Alanus ab Insulis: "Omnis mundi creatura / Quasi liber et scriptura / Nobis est, et Speculum" (Migne, *PL* 210. 579a). On the *topos* "book of nature" cf. above I.C.n.3. On the symbolic propensity see M.-D. Chenu, *Nature, Man, and Society in the Twelfth Century*, pp. 99–145; Dronke, *Fabula: Explorations into the Uses of Myth in Medieval Platonism*, pp. 32–47 (Image, Analogy, Enigma). To the classical expressions of enigma one could add Nock, ed., *Sallustius Concerning the Gods and the Universe*, p. 4.9–11: ἔξεστι γὰρ καὶ τὸν κόσμον μῦθον εἰπεῖν, σωμάτων μὲν καὶ

not only through God's acts, in and through them, but through his participatory-symbolic presence in them.

3. Analogy of Being and the Understanding of Ubiquity until Thomas Aquinas

Sound theology employed the symbolizing propensity of medieval Christianity, but it also had to put some strictures on it. If nature is to signify God's presence, then not in equal measure or degree; he could not be made simply into the essence of all things. Medieval Scholastic theology sought at first a balanced, restricted expression for the symbolic sense of nature and history. Thomas Aquinas offered such a balanced theory with his doctrine of the analogy of being. He brought it to bear on one special question at hand, the understanding of God's ubiquity, of God's "existence in things." Three problems were linked to the theological meaning of "inexistence" and "place": the presence of God at all places, the presence of angels in some places, and the presence of Christ's body in the Host. Thomas Aquinas believed that God's ubiquity was an analogical mode of speech. The presence of angels at a place is an equivocal mode of speech. The presence of Christ's body in the Host must be taken literally; it is there with extension (*quantitas dimensiva*).

From the *Glossa ordinaria* through Peter the Lombard, whose book of sentences secured the attribute of ubiquity a choice status among Scholastic questions for generations of incipient theologians, Scholastic theology inherited a formula that determined which aspects are involved in God's *existentia in rebus*. "It is said that God exists in all things through essence, power, and presence."[29] God, according to Thomas Aquinas, is in all things by his power inasmuch as all things are subject to his power; by presence, inasmuch as God knows everything without mediation; by essence, inasmuch as he is the *causa essendi* of all things, inasmuch as he gives them being.[30] *Esse dare* was a key term in Thomas's thought from the very

---

χρημάτων ἐν αὐτῷ φαινομένων, ψυχῶν δὲ καὶ νῶν κρυπτομένων. The world itself is a myth because it inhabits hidden entities.

[29] Thomas Aquinas, *Summa theol.* 1 q.8 a.3; Peter Lombard, *Sententiae* 1 d.37 c.1, ed. Quaracchi, pp. 229–30. For a long time it was mistakenly thought to be a quotation from the Job commentary of Gregory the Great.

[30] Thomas Aquinas, *Summa theol.* 1 q.8 a.3, resp. He also offers an interesting historical remark: *per potentiam* is an antidote against Manicheans, who believed the world to be *potestati principii contrarii*. *Per presentiam* is an antidote against all those who deny God's

beginning of his literary career. On the one hand, Thomas radically separated essence and existence (*esse*); on the other hand, he envisaged a hierarchy of beings according to how much their essence implies, or demands, being. Form is a principle of actuality; inasmuch as a thing has form, it has being.[31] This is least true of corporeal substances in our immediate (sublunar) vicinity, whose matter is generable and corruptible; since their form needs for its instantiation, or individuation,[32] perishable matter, the individuals must perish while the species is maintained through repeated generation or mechanical constellations. It is truer of celestial bodies made of matter which, of itself, is neither generable nor corruptible. It is truer yet of man, whose soul is immortal. It is truest of incorporeal substances, whose essence implies being as a circle implies roundness.[33] But to imply being is not the same as to be: God still has to *give* existence to them, their substrate is an *esse participatum*. They are distinguished according to the measure to which their actuality exceeds their potentiality.[34] Leibniz stated of his substances that they have an *exigentia existentiae*. This is certainly true of Thomas's separate forms, of which each one is a species in and of itself. The more similar or closer to God they are, the more their essence demands existence; God alone is "pure act," his essence is his being. "And in this manner all things that are from God become similar to him, the first and universal principle of all being, *inasmuch* as they are beings";[35] wherefore all things are "in God."

This hierarchy of similitudes is the *ontological* sense of the analogy of being.[36] As a doctrine of knowledge, it takes its cue from Aris-

---

providence (Epicurus?). *Per essentiam* against those who deny that God created everything without the mediation of secondary causes (Averroists?).

[31] Thomas Aquinas, *Summa contra gent.* 2.55: "esse autem per se consequitur formam." Cf. Überweg and Geyer, *Grundriss* 3:435.

[32] On the principle of individuation see below III.B.3.

[33] Thomas Aquinas, *Summa contra gent.* 2.55: "Substantiae vero quae sunt ipsae formae, numquam possunt privari esse, sicut si aliqua substantia esset circulus, numquam posset fieri non rotundum." Cf. below II.H. nn. 5,7 (later medieval interpretations).

[34] Thomas Aquinas, *De esse et essentia* c.4, in *Le "De ente et essentia" de S. Thomas d'Aquin*, ed. Roland-Gosselin, p. 36.10–15: "Est ergo distinctio earum ad invicem secundum gradum potentie et actus. . . . Et hoc competur in anima humana etc."

[35] Thomas Aquinas, *Summa theol.* 1 q.4 a.3: "Et hoc modo illa quae sunt a Deo, assimilantur ei inquantum sunt entia, ut primo et universali principio totius esse."

[36] Ibid.: "Non dicitur esse similitudo . . . propter communicationem in forma . . . sed secundum analogiam tantum, prout scilicet Deus est ens per essentiam, et alia per participationem."

totle's repeated insistence that "being," even though it can be understood "in many ways," is not always "equivocal by accident," as
when two meanings of "to be" share a common referent. Boethius
added analogy as a permissible mode of equivocation.[37] Thomas
urges us not to confuse the indeterminate and equivocal sense of
being (*esse tantum*) with its precise meaning that is specific to each
and every *ens*: such a confusion led Amalric of Bena to understand
the doctrine that God is pure being in a pantheistic vein.[38] The only
chance to grasp the precise sense of being, let alone God's *esse per
essentiam*, in its identity and difference is by analogy to the modes in
which essence demands being, that is, actual existence. (That a faint
pantheistic flavor remains nonetheless can be seen in the formulations of Meister Eckhart.)[39] And since comprehending being thus
amounts to comprehending God, analogy of being becomes the ve-
hicle by which to construct a legitimate discourse about divine
predicates.

As a *theory of signification* of divine names, the doctrine of analogy
is grafted onto the Aristotelian distinction between meaning and similitude, *significatio* and *representatio*. Words name (or refer to) objects, but acquire their meaning through the mediation of concepts.[40] A concept represents the object of which it is a concept by

[37] Aristotle, *Metaphysics* z.1030a33–1030b2; cf. Owens, "Analogy as a Thomistic Approach to Being," pp. 303–22. For the understanding of the genesis and the different
modes of Thomas's analogy of being, the book of Lyttkens, *The Analogy between God and
the World: An Investigation of Its Background and Interpretation of Its Use by Thomas of Aquino,*
is still indispensable. It seems that Thomas, between the *Sentences* commentary and the
Summa, shifted emphasis from the analogy of proportionality to the analogy of attribution; the former, though more elegant, is also more problematical.

[38] Thomas Aquinas, *De esse et essentia* c.5, pp. 37–38: "Nec oportet, si dicimus quod
Deus est esse tantum, ut in illorum errorem incidamus, qui Deum dixerunt esse illud esse
universale, quo quaelibet res est formaliter." In the *Summa theol.* 1 q.3 a.8 Thomas distinguished what we may call three forms of pantheism: the error of those who equate God
with the *anima mundi*, those who say "Deum esse principium formale omnium rerum. Et
haec dicitur fuisse opinio Almarianorum. Sed tertius error fuit David de Dinando, qui
stultissime posuit Deum esse materiam primam." Thomas, of course, does not speak of
pantheism, but of "introducing God into the composition of things."

[39] "Unless safeguarded by limiting dogmas, the theory of Immanence, taken alone, is
notoriously apt to degenerate in pantheism" says Underhill, *Mysticism: A Study in the Nature and Development of Man's Spiritual Consciousness,* p. 99; Thomas's doctrine of divine
presence: ibid., n. 3 (relying on *Summa contra gentiles* 3.68); Eckhart: ibid., p. 101; her
contention, however, that Christianity is—historically and in principle—closer to mystical experiences and better capable of expressing them confounds, I fear, what she
wished to separate—mysticism and "mystical theology."

[40] Aristotle, *De interpretatione* 1.16a3–9; Thomas Aquinas, *Summa theol.* 1 q.13 a.1.

virtue of its resemblance (*similitudo*) to it. To the degree in which the essence of a thing is known to us, our concept of it represents that thing, is a true picture of it. But things themselves resemble God to the degree in which they have perfections, because each perfection in them, however incomplete, reflects—and therefore represents— its paradigm in God, the source of all perfections.[41] The various perfections (essences) constitute a coherent mutual order that again reflects and represents—however vaguely—the unity and simplicity of God. Each thing, inasmuch as it is, can be thus said to be in God's image; and all things as a unity within a multiplicity represent God in an additional sense. The whole world is an *imago Dei*. Our concepts of God are, so to say, pictures of pictures, representations of representations.[42] Our self-knowledge is, therefore, the image of God closest to us, because it is the least mediated: in this fourth, *psychological* foundation of the doctrine of analogy, Thomas followed Augustine.[43]

All divine attributes must be understood analogically.[44] But because the most eminent sense in which things are said to resemble

[41] Thomas Aquinas, *Summa theol.* 1 q.13 a.2: "Significant enim sic nomina Deum, secundum quod intellectus noster cognoscit ipsum. Intellectus autem noster, cum cognoscat Deum ex creaturis, sic cognoscit eum, secundum quod creaturae eum repraesentant . . . cum igitur dicitur 'Deus est bonus' non est sensus 'Deus est causa bonitatis' vel 'Deus non est malus' [Maimonides' attributes of action and negative attributes; see below n. 44], sed id quod bonitatem dicimus in creaturis, praexistit in Deo."

[42] Ibid. 1 q.13 a.4: "Sicut igitur diversis perfectionibus creaturarum respondet unum simplex principium . . . ita variis et multiplicibus conceptibus intellectus nostri respondet unum omnino simplex, etc." The Wittgenstein picture metaphor is, I believe, helpful, even if Thomas never precisely formulated or intended it. The "similarity" between concepts and things is a structural, not a material, one: the various concepts of things correspond to their aspects and relations. A picture of a picture is a transitive, but not necessarily a symmetrical, relation; it need not include all that constitutes the original picture.

[43] On the reflexive meanings of analogy see Lonergan, *Verbum: Word and Idea in Aquinas*, pp. 183–220.

[44] Alexander of Hales demanded of divine attributes that they be understood "non aequivoce, sed analogice": *Summa theologiae* p.1 inq.1 t.4 q.1 c.11, ed. Quaracchi, 1:203 and *passim*; cf. Lyttkens, *Analogy*, pp. 123–31. Thomas's doctrine is far more comprehensive, yet both reacted *expressis verbis* against Maimonides' so-called doctrine of negative attributes. Maimonides permitted, as proper modes of locution about God, only "attributes of action" and "negations of privations" (*The Guide of the Perplexed* 1.52–58, trans. Pines, pp. 114–37). Thomas Aquinas, *Summa theol.* 1 q.13 a.2, criticizes the theory of negative attributes because it does not permit any significant attribution at all; saying of God that he is "not weak" is no more meaningful than saying of him that he is "not colorless." What, however, saves Maimonides' theory from an indiscriminate, infinite enumeration of negations is his seldom understood theory of generative construction of divine attributes (1.58) in analogy to the defining construction of the concept of a being.

God is in their mode of being, the attribute of ubiquity—of the divine existence in things—actually represents, more than others, the ways of analogy. Thomas began his explication of ubiquity with a simile that was to prove misleading in his time and in ours. He said that, in the same way in which a mover or an agent must be contiguous to that which is moved or acted upon, God should be at or in his creation. The most immediate sense in which God acts on things is that he gives them being. God is in things—including demons—to the measure in which they participate in being.[45] The immediacy or closeness of God to things is measured by their mode of being; God is closest to himself as *ipsum esse per suam essentiam*. The physical analogy ought not to have been taken by Duns Scotus or, more so by Ockham, literally. True, Thomas did say that no agent, however powerful, can act at a distance. He does not say, as Ockham inferred, that God cannot do so.[46] He merely says that, by analogy, while physical agents need a medium if distant, God does not because he is never distant from that upon which he acts. And Thomas hastens to add that by "distance" or "proximity" he means the similitude of things to God inasmuch as they participate in being.

The term "ubiquity" captures the various analogical meanings of in-existence on many levels. Places (in Aristotle's sense) have accidental and essential properties. It is true essentially of a place, by definition, that it be filled with a body. It is true only of the earth that heavy bodies fall toward it; the *virtus locativa* of a place qua place is an accidental property. That God gives places (which are also bodies of sorts) being and *virtus locativa* corresponds to the meanings of ubiquity "by power, by presence, by essence." But that a place be always filled with a body represents and corresponds to the sense in which it is God's essential property, and his only, to be at any

---

[45] Thomas Aquinas, *Summa theol.* 1 q.8 a.1, resp.: "Quamdiu igitur res habet esse, tamdiu oportet ut Deus adsit ei, *secundum modum quo esse habet.*" (Demons: ibid., ad 4.)

[46] William of Ockham, *Scriptum in librum primum sententiarum ordinatio* (henceforth *Ordinatio*) 1 d.37 q.1, in *OT*, ed. Etzkorn and Kelley, 4:563: "Istud ultimum *probant* ipsi . . . secundum Philosophum," i.e., Thomas did not present a proof, but an analogical construct. It is true, however, that in his *Sentences* commentary, Thomas was more positive, and that his reference to Aristotle there could—but need not—be construed as a proof: "Ad cuius evidentiam oportet tria praenotare. Primo, quod movens et motum, agens et patiens, et operans et operatum, oportet simul esse, ut in 7 *Physica* [text 20] probatur. Sed hoc diverso modo contingit in corporalibus et spiritualibus." Thomas Aquinas, *Commentum in quatuor libros sententiarum* 1 d.37 q.1 a.1 (ed. Parma 6:298). The analogical character is emphasized less.

place—yet not dimensionally as a body.[47] If the meaning of ubiquity was dimensional, even a grain of sand would fill the whole universe provided it were the only thing existing in it. God is in all places substantially, everywhere as a whole and not in parts, because it is his essential property to be in and by himself even when there was no place. And Thomas adds, in a speculative vein: Were we to assume infinitely more places than there are, God must be in all of them because he is essentially everywhere.[48] Angels, in contrast, are said to be somewhere only equivocally, *per contactum virtutis*; that they cannot occupy two places at once has the comforting meaning that even an angel cannot do more than a thing at a time. Christ's body is in the Host extensionally (with a *quantitas dimensiva*), that is, univocally.[49]

If, as we claimed, Aristotle's science wanted to be as univocal as language permits, how could the Thomistic view of nature as a symbol and analogy of God profit from it? Because of the balance it maintains between the analogical and unequivocal meaning of concepts; because it emphasizes that the latter is to us the primary meaning which should carefully be detached from the former. Thomas's doctrine of analogy did as much to restrict the medieval sense of God's symbolical presence as it did to promote it. The same considerations apply to God's (or Christ's) symbolic vestiges in history which were found in the rich systems of prefigurations and their fulfillment, systems that proliferated more than ever before in the twelfth and thirteenth centuries.[50] The distinction between *significatio* and *representatio* is also at the heart of Thomas's exegetical theory. In contrast to a long preceding tradition, Thomas articulated the new, revolutionary exegetical understanding initiated a century earlier by the Victorines. He emphasized that the language

[47] Thomas does not use the distinction, drawn from Peter Lombard and employed by Alexander of Hales, between *esse in loco per definitionem* and *per circumscriptionem*. Alexander of Hales, *Summa theol.* p. 1 inq. 1 t.2 q.3 tit.2 c.1, 1:64; Peter Lombard, *Sent.* L. 1 d.37 c.6, p. 236.

[48] Thomas Aquinas, *Summa theol.* 1 q.8 a.2, resp.; a.4, resp; cf. 1 q.46 a.2 ad 8 (*locum imaginatum tantum*). We obtain a double analogy: (i) accidents of places:places::God: places; (ii) bodies:places::God:places (qua bodies). The imagination of "more places" confirms, of sorts, Grant's conjecture (*Much Ado*, p. 146), though Thomas speaks of places, not of a void.

[49] Thomas Aquinas, *Summa theol.* 1 q.52 a.1, resp. (angels); 3 q.77 a.1; cf. Martin, *Ockham*, pp. 72–75.

[50] Below IV.D1,2.

of the Scriptures must be understood on a plain, literal level only. The *sensus litteralis* includes, however, similes and parables if they were the "intention of the author"; previously those were allocated to the spiritual sense. But the words of the Scriptures have no deeper sense. The things, events, persons, or prophecies of which the Scriptures talk—and they only—have a deeper, symbolical or mystical meaning, a *sensus spiritualis*.[51] Neither our discourse about nature nor about history need be equivocal. The hidden significances of things can and must be separated from the immediate meanings and references of words. Theology and physics can be separated.

Like Aristotle's, the material universe of Thomas can be described with precision. Again like Aristotle's, his was in no way a homogeneous universe. It consisted of a hierarchy of forms perfectly attuned to each other. God can be said to "express" himself in them, and in the harmony of the whole. The unity-within-the-multiplicity is the reason why God did not create the monochromatic universe that he could have created. Origenes, Thomas says, assumed that "in the beginning all things were created equal by God."[52] God created homogeneous spiritual beings. Some turned toward God, others away from him; the latter were driven into material bodies—each according to its sins. To the Angelic Doctor this was not only metaphysical nonsense,[53] it also offended his perception of God's goodness. Had it been so, he argues, then all material things would be existing as punishment for sinners, not as so many expressions of God's goodness. Yet "God saw everything that he did, and it was very good" (Gen. 1:31). A universe with only one degree of perfection would not be truly perfect. The equality in the

---

[51] Thomas Aquinas, *Summa theol.* 1 q.1 a.10: ". . . auctor sacrae Scripturae est Deus, in cuius potestate ut non solum voces ad significandum accommodet (quod etiam homo facere potest), sed etiam res ipsas. . . . Illa ergo prima significatio, qua voces significant res, pertinet ad primum sensum, qui est sensus historicus vel litteralis. Illa vera significatio qua res significatae per voces iterum res alias significant, dicitur sensus spiritualis; qui super litteralem fundatur, et enim supponit." Lubac, *Exégèse médiévale: Les quatre sens de l'écriture,* 2.2, pp. 272ff., 285ff.; Smalley, *The Study of the Bible in the Middle Ages,* pp. 303ff. (Her example for the exegetical revolution is less exegetical than historical—the reception of Maimonides' theory of "reasons for the precepts." About it see below IV.B).

[52] Thomas Aquinas, *Summa theol.* 1 q.47 a.2; Origines, *De principiis* 1.6–8, ed. Kötschau, pp. 78–105.

[53] Because spiritual beings, individuated by themselves (i.e., by their form), can be only one of a kind: see below III.B.3.

universe is also its inequality: the *equalitas proportionis*, the relation of things to each other and to God which makes them similar to him; similar especially if they can reflect on themselves and on him, and to the measure in which they can do so.

D. LATE MEDIEVAL NOMINALISM AND
RENAISSANCE PHILOSOPHY

1. *Sacrificing Physical Connotations*
for Univocation: Scotus

If Thomas conceded the need for some equivocation in our theological discourse, Duns Scotus and, even more so, the so-called Nominalists of the fourteenth century objected to any equivocation, whether restrained or not. About that which one cannot speak without equivocation they preferred to remain silent. They aimed at an absolute transparency of the language of every science. Our terms, they insisted, must be either denotative or connotative. The former name singular, discrete entities (subjects) or their absolute properties directly, the latter indirectly (*in obliquo*). Only the former can be said to refer to existing entities that do not need the mediation of universals either for their existence or for our immediate cognition of them (*notitia intuitiva*). Connotative terms, such as relationships, or terms of quantity, extension, and motion are valid only insofar as they are coextensive with a set of singular entities; no terms with different meanings could be coextensive. If one wishes to speak of research programs, theirs was a program of semantic reduction—simplification—of our world view in the name of both logic and theology. In a world of independent entities, each of which could exist, had God so willed, entirely by itself (*toto mundo destructo*), connotative terms refer to contingent constellations of things.[1] The Terminists could not but object to any attempt to see God symbolized in nature because the order of nature was, in their eyes, so utterly contingent upon God's will. Not only was the physical order of things in relation to each other (*ordo ad invicem*) changeable at any time through God's absolute power (*de potentia Dei absoluta*), even the order of salvation was in no way necessary.

---

[1] Logic, univocation: above II.A.2–3; below II.H.7; V.C.3–4. See also Langston, "Scotus and Ockham on the Univocal Concept of Being," pp. 105–29. Principle of annihilation: below III.B.3, C.3, D.3; V.C.3.

Had God wanted, he could place the whole universe in a (logically possible) infinite space and let it move there indefinitely in a straight line. Or, he could have created many universes like ours;[2] he could even, rather than sending his son to save us, have assumed the nature of a stone or a donkey—*aut lapis, aut asinus.*[3] Of the many logically possible universes, ours is neither the best nor otherwise the product of a particular, discernible aim representative, thereby, of God's image. The Nominalists had to reject the doctrine of analogy because they had already desymbolized the universe (as well as history) almost completely.

In view of such an obsession with the precise usage of terms, how did the predication of ubiquity fare? Those who, since Duns Scotus, rejected the analogical status of being could not admit an analogical construction of "being in" either. For Scotus, being and essence are not closer to each other in separate intelligences than they are in material substances. Existing as well as possible essences have *esse*, and even man can create or give "being," though not *ex nihilo*. Essences cannot be measured according to the degree they *imply* being; no possible or actual substance does. All substances are individualized already as possibles and without matter—by formal criteria, as Scotus thought, or simply because they are, as Ockham insisted. There can indeed be more than one angel of a kind. God knows all possible beings, and actualizes but a few of them. Potentiality is not "mixed in" with actuality in degrees; actuality is always added.[4] Scotus and Aureoli therefore distinguished much more sharply than Thomas between God's immensity and omnipotence, between ubiquity "by essence and presence" and ubiquity "by power."[5] These predicates are not coextensive because they do not imply each other. The new

---

[2] Below III.B.3.

[3] *Centiloquium theologicum* conc. 6, 7a, ed. Boehner, p. 44. The *Centiloquium* is probably not an authentic work of Ockham. Cf. Iserloh, "Um die Echtheit des *Centiloquiums*," pp. 78–103, 309–46; Boehner, "On a Recent Study of Ockham," in: *Collected Articles* pp. 33–42; and Baudry, *Guillaume d'Occam: Sa vie, ses oeuvres, ses ideés sociales et politiques*, pp. 270f., 286. But Ockham, like Scotus before him, concedes the possibility that God may save an unrepenting Judas Iscariot, or destroy the just. Ockham differs from the author of the *Centiloquium* rather in other matters, below III.B.3

[4] Below, III.B.3.

[5] Johannes Duns Scotus, *Ordinatio* 1 d. 37 q.u., in *Opera*, ed. Balič et al., 6:299–302; *Lectura* 1 d. 37 q.u., in *Opera* 17:477–79; Johannes de Ripa, "Jean de Ripa I sent. dist. XXXVII: De modo inexistendi divine essentie in omnibus creaturis," ed. Combes and Ruello, pp. 161–267, esp. 264–65 and nn. 88–89 (Aureoli, Scotus); Grant, *Much Ado*, p. 146 and n. 137.

ethos of demonstration comes to the fore in Scotus's rejection of
Thomas's argument for God's ubiquity from the impossibility of
action *in distans*. Thomas, we saw, meant it analogically. Duns Scotus and his generation insisted that it either be, or not be, a strict
proof. God could act on things or create them even if he were distant. Angels can indeed be at one and the same place simultaneously
because they are incorporeal. Christ's body is in the Host in a locative, dimensional sense because God can place the same body at different locations simultaneously, in heaven and in many places on
earth.[6]

Later, Duns Scotus sharpened the distinction between immensity
and power by imagining an extramundane, empty space. The void
was recently revived also by Tempier's condemnation list, which
branded as heretical the opinion that God could not move the heavens in a straight line because he would leave a void.[7] It is more likely
that Scotus was moved by Thomas's speculations about extramundane places in the same context.[8] Since God can act without being
present, *in distans*, he could be totally absent from this putative space
yet act in it. Thomas understood the coincidence between place and
reality analogically. Without analogy, there was no ground to assume the coincidence at all, though Scotus held to it by faith only.
His definition of presence "by essence" shows once more that it is
but secondary to God's omnipotence. The latter is defined positively, immensity but privatively: *omni rei illabitur ratione suae illimi-
tatae immensitatis*.[9] In constructing his theory of divine attributes,
Scotus wanted first and foremost to secure God's absolute and infinite free will—*libertatem salvare*.

2. Ockham and Fourteenth-Century Scholasticism

Ockham joined Scotus in the critique of Thomas's simile of motion, which he explicitly took to be an intended and illicit proof. He

[6] Scotus, *Reportata Paris.* 4 d.10 q.3, ed. Wadding, II:636–45; *Ordinatio* 2 d.2 p.2 q.3,
ed. Balič, 7:268ff. Martin, *Ockham*, pp. 75–78; Seeberg, *Lehrbuch* 3:522, 526–27.

[7] Denifle and Chatelain, *Chartularium* 1277 no. 49, 1:546. Combes and Ruello ("Jean
de Ripa," p. 265 n. 89) remark rightly that, in view of Scotus's and Aureoli's ultimate denial of a vacuum so imagined, the importance of the condemnation, at least on this issue,
seems dubious. It gave, I believe, ammunition to those who wanted it, but it was by no
means binding, though very often quoted verbatim. For a different opinion see Grant,
Much Ado, pp. 108–11, and below n. 20.

[8] Above p. 55 n. 48.

[9] Scotus, *Reportata Paris.* 1 d.38 q.2 schol. 3, *Opera*, 11:217 (Wadding); Gilson, *Scot*,
pp. 390–97 (quotation on p. 391).

also emphasized that God can act *in distans*, all the more so since natural agents can. But unlike Duns Scotus he offers a proof from physics instead, hardly a more convincing one. The heart of the question, he recognized, was neither inexistence by power nor by knowledge *a parte cognoscenti*, but omnipresence in its essential, and locative, sense. If Scotus went a good part of the way to disarm the locative connotations, Ockham went almost all the way. He formalized the relation of "being in by essence" to a point that it did not need to be understood spatially at all; its minimal meaning is "to be in something else and not distant from it or from anything in it wherever it or something of it is."[10] It is a formulation worthy of Leibniz. Extension is, for Ockham, a connotative, relative notion by which a thing is recognized to have "parts outside parts" or parts separate from, yet together with, each other. Therefore a body can be thought of without this relation, so to say contracted to a point that is not "somewhere," and still be a body—such as the body of Christ.[11] Ockham now exchanged Thomas's alleged physical proof for another. Ockham commits God to "be in" at least one thing besides himself, because there is no real thing that is distant from *all* other real things in the universe, and God is a real thing. In this seemingly arbitrary postulate he could, of course, be thinking of the presence of God in Christ. But if God were to be in one place and not another, he would have to move to the other if he wished to be there—another physical necessity, no less so than Thomas's. Therefore God must be in all places.[12] Even if we assume that by "moving" Ockham meant "to undergo change of distance" in the above-mentioned formal sense, we are still left with the first commitment—that God must be in some things, and therefore in all. In short, Ockham returned to Thomas's claim that God is in all places, except that now we know even less what "place" means; it is left

[10] William of Ockham, *Sent.* 1 d.37, pp. 567–68: "Sed 'esse in' aliquo per essentiam, est esse in aliquo et nec ab eo nec ab aliquo sibi intrinseco distare quin sit ubicumque est ipsum vel aliquid ipsius."

[11] William of Ockham, *De sacramento altaris*, ed. Birch, pp. 148, 348, 466; *Summa totius logicae* 1.44; Martin, *Ockham*, pp. 78–87; Stump, "Theology and Physics in *De sacramento altaris*," in *Infinity and Continuity in Ancient and Medieval Thought*, ed. Kretzmann, pp. 207–30, esp. 215–16 (denial of quantity to Christ's body in the Host). It is noteworthy that on this issue Gabriel Biel follows Scotus rather than Ockham.

[12] William of Ockham, *Sent.* 1 d.37, p. 569. Ockham's proof rests merely on the assumption that "x is distant from y" (x/y) and "x is in y" define each other so that $\sim (x/y) \equiv x \cdot y$. We need not even invest ($x \cdot y$) with a meaning—which is precisely Ockham's intention.

deliberately uninterpreted. Thomas of Strassburg concluded this
development and freed God's omnipresence from all locative connotation whatsoever. He suggested that God cannot be said to be in
any place locally in either a circumscriptive or a determinative
sense, only *attinctive et conservative*.[13] Locality is but a metaphor, and
yet to be "simply everywhere" is the property of God alone. God
was driven out of all places in the same way that Gassendi and Descartes were to adopt later. Out of concern with univocation, they
turned God's omnipresence into an altogether equivocal attribute.

This movement toward a minimal construction of God's presence competed with a countermovement that sought a maximal
construction in an ever more literal sense. It, too, was encouraged
by Thomas's and Scotus's speculations about infinite extramundane
places or spaces. More and more theologians insisted that God can
create actual infinite magnitudes;[14] those among them who were
reared in the newly developed mathematical-logical techniques of
the Calculators also learned how to construct a one-to-one correspondence between different denumerable sets or different continuous magnitudes, though none of them tried to construe a proof of
nondenumerability. In this way, a velocity could be imagined to increase infinitely in one hour. And in this very same way one could
not only imagine an extensive infinite space, but also fill it with infinitely many coextensive infinite bodies of different intensities. Edward Grant, in his recent comprehensive study of medieval and
early modern concepts of space, has traced all the directions in
which the vigorous development of "imaginary space" went since
the fourteenth century.[15] Thomas Bradwardine almost equalled

---

[13] Thomas of Strassburg, *Commentaria in IV libros sententiarum* 1 d.37, fol. 106vb: "secundo dico quod deus est ubique . . . quia sicut se habet magnitudo molis infinita ad omnem locum occupative, sic se habet magnitudo virtutis infinita ad omnem locum attinctive et conservative." Angels, on the other hand, are in places "non circumscriptive, sed
diffinitive"; material bodies in both senses (fol. 108ra). The terms *circumscriptive* and *diffinitive* now became fashionable (Grant, *Much Ado*, p. 130). But, habitually, God was permitted to be in places at least *definitive*.

[14] On the "infinitists" in the fourteenth century see A. Maier, *Die Vorläufer Galileis im
14. Jahrhundert*, 1:196–215; *Metaphysische Hintergründe der spätscholastischen Naturphilosophie*, 4:381 n. 9 (here the word "infinitists"); Murdoch, "*Mathesis in philosophiam scholasticam inducta*: The Rise and Fall of the Application of Mathematics in Fourteenth-Century Philosophy and Theology," in *Arts littéraux et philosophie au moyen âge*, pp. 215–54, esp.
pp. 215–24; Breidert, *Das aristotelische Kontinuum in der Scholastik*, pp. 33–40. Rimini's
famous demonstration how God could create infinitely many angels in a finite time still
fascinated the generation prior to Galileo: Benedictus Pereira, *De communibus* 10, p. 593.

[15] Grant, *Much Ado*, pp. 116–47 and *passim*.

God's immensity with this imaginary space.[16] Johannes de Ripa objected to this identification, but spoke deliberately of God's "real presence" in space and in a superexistent location as a precondition for the putative existence of infinite bodies of different intensities in space.[17] Oresme, still speaking of a putative infinite space only, actually equated it with God's immensity: "Item, ceste espasse dessus dicte est infinie et indivisible et est le immensité de Dieu et est Dieu meismes, aussi comme la duracion de Dieu appellee eternité est infinie et indivisible et Dieu meismes." The statement apparently seemed to one scribe so horrendous that he omitted the words "and is God himself."[18]

How close were these and similar views to the early modern exaltations of space? Let Nicole Oresme be our guide. His space, like de Ripa's, is conceived as a precondition for possible, counterfactual states and worlds that God can create *de potentia eius absoluta*; it is not the precondition of our world. In more precise (Scotistic) terms, space was neither actual nor a mere logical possibility, but—so it seems—a *possibile realis* in a sense to be discussed in the following chapter.[19] More's space and Newton's was actual: because it was real but lacked a subject, it was predicated of God. Oresme recognized, more clearly than any other medieval author I know, the absolute nature of space as a precondition for absolute motion in it. The *espace ymaginé*, he says, must be "infinie et immobile."[20] To say the con-

---

[16] Thomas Bradwardine, *De causa Dei contra Pelagium*, ed. H. Savil, pp. 177–80, trans. E. Grant, *A Sourcebook in Medieval Science*, pp. 555–68; Koyré, "Le vide et l'espace infini au XIVᵉ siècle," pp. 45–91, esp. pp. 83–84.

[17] Combes and Ruello, "Jean de Ripa," p. 233. De Ripa's complex theory resembles Oresme's theory of representation of qualitative change (below V.B.3) in that, for each point in the infinite space, he assumes a range of intensities to infinity; "God" almost fulfills the role ascribed by Oresme to the missing fourth dimension.

[18] Nicole Oresme, *Le livre du ciel et du monde* 1.24, ed. Menut and Denomy, p. 176. The context of the discussion is the plurality of worlds (see also below III.B.3). Oresme, who suggests not only the possibility of worlds side by side but also one inside the other (another universe within the earth), shows awareness in his discussion of the *relativity* of spatial measurements, which reminds one of the arguments for the undetectability of the Fitzgerald contraction (p. 168).

[19] Below III.B.4; Grant, *Much Ado*, pp. 119, 132f., takes "reality" to mean "physical reality" and cannot reconcile it with the designation of space as "imaginary." By interpreting reality as *possible realis* one can, I believe, remove the ambiguity. If so, then "imaginary" need not, as Koyré ("Le vide et l'espace," p. 52) assumes, refer to an "independent" space only.

[20] Nicole Oresme, *Le livre du ciel* 2.8, pp. 368–70. Oresme argues the possibility of the indefinite rectilinear motion of the universe in space. See E. Grant, "The Condemnation

trary is to assert an opinion "condemned in Paris"; without an absolute space God could not, if he so wishes, move the universe in a straight line without creating first another reference—body. Space is the precondition of absolute motion, but not—as to the seventeenth century—the precondition for the absolute distinction between motion and acceleration. Only the latter made space into a physical reality; without absolute change of motion there can be no real *forces*. None of the medieval speculations in the fourteenth century introduced space to solve real physical problems. A different physics was needed for that, a physics committed not only to the ideal of precision in our concepts, but also to the ideal of homogeneity of matter and the existence of real forces.

Franciscus Suarez summed up the medieval discussion about God's omnipresence with his typical fairness to all points of view. He, too, believed that Thomas attempted to prove God's ubiquity by physical argument. He concedes that the impossibility of action *in distans* even in nature cannot be proven rigorously.[21] He defends Thomas on the grounds that the impossibility of action *in distans* can never mean that an agent can only act on that which is near him, but it can also mean that something cannot be acted upon except directly or by another cause: in this second sense, he believes, Thomas was right in inferring that God is always in contact with that which he acts upon, whether in this world or in imaginary spaces. New in all this is that Suarez is so eager to develop a *theologia naturalis* in which even the immensity of God could be proven *mere naturalibus*, that he reads Thomas like a Nominalist even while defending him.

3. Homogeneity: Cusanus and Telesio

The universe of the Scotists or Terminists was unequivocal to the extreme. But was it also homogeneous?[22] On the contrary—for one

---

of 1277: God's Absolute Power and Physical Thought in the Middle Ages," pp. 211–44, esp. p. 230. Grant exaggerates, perhaps, the impact of the condemnation; cf. above. n.7.

[21] Franciscus Suarez, *Disputationes Metaphysicae* dis. 30 sec. 7, 3–4, 11, in *Opera omnia*, ed. Berton, 16:95, 96, 98.

[22] To some extent, the metaphysics of light, nourished from Neoplatonic sources, also supplied a model for the homogeneity of matter—at least primordial matter. It developed most strongly in the Middle Ages in two independent traditions. Grosseteste's metaphysics of light was not forgotten: Crombie, *Robert Grosseteste and the Origins of Experimental Science 1100-1700*, pp. 128–34. The notion of *species* as forces (*propagatio specierum,*

thing, most of them accepted Aristotelian physics as a matter of fact, only to add that it is not logically necessary. Moreover, Ockham's universe was split into as many possible orders as there are entities, each of which can exist without the others: *omnis res absoluta, distincta loco et subiecto ab alia re absoluta, potest existere alia re absoluta destructa* (or even, *toto mundo destructo*).[23] Against the conceptual reduction of our language about nature attempted by the Terminists there stands the speculative reduction of nature itself by Renaissance philosophers of nature; Telesio, Cardano, Campanella, and Bruno did not share the Terminists' obsession with precision of language. Their philosophy, so they believed, was not the Scholastic preoccupation with words and definitions,[24] but a *philosophia realis*, a turning to nature itself. What most of them really turned to, or returned to, was the universe of the Stoa.

We shall limit our discussion of Renaissance philosophies of nature to two examples almost a century apart: Nicolaus Cusanus and Bernardino Telesio. For very different reasons, both asserted the fundamental homogeneity of the universe; both exchanged the Aristotelian forms for forces. Both also returned to symbolical readings of the universe of the kind that Scholastic philosophers so labored to minimize. For Cusanus, the homogeneity of nature and the perpetual *imprecision* of our scientific language followed from the very same epistemological and ontological premises. Though he was, by his own admission, an heir to a long tradition of negative theology,[25] his speculative originality is manifest already in the bold

---

formarum) (ibid., pp. 104–16) may have generated a new conception of laws of nature (Schramm, below III.A.n.22). At times, the attributes of omnipresence seem transferred to light—together with an active sense of all-efficacy: Witelo, *Liber de intelligentiis* 7.1; 8.1–4; 9.1–2 ("natura lucis est in omnibus"), in *Witelo: Ein Philosoph und Naturforscher des XIII. Jahrhunderts*, ed. Baeumker, pp. 7–14. On the other hand, the Kabbala nurtured the image of primordial, undifferentiated light as symbol of the "unlimited" (*En sof, 'or en sof*). Indeed, the Neoplatonic "prime matter" is that light the Psalmist said God wears as clothing ('ote 'or kesalma): Nachmanides, *Perush hatora* to Gen. 1. While not unaware of these traditions, and their later fusion with new conceptions of space, they seem to me at best an added mode to express a sense of the homogeneity of matter once the need for it was felt later.

[23] William of Ockham, *Quodlibeta* 6 q.6, *Philosophical Writings*, ed. Boehner, p. 26; *Sent. prol.* q. 1, *OT* 1.1:38. E. Hochstetter, *Studien zur Metaphysik und Erkenntnislehre Wilhelms von Ockham*, pp. 56–57.

[24] Above I.C. nn. 2–3. (Gassendi, Galileo).

[25] Nicolaus Cusanus, *De docta ignorantia* 1.27, in *Werke*, ed. Wilpert, pp. 34–36 (pp. 27–28 in the Strassburg edition): "Hinc omnis religio in sua cultura necessario **per theolo-**

manner in which he turned this tradition on its head. True, God—
the absolute infinite—can be spoken of only negatively; the principle of noncontradiction does not apply in the domain of the divine,
which unites the opposites just as a circle with an infinite diameter
is *eo ipso* also a line.[26] But the universe, too, can never be totally captured by concepts. Our conceptualizations are mere approxima-
tions:[27] the universe is neither finite nor infinite, neither discrete nor
continuous, neither at rest nor absolutely in motion. God is present
in it "everywhere and nowhere" (*undique et nullibi*).[28] Because the
universe is not absolute, all our concepts of nature are likewise relative, and rest, for their validity, on analogy and similitude. No
place in the universe can be said to be absolutely at the center, and
no place can be said to be absolutely preferred to, or different from,
another. Translated into terms of the material composition of the
universe, it means that every material body contains or reflects
every other, and all are in God; as God is, by contraction, "in all
things."[29] Cusanus recognizes only four, the Aristotelian sublunar,
elements, of which also celestial bodies are made. These bodies,

---

giam affirmativam ascendit. . . . et ita theologia negativa adeo necessaria est quo ad aliam
affirmationis ut sine illa deus non coleretur ut deus infinitus, sed potius ut creatura, et talis
creatura idolatria est. . . . et hoc quidem quia verissimum verius per remotionem et negationem de ipso loquimus, sicut et maximus dyonisius . . . quem rabbi salomon [ibn
Gebirol] et omnes sapientes sequuntur." Unlike the Neoplatonic tradition, and unlike
Maimonides, the propelling notion of Cusanus's negative theology is not God's oneness,
but rather God's *infinity*.

[26] Ibid. 1.13, pp. 15–16 (pp. 13–14).

[27] On Cusanus's epistemology see Cassirer, *Das Erkenntnisproblem in der Philosophie und
Wissenschaft der Neueren Zeit*, 1:21–61, esp. 25–31 (*approximatio, similitudo*). I am not
aware of a study of Cusanus's theory of language. Cusanus insists that all our terms and
names are *imprecise*, and win *praecisio* only from the notion of God: e.g., *Idiotae de mente*
3, *Werke*, p. 242 (p. 172): "Nam deus est cuiuscunque rei precisio" etc.

[28] Finitude, motion: Nicolaus Cusanus, *De docta ign.* 1.11, *Werke*, pp. 61–63 (pp. 45–
46). Cf. Koyré, *From the Closed World to the Infinite Universe*, pp. 6–24. Ubiquity: Cusanus, *De docta ign.* 2.12, *Werke*, p. 63 (p. 47): "unde erit machina mundi quasi habens
undique centrum, et nullibi circumferentiam, quoniam circumferentia et centrum deus
est qui est undique et nullibi." The origin of the metaphor is the "Book of the xxiv Philosophers" (cf. Harries, "The Infinite Sphere: Comments on the History of a Metaphor,"
pp. 5–15). Cf. also Cusanus, *Apologia doctae ign.*, *Werke*, pp. 110, 115 (pp. 79, 82).

[29] Nicolaus Cusanus, *De docta ign.*, 2.4–5, *Werke*, pp. 44–48 (pp. 34–38), summarily at
p. 47: "Subtili intellectu ista altissima clare comprehenduntur: quomodo deus est absque
diversitate in omnibus, quia quodlibet in quolibet, et omnia in deo quia omnia in omnibus, sed cum universum ita sit in quolibet quod quodlibet in ipso, et universum in quolibet contracte id quod est ipsum contracte etc." Cf. *De visione Dei*, *Werke*, pp. 305–307
(pp. 219–21).

too, are corruptible.[30] Each physical body contains all four elements. Each is, in a sense, animated; Cusanus explicitly wished to mediate between the *veteres Stoici*, who claim that matter includes in itself, *actualiter*, all possible forms, and the "Peripatetics," who allow matter to bear forms only as mere possibility.[31]

The difference between the negative theology and the negative cosmology is the difference between the absolute and the relative. Infinity, when attributed to God, is a *negative* predicate; when attributed to matter it is only *privative*, that is, subject to qualifications.[32] The boundlessness of the world, though strictly speaking incomparable with God's simple, absolute infinity, is nonetheless an image of it, an analogy; and every order in the world likewise symbolizes the divine, because the world *is* a "contracted" God, just as, for the Scotists, the individual instantiation of a quality was a contracted form.[33] Cusanus offered a genuine speculative synthesis of different and disparate traditions of thought: negative theology, the doctrine of *analogia entis*, the late Scholastic enthusiasm for actual infinities and for mathematical arguments in theology. In the context of this discussion I want only to establish that Cusanus, much as Leibniz later, held both to the homogeneity of nature and—for the same reasons—to the necessary imprecision, if not equivocation, of our scientific language.

Different considerations and a different philosophical temperament led Telesio to assert the homogeneity of nature, not the obsession with the infinite, but the sense of universal sympathy. His system is, in many ways, a *tertium comparationis* between the Stoic universe and the universe of Henry More.[34] Like the Stoics, he

---

[30] Nicolaus Cusanus, *De docta ign.* 2.13, *Werke*, pp. 67–69 (pp. 50–51); *De coniecturis 2*, *Werke*, pp. 152–54 (pp. 110–11).

[31] Nicolaus Cusanus, *De docta ign.* 2.7, *Werke*, p. 53 (p. 40): "Unde aiebant veteres stoyci formas omnes in possibilitate actu esse: sed latitare et per sublationem tegumenti apparere. . . . Peripatetici vero solum possibiliter formas in materia esse dicebant, et per efficientem educi. Unde istud verius est: . . . forme non solum sunt ex possibilitate sed efficiente."

[32] Nicolaus Cusanus, ibid. 2.1, p. 39 (p. 30): "Solum igitur absolute maximum est negative infinitum . . . universum vero . . . privative infinitum." Cf. Blumenberg, *Die Legitimität der Neuzeit*, p. 474.

[33] E.g., Nicolaus Cusanus, *De docta ign.*, 2.4, *Werke*, pp. 44–46 (pp. 34–35); cf. above n. 25.

[34] Bernardino Telesio, *De rerum natura juxta propria principia*, ed. Spampanato. On him, see Cassirer, *Erkenntnisproblem*, pp. 232–40. A similar link between sensualism, materi-

wished to replace the Platonic or Aristotelian "forms" (i.e., essences) with real *forces*, all of which are reducible to attraction and repulsion, or to heat and cold (as had Empedocles at one time). These forces operate on one, homogeneous, actual yet passive matter; the actuality of matter was already stressed by the Nominalists. Each body represents a balance of these forces and seeks to preserve that balance; each body possesses an instinct for self-preservation.[35] The center of heat is the sun, the center of cold the earth; the harmony of the universe results not from one preestablished goal but rather from the activity of each animated body—indeed, all bodies are animated by force—in its own self-interest. This is one of the earliest occurrences of an antiteleological, political, ethical, as well as natural, principle of an "invisible hand of nature."[36] The soul is only a finer, more subtle matter than the rest of the body, hence reducing epistemology to a sensualistic account and eliminating the *species intelligibiles* of the schools. And, since matter is not mere potentiality, time and place must be distinguished from it as absolute receptacles. The war between the Aristotelian adherents of form and the new adherents of force was fought in Italian universities in the generation before Galileo, at times with bare fists.

The animated universe of many natural philosophies in the Renaissance was homogeneous in the sense that the absolute distinction between celestial and terrestrial matter was eliminated and the number of elements reduced to two or less. It was also a universe that, like its Stoic ancestor, was much more ambiguous than the Aristotelian-Scholastic one. The very notion of a "force" in Telesio's system expresses little more than the affinity of like bodies for each other, their "similitude"; in a universe that is held together by bonds

---

alism, and dynamism characterizes many Italian new philosophies of nature: Cardanus, Campanella, Bruno. Bacon, we are reminded by Randall, called him "primus novorum virorum"—first of the moderns: *The Career of Philosophy* 1:202. On his concepts of space see Grant, *Much Ado*, pp. 192–94, with whom I agree about vestiges of Stoic physics— not so much because of the concept of space, as because of the concept of forces. Much as I would wish them, I do not find traces of Philoponus in Telesio.

[35] Telesio, *De rer. nat.* 4.xxiv, p. 728; this, too, is a Stoic doctrine. See also Höffding, *History of Modern Philosophy*, pp. 92–102.

[36] On its early history in modern social thought see below IV.A.I. This, too, has Stoic origins: Kristeller, *Eight Philosophers of the Italian Renaissance*, p. 102 (self-preservation). Less convincing is Kristellar's derivation of the triad, hot, cold, and matter from the Aristotelian triad (above II.B.2) form, matter, and privation. He rejects—without specifying why—the analogy to the Empedoclean love, hate, and matter.

of sympathy, everything becomes a "sign" of something else and the world is full of hidden connections. In one critical respect, however, Telesio deviates from his Stoic model: bodies act only according to their own forces and their own self-interest. Telesio opposes final causation without any attempt to de-animate nature: his universe shows no trace of goals, of a grand design. A century later, even those who saw in nature a divine design opposed final causes *because* they animated nature. Put differently, the ideal of monocausality was a critical rather than a constructive ideal, and ambiguous at best. Like the Stoics, the Renaissance developed a passion for minute and endless details, with or without a theory.

4. Homogeneity and Infinity: Copernicus

Evidently, then, the new sense of the homogeneity of nature—and with it the shift from forms to forces—*preceded* the Copernican theory. It continued to inform many Renaissance philosophies of nature, whether geocentric (like Telesio's) or heliocentric, even after the publication of *De revolutionibus*. Yet the Copernican theory had a crucial role in the promotion of this new sense, similar in many ways to the way in which it promoted the sense of infinity. In due time, the heliocentric planetary system was embedded in a universe that was both infinite and homogeneous. And while it is true, theoretically as well as historically, that all three tenets can be held independently and, indeed, were at times so held, they have, nonetheless, an affinity for one another: *once clearly defined*, they are more easily held together than put asunder.

"Affinity" is a vague term; I use it in a weaker and stronger sense. A heliocentric system does not demand an infinite universe—only an immensely big one—to account for the imperceptibility of the parallax of the fixed stars. This was clearly recognized already in Antiquity.[37] Even a geocentric universe could be thought of as infinite—as the many medieval discussions about the hypothetical plurality of worlds or the imaginary extramundane space prove—but it may nonetheless be said, in a weak sense, to be easier to declare a heliocentric universe to be not only immense, but virtually infinite. Copernicus himself refused to commit himself.[38]

---

[37] *The Works of Archimedes*, trans. and ed. Heath, pp. 221–22.

[38] Nicolaus Copernicus, *De revolutionibus orbium coelestium* 1.1 c.8. If Copernicus believed in a finite, spherical, material universe embedded in infinite space (as his remarks

At the time of Copernicus, the infinity of the universe had long ago ceased to be seen as an absurd proposition. To the contrary, a casual remark of Nicole Oresme reveals, more than many medieval formal discussions about the possibility and nature of space, an important change in the climate of opinion induced by the "infinitist." It is our natural inclination, he says, to conceive of the universe as infinite; only science—he means, of course, Aristotle's cosmology—teaches us that it is not so.[39] The finitude of the world has hitherto been a common-sense if not self-evident proposition; it now became counterintuitive, at least in some quarters, long before the spread of the Copernican revolution. Natural philosophy in the fifteenth and sixteenth centuries was unable to say which truths about space are "intuitively" true.

The image of a homogeneous universe, we saw, was likewise independent of the dimension of the universe. It was by no means necessary, as Cusanus proves, even to ascribe an absolute center, let alone preferred, "proper places" to a finite universe. But did not the heliocentric hypothesis call for the collapse of the distinction between "celestial" and "terrestrial" bodies and regions? Not necessarily—it is possible that Copernicus himself still upheld that distinction, though I do not believe so.[40] Heliocentricity and homogeneity have, I believe, a much stronger affinity than heliocentricity and infinity. Special reasons must be sought to explain why and in what sense a heliocentric universe still involves an absolute separation of celestial and terrestrial regions. In fact, empirical evidence for the homogeneity of nature—such as evidence for the corruptibility of celestial bodies (comets) or the irregular, earthlike features of the moon—were advanced as supportive evidence for the Copernican theory; and all that Cardinal Bellarmine could

tend to suggest), we again may suspect Stoic influence. Koyré, *From the Closed World*, pp. 31–43, reminds us that Copernicus's universe is "still hierarchical"; cf. also his *The Astronomical Revolution: Copernicus-Kepler-Borelli*, trans. Moddison, p. 72, n. 7.

[39] Nicole Oresme, *Livre du ciel* 1.24, p. 176: "Je respon, et me semble premierement, que entendement humain aussi comme naturelment se consent que hors le ciel et hors le monde qui n'est pas infiny est aucune espace quelle que elle soit, et ne peut bonnement concevoir le contraire." (The same is true of eternity.)

[40] Guerlac, "Copernicus and Aristotle's Cosmos," pp. 109–13. It is not a convincing argument. Whether Copernicus believed in material, even rigid, spheres is likewise hotly debated. Cf. N. Jardine, "The Significance of the Copernican Orbs," *Journal for the History of Astronomy* 13 (1982): 168–94.

do was to stress that it was not demonstrative evidence.[41] In short, the heliocentric theory, and the infinity and the homogeneity of the universe, once clearly defined, were more easily held together than apart; eventually they seemed to corroborate one another. The Copernican theory enhanced the sense of the uniformity of nature—like the infinity of the universe—without initiating or necessitating either theoretical position. It was a catalyst to the new sense of homogeneity, not its author.

5. Protestant Interpretations of God's Presence

The new cosmologies of Telesio, Cardano, Campanella, and Bruno are marked by a penchant for speculative reductionism. The Scholastic obsession with precision of terms, with unequivocation, was set aside by a generation which believed that "real philosophy" ought to be concerned with nature itself, not with words. How much more easily such a universe became susceptible to pantheistic reading we see not only in Ficino or Bruno, but already in the thought of Cusanus: the world is an explication, a self-expression of God—God contracted himself into the world. God, it seems, began to regain a body. The world recovered its symbolic meanings. We may safely assume that, with the exception of a few medieval intellectuals, it never lost it altogether.

Did the Reformation help God to regain a body? I argued earlier that Christian fears of pantheistic doctrines derived not only from the fear of deifying nature, but, more specifically, from the fear of diluting the meaning of Christ's particular, selective, real presence in the Host as managed by the priestly hierarchy. Protestant theol-

---

[41] Galileo, *Opere*, 12:171ff. A recent biography aptly summed up the methodological difference between Bellarmine and Galileo: "Galileo also sprach von 'tausend Beweisen', während Bellarmin auf den einen 'zwingenden Beweis' wartete. Hinter diesem Unterschied verbirgt sich die radikale Differenz der Erkenntnishaltung der neuen Wissenschaft einerseits und der scholastischen Philosophie andererseits": Fölsing, *Galileo Galilei: Prozess ohne Ende, Eine Biographie*, p. 322. But Fölsing errs in generalizing "Scholastic philosophy." True, the criteria of *demonstration* have been heightened since the fourteenth century to match the Aristotelian ideal of demonstrative science in the *Posterior Analytics*; this was the reason, e.g., why Scotus and Ockham rejected Thomas' physical "proof" of ubiquity (above II.D.1–2). But Ockham and his generation, as once Aristotle, likewise encouraged proofs from probability: A. Maier, "Das Problem der Evidenz in der Philosophie des 14. Jhs.," in *Ausgehendes Mittelalter: Gesammelte Aufsätze zur Geistesgeschichte des 14. Jhs.*, 2:367–418. On Galileo's employment of the exegetical principle of accommodation, below IV.B.2. See also Wallace, *Galileo and His Sources*, pp. 99–178.

ogy lost this fear. Even in its doctrines of the sacraments it could pursue, to the extreme, the utterly transcendent or utterly immanent image of the divine, claiming in either case that it is true to the Scriptures. To Zwingli (and to Karlstadt and others earlier) the words "This is my flesh" carried a symbolic meaning only.[42] Luther held, from the onset, to the real presence of Christ in the Host. His preference of the doctrine of consubstantiation over the doctrine of transubstantiation, though it relied on a minority tradition in Scholastic thought, may have been informed by the new sense of nature discussed above. Imagining interpenetrating substances was, to the Middle Ages, no less a conceptual problem than conceiving accidents without their proper subject. To a more Stoic-oriented sense of nature, the complete interpenetration of bodies, *velut ferrum ignitum*,[43] became much less repugnant to common sense. Luther's preference certainly had no better grounding in the Scriptures. But unlike the medieval predecessors of either doctrine, Luther could never acquiesce to the strong locative sense of the real presence. Christ's body "to the right hand of God" is not a distinct thing in place like "a bird in a tree." The right side of God stands for his omnipotence.[44] Christ's body was, and always is, permeated through and through by his divine nature. And, like God's power and essence, even Christ's body is everywhere (*ubique*) at all times. The communion is only the occasion at which Christians are instructed by the word of God where to concentrate on finding Christ's pres-

[42] *Credere est edere*: Ulrich Zwingli, letter to Alber (November 1524), *Sämtliche Werke*, ed. Egli and Finsler 3:341; cf. Potter, *Zwingli*, pp. 156–57; Bizer, *Studien zur Geschichte des Abendmahlstreits im 16ten Jh.*, pp. 40ff.; Seeberg, *Lehrbuch*, 4.1, pp. 396–407, 458–79; Ozment, *The Age of Reformation, 1250–1550*, p. 336.

[43] Luther, *Werke, Kritische Gesamtausgabe*, 6:510. It seems, though, that Luther never used the term *consubstantiatio*; Seeberg, *Lehrbuch* 4.1, p. 400. On the development of Luther's positions see Bizer, *Studien*, and Hausamman, "Realpräsens in Luthers Abendmahllehre," in *Studien zur Geschichte und Theologie der Reformation, Festschrift für Ernst Bizer*, pp. 157–73.

[44] "Sol er macht haben und regiern, mus er freilich auch da sein gegenwertig und wesentlich durch die rechte hand Gotts, die allenthalben ist": Luther, *Werke*, 23:145; cf. 23:159, 28:141; Seeberg, *Lehrbuch* 4.1, pp. 462–66, 469ff. (ubiquity); Bizer, "Ubiquität," in *Evangelisches Kirchenlexicon* 3:1530–32. There seems to be agreement that Iserloh, *Gnade und Eucharistie in der philosophischen Theologie des Wilhelms von Ockham, ihre Bedeutung für die Ursachen der Reformation*, has exaggerated Ockham's role and the Scholastic roots; how different Luther's doctrines are, in all of their phases, from those even of contemporary Nominalism can be seen from the precise analysis of Biel's positions in Oberman, *The Harvest*, pp. 275–76.

ence. Protestantism had much less to fear from pantheistic inclinations than Catholicism. Indeed, they occur more often. Jacob Boehme's thought may have been richer or deeper[45] than the vulgar pantheism of the poor village miller mentioned before, but the main difference between them was that Boehme was not burned at the stake. On the other hand, Protestant theology also encouraged, at least on the level of exegesis, unequivocation: it called for a return to the *sola scriptura*.

It seems to me that only in the seventeenth century did both trends converge into one world picture: namely, the Nominalists' passion for unequivocation with the Renaissance sense of the homogeneity of nature—*one* nature with forces to replace the many Aristotelian static natures. Protestant theology may have acted at times as a catalyst to the fusion. Once both ideals of science converged, the vision of a unified, mathematized physics could emerge, in which Euclidian space was the very embodiment of both ideals. Now, and only now, a clear-cut decision has to be made as to how God's ubiquity—to which the Lutherans added the ubiquity of Christ's body—had to be understood; to decide whether God must be placed within the universe, with or without a body, or outside it. Pascal's defense of a metaphorical-symbolical language of theology was the exception, not the rule.[46] Scientists and theologians in the seventeenth century spoke pure prose; but, unlike Mr. Jourdain, they knew it.

E. DESCARTES AND MORE

1. *Descartes's Dilemma*

The universe that became, in the seventeenth century, both unequivocal and homogeneous inspired a fusion between theology and physics to an extent unknown earlier and later. Theological and

---

[45] I am often reminded of Heine's remark: "Karl I hatte von diesem theosophischen Schuster eine so grosse Idee, dass er eigens einen Gelehrten zu ihm nach Görlitz schickte, um ihm zu studieren. Dieser Gelehrte war glücklicher als sein königlicher Herr. Denn während dieser zu Whitehall den Kopf verlor durch Cromwells Beil, hat jener zu Görlitz durch Jacob Böhmes Theosophie nur den Verstand verloren." *Geschichte der Religion und Philosophie in Deutschland*, in *Werke* (Berlin, n.d.) 8:62.

[46] Blaise Pascal, *Pensées*, fragments 383, 606 *Oeuvres complètes*, ed. Chevalier, pp. 1188, 1282. Cf. Goldmann, *Le dieu caché: Etudes sur la vision tragique dans les Pensées de Pascal et dans le théâtre de Racine*, pp. 57ff, 216ff, 264ff.

physical arguments became nearly indistinguishable. This circumstance brought with it advantages as well as disadvantages to both
theology and physics. Some of the most pressing theological problems of the Middle Ages dissipated with the commitment to new
physical systems as, for example, how to translate the immortality
of the soul or *creatio ex nihilo* into Peripatetic terms. The eternity of
the world and the conception of the soul as the organizational principle of the body were an integral part of Aristotle's physics and
psychology, yet were unacceptable theologically. Early modern
physics could easily sustain a genuine cosmogony and eschatology,
that is, a conception of the universe as coming to be and about to
pass away at the end of days. In exchange, however, for problems
solved, the seventeenth century faced new problems which grew
out of the need to invest certain theologumena with a precise physical meaning. Descartes's physics is a case in point.

Descartes's vision of a homogeneous material universe governed
always and everywhere by the same unequivocal, that is, mathematical, laws held in its spell even those thinkers who recognized its
flaws. The flaws in the system are innumerable; they result mostly
from the very same circumstance responsible for its fascination. In
his eagerness to mathematize physics thoroughly, Descartes recognized only one indispensable attribute of matter: extension. Bodies
are nothing but extended things (*res extensae*). Thus, space and matter had the same meaning: the material world is one infinite continuum, and in fact, all of matter is one substance. Motion means that
a body changes from the vicinity of some bodies to the vicinity of
others; it is an entirely relative concept.[1] Since all bodies move at
once, what does it mean that they move at all? And if a body in motion collides with another and both move in one direction, are they
not one body? Obviously, Descartes lacks a principle of individuation for single bodies as physical entities.

[1] René Descartes, *Principia Philosophiae* 2.25, AT 8.1, p. 53: "dicere possumus [motum]
esse translationem unius partis materiae sive unius corporis ex vicinia eorum corporum,
quae illud immediate contingunt et tanquam quiescentia spectantur, in viciniam
aliorum." In Descartes's view, only the whole, universal continuum of extended substance really deserves the name of a body and has a constant proportion of motion and
rest (*quantitas motus*). The single body can neither be said to have distinct boundaries nor
a distinct absolute motion; its inclination to retain a uniform rectilinear motion is a never
realizable "inclination." Cf. Kenney, *Descartes: A Study of His Philosophy*, pp. 200–15.
Descartes, then, lacks a physical principle of individuation; and Spinoza, Hobbes, and
Leibniz also constructed their theories so as to remedy this difficulty.

Furthermore, Descartes's famous rules of motion seem to assume the impenetrability of colliding bodies.[2] Now, motion of bodies in Euclidian geometry—say, in proofs of congruency—always assumes that bodies pass through each other (lines, areas) or coincide with each other. Impenetrability can certainly *not* be derived from the geometrical characteristics of bodies as extended things, even if we understand it in the minimal sense that one body cannot occupy the same place simultaneously and totally with another. Once Descartes, however, made this assumption, he was at a loss to explain the origin of various states of density or the phenomenon of elasticity. Moreover, the assumption of solid bodies speeding toward each other with no obstacle between them contradicts the image of matter as a continuum. For the very same reason, Descartes's two laws of inertia are also counterfactual conditionals: no body is separable from other bodies so that it can move uniformly and rectilinearly; wherefore Descartes speaks only of the "tendency" of bodies to so move "inasmuch as they can"—if "considered by themselves."[3] Indeed, the only sensible way to interpret Descartes's laws and rules of motion, including the postulate of impenetrability, is by taking them as counterfactual conditionals that function as limiting cases. Descartes's physics is altogether hypothetical, and so, by his own admission, is his cosmology: *if* we assume that God imparted a constant quantity of motion to the universe and then left it to its own devices, *then* matter will form a vortex from which a system of planets will emerge such as ours. To make things worse, God is not even needed; all we need is the quantity of motion—even from eter-

---

[2] Descartes, *Principia* 1.37–53 (AT 8:62–71). An explication of the rules in algebraic notation was tried by Aiton, *The Vortex Theory of Planetary Motions*, p. 36. As Huygens rightly remarked, "videtur corpus secundum Cartesium non differe a vacuo philosophorum." *Pièce concernant la question du "mouvement absolu," Oeuvres complètes*, 16:221.

[3] Descartes, *Principia* 1.37 (AT 8:62): *Quantum in se est*; (p. 63): *seorsim spectatam*. On the Lucretian origin of the term see I. B. Cohen, "Quantum in se est." Another possible source is ethical-theological: cf. Oberman, "Facientibus Quod in se est Deus non Denegat Gratiam," pp. 317–42. Why did Descartes split the inertial principle into two—one governing motion (i.e., the *quantitas motus*), the other governing direction? Leibniz thought that, in this way, Descartes hoped to secure a structure by which spirits could influence minute bodies, not by changing their motion but merely by changing their direction. But this is more likely to be an afterthought, a side benefit (if indeed it solves the difficulties of mind–body interaction, which I do not believe). Rather, $(m \cdot v)$ is, to Descartes, always a scalar; throughout the "rules of motion," change of direction obeys another logic than change of motion.

nity. Newton's outcry, "hypotheses non fingo," may be directed not only to a particular theory (the *vortices*) within Descartes's physics; it pertains to all of it.

In the next chapter I will argue that the mathematization of nature since Galileo presupposed the employment of counterfactual conditionals as limiting cases of reality. The strength and novelty of seventeenth-century science, both theoretical and experimental, was in its capacity to take things out of context and analyze their relations in ideal isolation. It was a new form of abstraction, or generalization; and it was recognized by many who employed it as new, as the source of the advantage of the new science of nature over the old. Only with the aid of mathematical models could the ideal of homogeneity and the ideal of unequivocation be joined. But precisely because this is so, the problem had to arise: what are the limits of mathematical abstraction? Until what point is the scientist permitted, even encouraged, to "disregard material hindrances" without abandoning the true understanding of nature? Indeed, Descartes was too faithful a mathematician to be a good physicist. He threw out the baby with the bath water. In his physics there is no room for elastic bodies or for real forces.[4] Motion and the direction of motion are entirely sufficient to determine the impact of one body on another—which consists of the translation of motion from one body to another according to laws of conservation ($m \cdot v$). If so, then in Descartes's physics there is no place for, and no meaning to, Galileo's great discovery that only acceleration, not the motion of falling bodies as such, is caused by a specific force, namely gravitation. It is ironic that Descartes, who first formulated the inertial principle properly, never made proper use of it, while Galileo, who never formulated it as a general law, employed it most fruitfully.[5] Kinematically, there is no absolute measure by which to distinguish uniform motion from acceleration. Take any two bodies that move uniformly away from each other from a given point in a parabola or a circle. Only from the vantage point of another, third body could we

[4] But see, against this more common interpretation, Gabbey, "Force and Inertia in the Seventeenth Century: Descartes and Newton," in *Descartes: Philosophy, Mathematics and Physics*, ed. Gaukroger, pp. 230–320. Descartes's specific term for force, he argues, is *determinatio*. Even if so, it is, I believe, a poor substitute for real forces. There is, however, much more attention given to forces and accelerations in Descartes's mathematical letters than in the *Principia*. See below V.B.5–6.
[5] Below III.C.4.

decide that they change their direction; to these two bodies their motions seem perfectly uniform and rectilinear! Accelerations are absolute only if caused by real forces. Descartes, in formulating a purely geometrical physics, evicted real forces from his universe.

If forces were removed from matter, all the more so were spirits and God. These are substances sui generis. The only attribute characterizing spirit is cogitation (what Descartes has in mind comes closest to the phenomenological "intention"). But, if so, how can spirits act upon matter—as indeed our soul acts upon our body? And how does God act upon matter? What does it mean, literally, that God implanted certain rules and a quantity of motion in matter? By what mode of causality is this conceivable? Matter-in-motion is conceived of by Descartes as devoid of any final cause or aim. But it is even difficult to see how a spirit could intervene in this closed system of causality by motion. If animals are pure *automata*, why not also all the actions of the human body? Finally, Descartes insists, as we shall see, on a voluntarism more radical than that of the most radical Nominalists. God is first and foremost omnipotent and self-caused; all his other attributes depend upon his will. If he so wanted, he could invalidate our "clear and distinct" ideas; even eternal truths are contingent upon his will.[6] Eventually, the sharp separation between matter (extension) and spirit (cogitation), the source of so many troubles in Descartes's system, is justified on the grounds that it constitutes a "clear and distinct" idea. Could God invalidate it too? Or could he create a world in which spirits are extended?

These and similar questions are theological as well as physical. They stem from the unique Cartesian fusion of theological and physical arguments such that the most fundamental laws—the inertial law and conservation of motion—are derived from the first law of physics: the constancy of God. God is constant, that is, he does not change without sufficient reason, because he is also good.[7] God's presence in the world is likewise understood unequivocally. It is neither material nor symbolical; it is metaphysical only in the sense that all other beings depend, at any moment of their existence, on God's will to preserve them. It is a relationship of logical impli-

[6] Below III.A.I, D.1–3.
[7] Descartes, *Principia* 2.36, AT 8.1, p. 61.

cation, the medieval *esse per potentiam* only. This idea served, ironically, Calvinist theologians in their refutation of the (Lutheran) notion of the ubiquity of Christ's body.[8]

2. More's Solution

Henry More developed his positions in a constant dialogue with and against Descartes. Much emphasis is given to his insistence on the absolute and infinite nature of empty space.[9] It seems to me that this is only a derivative concern of his. His fundamental, never modified or qualified position asserted against Descartes the extended nature of spirits. The presence of spirits (and ultimately of God) in the world was not only metaphysical, qua substances, but also physical: with bodies they share dimensionality. It seemed to More that most of Descartes's problems were solvable in this manner: the psychophysical interaction (*commercium mentis et corporis*), the assertion of absolute motions, and hence the introduction of real forces into matter—a way to deal with impenetrability as a real rather than as a hypothetical condition of bodies—in an equivocal divine mode of causation. Seen from this vantage point, More's concerns and solutions were not very far from those of Leibniz.

Spirits and bodies are *res extensae*. It is fair to say (though against More's objections) that spirits and solid bodies are bodies in most senses of the word: they occupy places and can interact among one another as well as with themselves. The difference between spirits and bodies lies in the nature of the forces they represent. Bodies, though breakable, are impenetrable; spirits, though indivisible, are penetrable. Spirits can penetrate bodies as well as each other; they can also contract and expand.[10] Upon reflection, we see the reason

---

[8] M. Heyd, *Between Orthodoxy and the Enlightenment: Jean-Robert Chouet and the Introduction of Cartesian Science in the Academy of Geneva*, pp. 72–80. Cf. also below II.H.6 (Leibniz).

[9] Notably Koyré, *From the Closed World*, pp. 125ff.; Jammer, *Concepts of Space*, pp. 26–32, 39–48; Grant, *Much Ado*, pp. 221–30.

[10] "But for mine own part I think the *nature* of a *spirit* is as conceivable, and easy to be defined as the nature of anything else. . . . As for example, I conceive the intire *Idea* of a *Spirit* in generall, or at least of all finite, created and subordinate *Spirits* to consist of these severall powers or properties, viz. *Self-penetration, Self-motion, Self-contraction* and *Dilation*, and *Indivisibility*; . . . I will adde also what has relation to another, and that is the power of *Penetrating, Moving* and *Altering the Matter*." Henry More, *An Antidote against Atheisme; or, an Appeal to the Natural Faculties of the Minde of Man, Whether There Be Not a God* 1.4 §3, p. 15. This is a summary, quoted also by Koyré (*From the Closed*

why More regards both properties as one: when spirits penetrate each other, their intensity grows; so also, if a spirit contracts. Stoic physics asserted very similar states for the Pneuma and used the analogy of rebounding waves. More called this property, perhaps under the influence of the Scholastic method of *latitudo formarum*, "spissitude,"[11] and added that it may be conceived of as a "fourth dimension." Bodies in and of themselves lack spissitude. But since, in nature, all bodies are permeated by spirit of some kind, the ability of a complex body to maintain its size indicates a certain, constant spissitude. Every change within a body is accounted for by the spirit penetrating it, which is only another way of saying that there are real forces that account for absolute motions. Spirits are forces. Forces, properties, spirits are often interchangeable terms. By mechanical power More means Descartes's "quantity of motion"; spirits possess "plastic power."[12] Once spirits are admitted into the realm of nature, one can ascribe to matter as such all the properties ascribed to it by Descartes: it is incapable of self-motion, has no force of its own, its motion as such is always relative, and it may

---

World, pp. 127–28), of *Enchiridion metaphysicum*, cc. xxvii–xxviii, also trans. in J. Glavill, *Saducismus Triumphans* (1681), pp. 99–179, under the title *The Easie, True, and Genuine Notion . . . of a Spirit*. More, then, goes about determining "clear and distinct" ideas in the same way as Descartes does.

[11] Henry More, *Enchir. met.*, cc. xxvii–xxviii; *The Immortality of the Soul* 1.2 § 11, p. 20. The "fourth dimension" of *spissitude* is different from Oresme's speculations about a fourth dimension (below V.B.3); Oresme meant the *representation* of qualities (intensive magnitudes), while More addresses their proper dimension. In fact, *spissitude* is Oresme's term for the third dimension. But there is enough of a similarity to wonder whether there is a connection. Koyré (*From the Closed World*, p. 132) compares the notion of spissitude to that of a field, much as the Stoic τόνος was compared to physical fields in recent literature (above II.B.3). Emphasis on the Neoplatonic rather than Stoic elements in the natural philosophy of the Cambridge Platonists led Cassirer to discard the theory of "spissitude" as an incidental "curiosity" within More's thought: Cassirer, *The Platonic Renaissance in England*, trans. Pettegrove, p. 150 n. 1.

[12] More, *Immortality of the Soul* 3.12–13, pp. 449–70, esp. p. 450, where "the spirit of nature" is described as "a substance incorporeal, but without Sense and Animadversion, pervading the whole Matter of the universe, and exercising a plastical power therein . . . as cannot be resolved into meer Mechanical powers." See also Cudworth, *The True Intellectual System of the Universe* 1.3 sec. xxxvii §3, p. 148: "Furthermore all such *Mechanists* as these, whether *Theists* or *Atheists*, do . . . but substitute as it were . . . *a Carpenters or Artificers Wooden Hand, moved by Strings and Wires, instead of a Living Hand*. They make a kind of Dead and Wooden World, as it were a Carved Statue, that hath nothing neither *Vital* nor *Magical* [!] at all in it. Whereas to those who are Considerative, it will plainly appear, that there is a *Mixture* of *Life* or *Plastick Nature* together with *Mechanism*, which runs through the whole Corporeal Universe."

therefore obey Descartes's geometrical rules. The distinction between fixed and changeable spissitude, or between plastic and mechanical forces, is not unlike the Leibnizian distinction between dead
and live force (*vis viva*)[13]—a circumstance of which Leibniz may
have been aware in his praise of More.

Such are the contours of the doctrine that More developed over
fifteen years. It shows its affinity with the Renaissance philosophies
of nature (notably with Telesio's) as well as with Stoic physics. Yet,
in contradistinction with both, he attends not only to the homogeneity of his (and their) animated universe; he also wishes the notions
of force, spirit, and matter to be "clear and distinct," that is, unequivocal. In so doing, he may have solved some of Descartes's
most pressing problems, but, in turn, he created new problems. He
may have solved, *more suo*, Descartes's psycho-physical dilemma,
but his God has difficulties of his own.

Like Descartes's God, More's is the Spirit-in-Chief. All other
spirits or forces depend on him. Some spirits lack reflection and
purposefulness, such as the (Stoic) *anima mundi*;[14] some have it; God
is the vertex in the hierarchy of spirits (or ideas). In the sense that
God is a spirit, More admits, though not without initial hesitations,
that he is extended; contrary to other spirits, however, his extension
is infinite—it is space itself. Now, this forces us to deny of God
what More ascribed to spirits—namely spissitude. God cannot expand or contract. He, like space, is always the same.[15] The only way
to avoid interpreting this circumstance as a deficiency or imperfec-
tion is as follows. Contrary again to Descartes, spirits and God are
not absolutely discrete substances. God is rather the spirit of all spirits, their source and place. More's is an emanational theology, not
unlike the Kabbalistic speculations he admired: God is veritably the
'*en sof* (*infinitum*), both identical with and different from the powers
(*sefirot*) of which he is the source. More's relation to the Jewish (and
Christian) Kabbala corroborates, I believe, my interpretation. He

---

[13] Below II.H.6–7. On More and Leibniz see Cassirer, *Platonic Renaissance*, pp. 150–56.
[14] Above n. 12.
[15] More, *Enchir. met.* I.vi.5, p. 42. Perhaps it would be more precise to say: God's spissitude is immense; he is "all-penetrating" and therefore the capacity to expand and contract, which other spirits have, cannot be ascribed to him anymore than the capacity of
self-annihilation. More, then, gives up the second half of the traditional formula that God
is "everywhere and nowhere." Cf. also ibid. I.27, p. 171: space is "confusior quaedam et
generalior representatio essentiae sive essentialis praesentiae divinae."

criticizes its anthropomorphic symbolism of God's body (*'adam kadmon*).[16] Nor does the "empty space" of Lurianic Kabbalism impress him: he criticizes it as being finite and a result of the contraction (*tsimtsum*) of the original infinite light, that is, God (*or 'en sof*).[17] Only derivative spirits, we remember, contract; God does not. But he is impressed with the emanational structure and process of the divine forces, with the vision of God as a balanced harmony of interacting and counteracting aspects. More's concept of the divine amounts to the concept of a harmonious sum total of all mechanical and purposive forces in the universe. Such a God could not but be reasonable.[18] He is the very embodiment of πρόνοια, much as was the Stoic Pneuma. He contrasts again with Descartes's God, in whom will had primacy over reason.

F. HOBBES, SPINOZA, AND MALEBRANCHE

1. From Two Substances to One: Spinoza

Hobbes's scattered remarks about the impossibility of conceiving incorporeal substances are part of his systematic theory of language, which I shall discuss in a later chapter. In the reconstruction of a precise and unequivocal language of science he saw at times the very essence of science, not only its instrument. From the beginnings of his literary career he also believed that, ultimately, all phenomena

---

[16] Copenhauer, "Jewish Theologies of Space in the Scientific Revolution: Henry More, Joseph Raphson, Isaac Newton, and Their Predecessors," pp. 489–548, esp. pp. 515–29; *'adam kadmon*, ibid., pp. 527–29; *tsimtsum*, ibid., pp. 523–26 (see next note). By contrast, Anne Conway, whose God was not extended, did approve of the divine contraction: Conway, *The Principles of the Most Ancient and Modern Philosophy*, ed. Loptson, p. 65: "Diminuit ergo in Creaturarum gratiam (ut locus ipsis esse posset) summum illum intensae suae lucis gradum, unde locus exoriebatur, quasi vacuus circularis, Mundorum spatium." Cf. *Kabbala denudata* (Sulzbach, 1677), 2:150: "Deus creaturos mundos contraxit praesentiam suam."

[17] It is worthwhile to remember, in addition to Copenhauer's arguments against Jammer (nn. 9, 16) that, even among Jewish Kabbalists who followed the Lurianic radical reinterpretation, it was fiercely disputed whether to understand "contraction" (*tsimtsum*) literally (*kifshuto*)—and conclude therefore that God's omnipresence, his "filling all the worlds" (*memale kol almin*) cannot be taken literally—or whether *tsimtsum* should be understood metaphorically (*shelo kifshuto*) so as to save God's real presence in the world. A history of this important dispute has yet to be written, but see Teitelbaum, *Harav miljadi umifleget habad*, 2:62ff., 78.

[18] Lichtenstein, *Henry More: The Radical Theology of a Cambridge Platonist, passim*. In this respect also, More's main tenets come close to those of Leibniz.

could be scientifically reduced into terms of matter-in-motion. He was, in this sense, a "mechanical philosopher," crusading against final causes, substantial forms, sensible or intelligible species, in short, anything that came from the despised vocabulary of Scholasticism. Geometry was to be the model of the new physics, in a sense also of his "new method" of political theorizing: in geometry, more than in any other science, man constructs the objects of our knowledge himself—and Hobbes believed that "truth" and "the constructable" are synonymous.[1]

Spinoza was the only seventeenth-century thinker who attributed a body to God, explicitly and unequivocally. It took him, however, some time to reach this conclusion. He rejected it in the beginning of his philosophical career. Spinoza's theory of God's ubiquity is one of the rare instances in which we can pinpoint not just a change from an opaque or confused position to a more articulate one; we can pinpoint a *genuine* change of position on a central issue.

The only book that Spinoza ever published under his name appeared in 1663.[2] It was an exposition *more geometrico* of Descartes's "Principles of Philosophy," to which Spinoza attached an appendix "containing metaphysical thoughts." Spinoza apparently intended the exposition itself to include only those propositions that he could either attribute directly to Descartes, or claim with good conscience that they clarify the latter's positions or make them more consistent. Some corrections to Descartes's physics were only intended to give it more consistency. In the appendix he intended, without saying so, to develop those consequences of Descartes's system that he knew to be considerably at odds with Descartes's positions. They may not all indicate his own positions at the time. Yet some of them are defended so vigorously, even though they are (or seem) at odds with his monistic commitments, that one wonders whether Spinoza's development toward the *Ethics* may not have taken a dialectical detour. In the exposition he did not treat the mind-body relationship—it was the Achilles' heel of Descartes's system in the eyes of

[1] Cf. below v.c.3–4.

[2] The *Theological-Political Treatise* appeared anonymously in 1673. It was the only other book of his published during his lifetime. The *Ethics* circulated for many years among friends and acquaintances; Leibniz, while in Holland, read its first part. For our discussion it matters little whether the *Cogitata* was composed before the *Principia*—as Curley (*pace* Freudenthal) believes: *The Collected Works of Spinoza*, 1:222–23.

friends and foes alike—or the doctrine of the mind. In the *Cogitata* he treats all that, and always in reference to its theological implications. It is, indeed, a testimony to his acquaintance with the later Scholastic terminology, a terminology used profusely here but dropped almost entirely in the later *Ethics*.

Descartes conceded that all of matter is one substance. Spinoza in the *Cogitata* hints at a theory that would solve Descartes's mind-body problem. That theory is not stated directly and unambiguously for the same reasons as Spinoza's reluctance to publish any of his true opinions: he was a cautious man. Nor did he yet develop to perfection the technique of the *Theological-Political Treatise*—namely using traditional terms but giving them an almost opposite meaning. Spinoza suggests, in effect, recognizing only two substances: mind and matter, the one consisting of ideas only, the other of extension; "and except for those two we know no other."[3] And while he distinguishes clearly between God and matter, the distinction between God's thought (which is also volition) and ours is ambiguous and relative. Whatever clear and distinct ideas we have, Spinoza has proven before, we share with God; we are in him as his objects of thought.[4] But God, he has already stated with Descartes, is incorporeal. Here he adds: matter owes its existence to God, but is a distinct being, and there need be no communication between matter and form except that God has—and sometimes we, too, inasmuch as we have adequate ideas—knowledge of all material constellations.[5] Ubiquity, or the immensity of God, is "commonly" (*vulgo*) understood as the spatial omnipresence of God. "If God, they say, . . . would not be everywhere, either he could not be

---

[3] Baruch Spinoza, *Cogitata metaphysica* 2.12, Van Vloten-Land, 4:231: "Substantia vereo extensa iam antehac [i.e., in the exposition of Descartes's theory] satis locuti sumus, *et praeter has duas nullas alias cognoscimus* [my italics]." Cf. ibid., p. 225: "Transeundum iam est ad substantiam creatam, quam in extensam et cogitantem divisimus. Per extensam, materiam . . . intelligebamus. Per cogitantem vero, mentes humanas *tantum* [my italics]." It means that all minds (ideas) are one. The difference between the divine and human cogitation is that the former is one, the human many (ibid. 2.7 §8, 4:215). Cf. also ibid. 2.1 §1, and 2.10, Van Vloten-Land, 1:221, where the separate existence of matter is unequivocal; God's cogitation, however, is only different from ours because it is spontaneous. But cf. ibid. 2.12, 4:228.

[4] Ibid. 1.2, 4:192.

[5] Ibid. 2.6 §1, 4:212: ". . . quia ostendimus [in the *Principia phil. Cartesianae*], in materia nihil praeter mechanicas texturas et operationes dari." But Spinoza does not yet develop the later maxim that the order of things is the same as the order of ideas.

wherever he wants to be, or he would by necessity [N.B.] have to move."[6] The source of the mistake is that they attribute quantity to God and therefore do not want it to be finite. Spinoza also rejects the distinction made by "some" between a threefold immensity "of essence, power, and presence" (*of* and not *by*). These are word-games, because essence, power, and presence must be absolutely convertible in God.[7] Ubiquity namely means that no thing exists unless created at every instant anew by God, that is, given or affirmed in its existence.

"Their arguments strive to affirm God's immensity from the properties of extension; nothing is more absurd." Now, even if Spinoza intended this passage merely to repeat Descartes's position, he would not have maligned the view that God is extended, calling it "common" and "absurd," if it were his own at that time. It was not. Moreover, the two-substances doctrine was not Descartes's either; Descartes admitted, besides the one extended substance, as many substances as there are souls, angels, and God. But Spinoza, in the *Cogitata*, seems to opt for *one* cogitative substance only, of which souls are presumably just so many modifications. His position could be interpreted as Occasionalistic, had we not known, from the "Short Treatise" (*Korte Verhandeling*), that his commitment to monism was earlier than the *Cogitata*. To complicate matters even more, there the doctrine of two substances is put into the mouth of "lust" (*Begerlijkeit*), not "love" or "understanding":[8] this comes close to calling it "vulgar." Either Spinoza, in the *Cogitata*, tries to eradicate the traces of his own views to the point of ridiculing them openly, in line with the interpretation of Strauss, or else we ought

---

[6] Ibid. 2.3, 4:207. The N.B. is Spinoza's: "Si Deus, aiunt, actus est purus, ut revera est, necessario est ubique et infinitus; nam si non esset ubique, aut non poterit esse, ubicumque vult esse aut necessario (N.B.) moveri debebit; unde clare videre est, illos *Immensitatem* Deo tribuere, quatenus ipsum ut *quantum* considerant." This, we remember, was the argument of Thomas and Ockham.

[7] Ibid. 2.3, 4:207–208. Note that the analogy of *per potentiam* with *potentia regum*, which Spinoza refutes, was already that of Thomas, *Summa theol.* 1 q.8 a.3: "Rex enim dicitur esse in toto regno suo per suam potentiam, licet non est ubique praesens."

[8] *Korte Verhandling van God, de Mensch, en deszelfs Welstand*, Van Vloten-Land, 4:15. Cf. Siegwart, *Spinoza's neuentdeckter Tractat von Gott, dem Menschen und dessen Glückseligkeit*, esp. pp. 110–34 (influence of Bruno). Indeed, Giordano Bruno seems also to have held a two-substances theory of sorts. We need not sort out the various layers of the *KV*, since it is evident that even the earliest of them demonstrate a clear monistic commitment.

to look for a different reading. Note that, in the *Korte Verhandeling*, Spinoza did not yet solve the oppressive mind-body problem. Perhaps, then, he saw for awhile (in the *Cogitata*) its solution in a doctrine of two substances which nonetheless constitute one *thing*. That they constitute one thing he refrained, in the *Cogitata*, from saying aloud. If he entertained for a short while such a view, it may have even been suggested to him by the doctrine of consubstantiation we discussed earlier.

The two substances became one in the *Ethics*, and Spinoza did not even shy away from the explicit conclusion that God is, or has, a body. "All who contemplated in some way the divine nature denied that God is corporeal: which they prove best by this, that by a body we conceive some quantity that is long, broad, and deep that defines a figure, which is most absurd if predicated of God, an infinite being."[9] This refuted position was exactly the one defended so vigorously in the *Cogitata*. That God, as he now maintains, is a corporeal substance follows from his definition of substance and attributes, which leaves him with only one substance having two attributes—*cogitatio* and *extensio*. Does this not mean that God's body is divisible? By no means: inasmuch as it is *substance*, it is indivisible.[10] The modifications and configurations within matter infringe no more on its substantial unity than the modifications and configurations of thought on the oneness of the divine mind, or on the attribute of thought. Thought and matter do not act on each other; rather, the order of ideas and their configurations is the same as the order and connection of things;[11] they are two modes of expression that stand in a one-to-one correspondence. Unequivocation and homogeneity became two exactly matching aspects of nature.

The difficulties in this theory are not greater or smaller than in other attempts to prove a one-to-one correspondence of mind and

---

[9] Spinoza, *Ethica ordine geometrico demonstrata* 1 prop.15, schol., Van Vloten-Land, 1:48–49.

[10] Ibid.: "substantiam corpoream, quatenus substantia est, non posse dividi"; cf. id., *Cogitata met.* 2.12, 2:226: "Clare enim concipimus, ubi ad humani fabricam attendimus, talem fabricam posse destrui; at non aeque, ubi ad substantiam corpoream attendimus, concipimus ipsam annihilari posse." Cf. also ibid. 2.7, 2:215: "Denique si ad analogiam totius Naturae attendimus, ipsam ut unum Ens considerare possumus, et per consequens una tantum erit Dei idea sive decretum de Natura naturata."

[11] Above II.A.n.6.

matter. In particular, it is hard to see exactly, given the cluster of ideas corresponding to a certain material configuration—in short, given an idea of a body—what it means to have an idea of that idea, or how the idea of an idea can be said to refer to any other additional material constellation. But Spinoza's theory had also many advantages over Descartes's beyond the overworked mind-body dilemma. Descartes cannot endow confused ideas with any reality at all; his *cogito*, as Malebranche also recognized, was too narrow a basis from which to reconstruct the world. Spinoza's theory allows confused ideas to match ill-defined material configurations.

The very idea of a body as an independent entity is objectively confused. Bodies are only relatively separated from others, namely through motion. Spinoza recognized clearly the deficiencies of Descartes's physics, in which motion was both relative and absolute. Instead of taking $(m \cdot v)$ to be ultimately a universal constant (with some arbitrary magnitude), he took it to be the signature of the relation of singular bodies to their environment.[12] The marks of Hobbes's influence on Spinoza's physics are evident. Simple bodies have all their parts moving with the same velocity. If they break, they break irreparably—their parts move differently as different bodies. They have motion. Complex bodies have internal motions of different velocities, the sum total of which is $(m \cdot v)$; Spinoza could thus explain much better the phenomenon of elasticity: as long as a part of elastic bodies does not break away, their system of motion tends mechanically to return to the previous balance, thus generating internal force. Even more complex bodies are a system of motions in motion such that, if one part breaks away, the others maintain nonetheless the same proportion of motion of the whole and

[12] Spinoza, *Ethica* 2 lemma 5, 1:87: "Si partes, Individuum componentes, majores minores evadant, ea tamen proportione, ut omnes eandem ut antea ad invicem motus et quietis rationem servent, retinebit itidem Individuum suam naturam ut antea, absque ulla formae mutatione." This is, I believe, his most basic physical proposition—as well as (below p. 338) a guiding image of his political theory. The section in the *Ethica* 2 between prop. 13 and prop. 14 (Van Vloten-Land, 1:85–89) is an inserted fragment of a "physics." Cf. Rivaud "La physique de Spinoza" and Lachterman, "The Physics of Spinoza's Ethics," in *Spinoza: New Perspectives*, ed. Shahan and Biro, pp. 71–112; p. 105, n. 19 lists pertinent literature. Lachterman exaggerates somewhat the importance of physics in Spinoza's thought. Cf. Gueroult, *Spinoza* 2:568. Both in the *Cogitata* and in the *Ethica*, Spinoza lays down the principles only—and refrains from the quantitative elaboration of Descartes's or his "laws of motion" (unlike, e.g., Malebranche). Spinoza merely uses physics for his psychology, ethics, and political theory.

are even capable of regeneration. But the whole universe can be
seen, in one way, as the most complex body (*facies totius universi*) of
which all other bodies are relative parts. If a single blood cell had
consciousness, it would believe that it is an independent entity
rather than part of the circulatory system.[13] Whether we call a body
one or many is a matter of point of view, albeit, if adequately conceived, a legitimate point of view.

Spinoza's account of the human personality and its emotions corresponds precisely to his (fragmentary) theory of bodies. There are
as many personalities in us as we have different ideas of the sum total
of our bodily constellations. Our mind—our idea of our bodily
configuration—is ever more clear the more we realize the relativity
of our "self," that its borders are relative both extensionally (as all
bodies are) and mentally: inasmuch as they have clear and distinct
ideas, all minds are one. Every being has a *conatus suum perseverare
motum*[14]—it is true for bodies and for thoughts. It is a law of nature
that everyone act according to his *self*-interest. But the self-interest
of the wise man, the more adequate (i.e., relative) his idea of his self
becomes, coincides with the interest of all, or the self-interest of the
whole. God, indeed, "loves himself in an infinite love."[15]

2. Malebranche on Intelligible Extension

It is not difficult to trace the lines of thought leading from Spinoza's
Cogitata metaphysica to his *Ethics*. Already in the *Cogitata* "creation,"
"conservation," and laws of nature (or *potentia Dei ordinata*) were

---

[13] Spinoza, *Ep*. 32, Van Vloten-Land, 3:119–23, esp. p. 121; cf. *Ethica* 2, lemma 7,
schol., 1:88. Cf. Sacksteder, "Spinoza on Part and Whole: The Worm's Eye View," in
Spinoza: New Perspectives, pp. 139–59. The *facies totius universi* (*Ep*. 64) should, perhaps,
be translated neither as "face" (which makes little sense) nor as "fashion or make," as
suggested by Hallet, *Benedict de Spinoza*, p. 30 and n. 39—which is even grammatically
odd—but rather as "character," "person," or "individual." I suspect that Spinoza borrowed it from the Kabbala (*partsuf*), yet avoided the term "person" because of its Christian, and anthropomorphic, connotations. His source may have been the *Sha'ar hashamayim*
of Abraham Cohen Herrera; on its possible influence on Spinoza see Scholem, introduction to the German translation: *Das Buch Sha'ar hashamayim oder Pforte des Himmels*, pp.
41ff. Cf. above II.A.n.7.

[14] Spinoza, *Ethica* 3 prop.6, 1:127. Note that, contrary to Descartes and Hobbes, Spinoza does not derive this principle from his physics. The physical notion of inertia is
rather derived from his peculiar version of the law of collision of bodies (ibid. 2 ax.2,
1:86: the angle of incidence equals the angle of refraction; if both are zero, a body simply
continues to move). On Spinoza's psychology see below V.C.5.

[15] Spinoza, ibid. 5 prop.35, 1:266: "Deus se ipsum Amore intellectuali infinito amat."

nearly synonymous.[16] But the *Cogitata* reveals another strand of thought, more in accord with the beginnings of an occasionalist position (La Forge, Cordemoy). Spinoza (like the Occasionalists) restricts God's omnipotence to the realm of existents—unlike Descartes—God neither creates ideas nor can God annihilate them.[17] Existing things—matter, perhaps souls—are through and through dependent on divine causation because they are only possibles; their non-existence does not entail contradiction. "Clear and distinct" ideas, even of possibles, are necessary, in the sense that they are necessarily valid even in God's mind. And Spinoza in the *Cogitata*, unlike in the *Ethics* (in which the order of things corresponds precisely to the order of ideas), believes that there is a surplus of *ideas* over *things*, of possible existents over actual existents.[18] Only the latter are dependent on divine causation, that is, creation and conservation. Extension, we saw, is identical with matter and in fact *one* substance. As such, it is not a divine predicate: the God of the *Cogitata* was incorporeal. But he "contains all the perfections of extension without its imperfections," without its divisibility:[19] there is no other way of interpreting this passage than attributing to God the idea of extension.

In a later chapter I shall discuss how and why Malebranche confined the mechanical construction of the universe to the realm of ideas or possibles only. His interpretation of God's ubiquity, however, ought to be mentioned here now, and with it the much maligned doctrine of the "intelligible extension" (*étendue intelligible*) that inhabits the divine intellect.[20] "The infinite intelligible exten-

---

[16] *Cogitata* 2.9, Van Vloten-Land, 4:217–19; p. 219: *potentia absoluta, ordinaria, extraordinaria.* On the history of the *potentia ordinata-absoluta* dialectics see below ch. 3.

[17] *Cogitata* 2.10, 4:219–21; cf. 1.3, 4:193–97; 1.2, 4:192. God's power is clearly confined to giving existence to the possible: "Deique omnipotentiam tantum circa possibilia locum habere" (4:217). In *Principia phil. Cartesianae* 1 prop.7, schol., Van Vloten-Land, 4:123, he even dismisses Descartes's own assertions as incompatible with his wit and his other words.

[18] But see the ambiguous statements in 2.7, 4:215 ("nam si Deus voluisset, aliam res creatae habuissent essentiam") and 2.9, 4:218 ("si aliter Deus decrevisset" etc.).

[19] *Principia phil. Cartesianae* 1 prop.9, schol., 4:133. It is doubtful whether Descartes would have spoken about the "perfections" of extended things in which God has a share. On the one-sidedness of Spinoza's presentation of Descartes see E. Gilson, *Études sur le rôle de la pensée médiévale dans la formation du système Cartésien*, pp. 299–315.

[20] Nichole Malebranche, *De la recherche de la vérité, Eclaircissements* 10, response to objections 2, 3, ed. Rodis-Lewis, in *Oeuvres complètes de Malebranche* (henceforth *OC*), 3:144–51; *Entretiens* 8.8, *OC*, 12–13: 184–88. Cf. *Defense against Arnauld, OC*, 6–7: 201.

sion is only the archetype of an infinity of possible worlds similar to our own. By means of it I only see certain determinate beings—material things. When I think of this extension I do not see divine substance, except insofar as it is representative of bodies and is participated in by them."[21] To perceive an object, to see it, is to see in it God, to participate in God's mind inasmuch as the idea of that object is "clear and distinct." There is no guarantee (except faith) that the object I see really exists even if my idea of it is clear and distinct. The intelligible world demands, for its reification, a continuous act of God's will. Existents depend, for their existence and interaction, on one cause only—the divine volition. This is as true of the interaction between bodies as it is true of the mind-body interaction (*commercium mentis et corporis*). Ideas, on the other hand, are in and of themselves valid; even God cannot invalidate them.[22] The precondition for all ideas of all possible material things is that of which they are a mere modification—namely extension. The idea of extension is therefore the idea of all possible things; it is the manner in which "God is in everything inasmuch as everything is in God," for God "possesses the perfections of all beings. He has the ideas of them all. He contains, therefore, in his wisdom all truths, speculative and practical."[23]

This doctrine was seen, by Mairan and others, as downright Spinozistic.[24] Since they had Spinoza's *Ethics* in mind, Malebranche's indignation was justified. His intelligible space, unlike Spinoza's, is not extension, but the idea of extension; and the idea of all possible material constellations, unlike Spinoza's *Ethics*, has an infinite sur-

---

The term hardly plays a role in the *Recherche* itself; it may have been developed against allegations of proximity to Spinozistic positions (below n. 25); it gained in importance in the *Entretiens* and later. That matter is one; that—save for the insistence of the Church—it would have to be judged identical with extension; and that its idea in God is immaterial—all this is already present in the *Recherche*. We need not look for any change in the overall theory to explain the emergence of the term except for the polemical exigencies just mentioned, as argued, e.g., by Connell, *The Vision in God: Malebranche's Scholastic Sources*, pp. 56–57, 322–55.

[21] *Entretiens* 2.3, OC, 12–13:52.

[22] Below V.A.1 and n. 11.

[23] *Entretiens* 4.14, OC, 12–13:98–99.

[24] Bayle, *Dictionnaire*, s.v. Leucippus; cf. (to the whole Arnauld-Malebranche dispute over extension and existence) ibid., s.v. Zeno of Elea n.H; cf. *Entretiens* 8.8 (next note). Letter to Mairan, 12 June 1714, OC, 19:882ff., quoted by Cassirer, *Das Erkenntnisproblem*, 2:569 n.1: "L'Idée de l'étendue est infinie, mais son ideatum ne l'est peut-être pas" (*idea, ideatum* are Spinozistic terms employed *ad hominem*).

plus of the possible against the actual. But we did see that many of Malebranche's positions were present, if vaguely and undeveloped, in Spinoza's *Cogitata metaphysica*—in particular, the doctrine of intelligible space. I do not think that Malebranche was, of necessity, aware of the *Cogitata*—he may not even have been aware, when writing his first work (*De la recherche de la vérité*), of Guelincx's Occasionalist doctrines. It becomes clear, however, that, prior to Malebranche, the most immediate solutions to a host of Cartesian problems were sought in the same direction—even by Spinoza.

Yet Malebranche's *étendue idéal* is much more than one of many divine ideas. It is even more than the archetypal idea of all possible material existents. It is the only "clear and distinct" idea of infinity that can convince us of God's existence.[25] Malebranche uses the idea of infinite extension in the same way in which Descartes used the idea of God:[26] the fact that we, as finite minds, find it in us is proof enough that it exists (as an idea) outside us; not necessarily as a material thing, yet undoubtedly as an aspect of God's mind, the only aspect fully revealed to us by the very fact that we are cognizant of the world around us.

G. NEWTON

1. An Unequivocal Theology

Of the fusion between the ideal of unequivocation on the one hand and the ideals of homogeneity and monocausality on the other we said that it was accompanied by a fusion of theology and physics into almost one science. Newton's philosophy of nature proves it. He demanded the unequivocation of theological terms no less than that of physical terms. In his interpretation of 1 John 5:7 he notes:

> In disputable places I love to take up with what I can understand. It is the temper of the hot and superstitious part of mankind, in matters of religion, ever to be fond of mysteries; and for that reason to like best

[25] "No finite mind can understand the immensity of God, or of any other attributes or modes of the divine. . . . Nothing, on the other hand, is clearer than intelligible extension," which, therefore, is not a divine attribute (*Entretiens* 8.8, *OC* 12–13:183–84). But it serves, as an idea, to prove God's existence, since a finite mind cannot be the author of an idea of infinity (*Entretiens* 2.1–2, *OC*, 12–13:49–52). I did not find this use of intelligible extension in the *Eclaircissements*.

[26] Descartes, *Meditationes* 3, AT 7:40–46.

what they understand least. Such men may use the Apostle John as they please; but I have that honor for him, as to believe that he wrote good sense; and therefore take that sense to be his, which is best.[1]

"Best" here means the translation of symbols and metaphors into unequivocal statements with the mediation of a consistent and precise code. The "dark side" of Newton was as rational as his bright side; one can easily understand why Newton doubted ambiguous theologumena such as the Trinitarian dogma.

Of God, Newton, like Descartes, More, or Spinoza, knows several things clearly and distinctly. Most of these matters pertain to God's activity more than they do to his essence. They add nonetheless to our knowledge of God, in the same way in which we may have precise knowledge of the attractive force between bodies without knowing its cause or essence. The attraction between bodies, Newton claims, is not an obscure quality, because we know precisely how it works: it obeys a universal, quantifiable relation. So also God: we do not know his nature, but we perceive his actions. Newton's God was first and foremost the *kosmokrator*, ruler over everything.[2] It can be shown that Newton needed space on both counts: to account for the reality of forces and for the reality of God's activity. In some sense, these are two aspects of the same thing.

2. *The Three Independent Physical Functions of Space*

Newton's physical concept of space carries a triple burden. It is attributed with homogeneity, absoluteness (immobility), and infinity, because it serves three different, though interdependent, functions. Space and time are always "equal to themselves," that is,

---

[1] Newton, *An Historical Account of Two Notable Corruptions of Scripture: In a Letter to a Friend, Opera*, ed. Horsley, 5:529–30; Buchholz, *Isaac Newton als Theologe*, pp. 36–40 and nn. 15 (literature on the *Comma Johanneum*), 17 (text); and Westfall, *Never at Rest: A Biography of Isaac Newton*, p. 490. Of both Hermeticism and Cambridge Platonism Samuel Parker spoke as "Conjectures of a Very Warm Brain," in *A Free and Impartial Censure of the Platonick Philosophie*, p. 107.

[2] Newton, *Philosophiae naturalis principia mathematica*, ed. Koyré and Cohen, pp. 528ff., 760–764. The "General Scholium" of Book III was added to the 1713 edition. McGuire, "Neoplatonism and Active Principles: Newton and the *Corpus Hermeticum*," in *Hermeticism and the Scientific Revolution*, pp. 95–142, pp. 106ff., has shown that Newton's emphasis on God as "Pantokrator" inclined him to oppose doctrines of *anima mundi* as well as related constructs that assumed intermediary, half-spiritual agents to account for forces, as indeed assumed by Cambridge Platonists.

without qualitative differentiations in their segments. Because they are *homogeneous* in all respects, nature can also be homogeneous; the same forces can act everywhere in the same manner, the same laws of nature can be valid everywhere. The famous "analogy of nature" is made possible only by the homogeneity of nature. That which is always equal to itself is one in all respects. Does it mean that space is divisible? Not in any real sense. If any real division of space were imaginable, its meaning would have to be more than merely dimensional: one segment of space would have, qua this or that space, properties of its own—say, motion (like the Cartesian bodies) or curvature (like our space). To Newton, all parts or locations of space are, qua spatial, equal to each other, and in this sense indivisible. In another sense, however, space is divisible *ad infinitum* because it is extended. The essence of dimensionality is the separation of places.[3] This dual meaning of the homogeneity of space does no harm to its theological connotations. Inasmuch as it is one, not even God can break it to pieces. Inasmuch as God knows that the earth and the moon are not at the same location in space, he has already divided it.[4]

Newton, somewhat like More, assumed a homogeneity of an order contrary to that of space—the absolute density of corporeal particles. His conception of different masses within an equal volume presupposes an unequal number of particles of equal volume which can, under no condition, penetrate one another. If they could, there would be no absolute measure for masses, and hence not for forces.[5]

[3] Indivisibility: Clarke, letter to Leibniz, in *Die philosophischen Schriften von Gottfried Wilhelm Leibniz*, ed. Gerhardt (henceforth GP), 7:368: "For *Infinite Space is One*, absolutely and *essentially indivisible*." Divisibility: Newton, *Opticks* 3.1, p. 403 (cf. below, III.E.1). On Newton's role in the formulation of Clarke's answers, A. Koyré and I. B. Cohen, "Newton and the Leibniz-Clarke Correspondence with Notes on Newton, Conti, and Des Maizeaux," pp. 69ff. I shall ascribe only those of Clarke's views of Newton that seem warranted by pronouncements of Newton himself.

[4] Clarke's fourth answer, GP, 7:383: "Parts, in the *corporeal* sense of the word, are *separable, compounded, united* . . . ; But infinite Space, though it may by us be *partially apprehended*, that is, may in our imagination be conceived as composed of *Parts*; yet Those *Parts* . . . being *essentially indiscernible* and *immoveable* from each other, and not *partable* without an express contradiction in Terms . . . Space consequently is in itself *essentially One*, and also *absolutely indivisible*." Cf. Grant, *Much Ado*, pp. 250–51.

[5] Newton, *Principia* I def. 1 (*quantitas materiae* = density · magnitude); 3 prop. 6, theorem 6, cor. 4, p. 404 (equal density = equal proportion of inertia and bulk); added to the third ed. *Id., Opticks*, query 31, p. 389: "All bodies seem to be composed of Hard Particles." Ibid., p. 400: ". . . it seems probable to me, that God in the beginning form'd Mat-

Particles are parts of space that are indivisible in an additional sense. But unlike space, the indivisibility of which is conceived in analogy to God and outside his power, elementary bodies are divisible, at least to God. This, I believe, is the sense of Newton's opaque remark about God's ability to divide space *ad infinitum*.

Secondly, space and time are *absolute*, and on that account enable unequivocal causality. Newton only stated and never argued the absolute nature of time, but it is clear why he needed it. If the temporal relation of two events were to be relative, so would also be the cause-effect relation. As for space, Newton recognized that without a preferred inert system, there is no way in which change of motion or direction—acceleration—could be identified. His own language is somewhat different: he believed in the existence of a point in space that is absolutely at rest.[6] Uniform motion is relative. If Newton's space were inhabited by only two bodies, and they were to move toward or away from each other uniformly, there would be no way to determine which of them really moves. Newton's first law assures us that there is no physical meaning to this question either, because no external forces are involved. If absolute forces exist, so does absolute acceleration. But absolute acceleration requires an absolute point at rest. Given such a point, Newton can identify both

---

ter in solid, massy, hard, impenetrable, moveable Particles . . . in such Proportion to Space, as most conducted to the End for which he formed them; and that these primitive Particles being Solid, are incomparably harder than any porous Bodies compounded of them; . . . no ordinary Power being able to divide what God himself made one in the first creation." Cf. also above, n. 3 and *Principia* 3 rule 3, p. 388, Cajori, p. 399. Mach, *Die Mechanik in ihrer Entwicklung*, p. 188, called this definition of mass circular; cf. Cajori ed., p. 638, who drew attention to Newton's Atomistic foundation of the notion of density; cf. also *Unpublished Scientific Papers of Sir Isaac Newton*, ed. Hall and Hall, p. 316 and n. 2, and recently, Freudenthal, *Atom und Individuum in Zeitalter Newtons: Zur Genese der mechanistischen Natur- und Sozialphilosophie*, pp. 36–40; but, as Cajori already noticed, the atoms may be of different sizes. On the development of Newton's concept of mass see Westfall, *Force in Newton's Physics: The Science of Dynamics in the 17th Century*, pp. 340–50, 448–56; cf. also I. B. Cohen, "Newton's Use of 'Force,' or Cajori versus Newton," pp. 226–30. Newton's difficulties result, in part, from the unaccounted for, precise identity of gravitational and inertial mass; cf. Cajori ed., p. 572.

[6] From its first edition onwards, Newton said no less than that the *principia* were composed to distinguish true from apparent (relative) motions: *Principia* (1686), p. 11. Absolute space, in the "Scholium," is synonymous with immobile space. That the center of the solar system is also the center of the universe, does not follow at all; it is a further hypothesis: *Principia* 3, hypoth. 1, p. 408 (hypoth. 4 of first ed., pp. 402, 417): "Centrum Systematis Mundani quiescere. Hoc ab omnibus concessus est, dum aliqui Terram alii Solem in centro quiescere contendant."

internal and external forces. The *vis insita* comes to the fore only at
the point of change of motion. Given the absolute point of rest, we
can say that even if only one body existed in the universe, it would
have inert force in the sense that it would resist change of velocity
or direction. Attractive forces assume at least two bodies. Without
a resting point in space we could not, however, distinguish between
uniform motion and acceleration: kinematically, the terms of one
are perfectly translatable into terms of the other. Newton calls this
point the "center of gravity"[7] not because it attracts bodies—only
bodies attract each other—but because it allows him to identify
gravity, that is, the rate of acceleration of bodies of different masses
toward each other. Gravity is, therefore, never the absolute property of a body (like the *vis inertiae*), but an absolute relation: a property of the world as a whole, rather than its parts.[8] The center of
gravity is nothing but a point in space; to say that it is at rest is to say
that the whole space is at rest, since space is homogeneous. While
Leibniz, as we shall see, believed that absolute forces have meaning
without absolute space, Newton did not. This permitted Newton,
in contrast to Leibniz, to conceive of force as a sequence of impulses
that can increase or diminish or stay the same. He did not need a
conservation law for forces, while Leibniz could not do without it.[9]
To prove the existence of this center of gravity, Newton devised the
experiment of the rotating bucket filled with water.[10]

Much has been said of Mach's famous critique of this and similar
experiments. It was preceded by an almost unbroken chain of critics
of the notion of absolute motion since Huygens, Leibniz, and

[7] Newton, *Principia* 3, prop. 11, p. 408 (Cajori ed., p. 419). In an unpublished fragment, Huygens once said of absolute as against relative motion: "non est mathematice
difficilis materia, sed physice aut hyperphysice": *Oeuvres complètes* (1927), 6:213.

[8] *Princ.* 3 rule 3, schol., Cajori p. 388: "The extension, hardness, impenetrability, mo-
bility, and inertia of the whole result from the . . . parts." (Newton extrapolates by virtue
of the analogy of nature: these are universal qualities because "they are not liable to di-
munition." McGuire, "The Origins of Newton's Doctrine of Essential Qualities," pp.
233–60.) Gravity is a universal phenomenon (in an unpublished marginal note, p. 402,
Newton calls it "the quality of all bodies"), and yet, unlike the *vis insita*, cannot be ascribed to each body individually: "Not that I affirm gravity to be essential to bodies: by
their *vis insita* I mean nothing but their inertia. This is immutable. Their gravity is di-
minished as they recede from the earth." The distinction between properties of all bodies
and properties of (single) bodies has been clearly shown by Freudenthal, *Atom und Individuum*, pp. 42–46.

[9] Below II.H.6.

[10] Newton, *Principia* 1, pp. 10f. (Cajori ed., pp. 10–12.)

Berkeley.[11] Why should we not assume that, if the bucket stood still and all the mass of the universe rotated around it, the water at its edges would likewise rise? Newton, it seems, did not *need* the construct of an absolute space; it was superfluous metaphysical baggage. This, however, is far from being the case without qualification. Leibniz, who abandoned the absolute distinction between uniform motion and acceleration, abandoned with it any kinematical clue to recognize forces; he simply decreed that they exist. Moreover, a relativistic theory of motion must explain the coincidence of inertial and gravitational mass, which an absolutistic theory can leave unexplained as a fortuitous circumstance, as a universal fact of nature. Given the mathematical tools of the seventeenth century, an explanation of this sort would confront insurmountable difficulties. Newton may not have been aware of them—neither was, for that matter, Mach. But a good intuition saved him from embracing a concept which, in order to be physically meaningful, required a non-Euclidian geometry. He was, with the state of the mathematical art at his time, much better off with an absolute space—with all of its metaphysical burdens.

Absolute space and God are both preconditions for action. If the universe consisted of one body only, then God alone could set it in motion, or stop it once it moved. If the universe consisted of two separate bodies, they would set each other in motion inasmuch as they would attract each other. But what is the subject of this absolute force, and how is it propagated? For awhile, Newton was willing to consider the aether as the subject and propagating medium of gravitation.[12] Even so, he would not have been able to account for the instantaneous effect of gravitation. Later on, when he took God's spatial omnipresence more and more literally, he could burden God, the source of all power, with its conservation and mediation.

---

[11] Mach, *Mechanik*, p. 226; Jammer, *Concepts of Space*, pp. 158–60; for Huygens, ibid., pp. 133–37; Freudenthal, *Atom und Individuum*, pp. 22–50, esp. pp. 34–35 (who tries to link Newton's redundant assumption of space, and his corpuscular theory, to the new reality of bourgeois society, very much like Macpherson, below v.c.n.16). For a more positive physical role of Newton's space see Toulmin, "Criticism in the History of Science: Newton on Absolute Space, Time, and Motion," pp. 1–29, 203–27.

[12] Newton, *Opticks* qq. 19–24, pp. 349–54; Westfall, *Force*, pp. 394–400; Koyré, *From the Closed World*, pp. 206–20. The conjecture of a universe with one, or two, bodies is mine. On Newton's speculations about forces other than gravitation, and their theological implications, see McGuire, "Neoplatonism" (above nn. 2,8).

Finally, space is *infinite*. Unless it were, it could not be literally predicated of God. Infinite space allows God, in his omnipotence, to create other worlds than ours (but I defer the discussion of this facet to the next chapter). There were also good physical reasons to assume the infinity of space. Newton's first law would not be valid whenever a uniformly moving body would reach the limit of the universe, given such a limit.[13] All bodies in space would collapse into one body, unless God held them apart: since holding them apart would have to be in proportion to the attractive force, this would presuppose an additional law of repulsion. The same problem also existed for Newton concerning the solar system in spite of the infinitude of space; but that was a partial problem that God could take care of by special dispensation, giving our planets an occasional additional impulse so that they would not be deflected by the sun's gravitation from their orbits. Newton's calculations probably induced him to belive that, during the five-and-a-half thousand years of creation, God had no need to add to the acceleration of the planets to keep them in orbit;[14] but his God, unlike Descartes's, was by no means a lazy gentleman who does not interfere in his creation. The very same calculations also could have served Newton in his objection to apocalyptic calculations of the end of the world, so common among radical theologians in his time. He argued against them that prophecies were not given to us to predict the future, but only to help us interpret and understand the course of past events; how else could free will be maintained? And since we do not know when or whether God shall refrain from correcting endangered planetary orbits, we cannot calculate the end of the world with the aid of science either.[15] Leibniz, who opposed many of Newton's

---

[13] Grant, *Much Ado*, p. 244.

[14] *Opticks*, p. 402 ("till this system wants a Reformation"); cf. Burtt, *The Metaphysical Foundations of Modern Science*, pp. 291–97; *Principia* 3 prop. 10, theorem 10 ("Motus Planetarum in Celis diutissime conservari posse"); *Opticks* 3 q.22, pp. 352–53 ("And so small a resistance would scarce make any sensible alteration in the Motions of the Planets in ten thousand Years").

[15] Newton, *Observations upon the Prophecies of Holy Writ* 21.8, *Opera*, 5:449: "The folly of interpreters hath been, to foretell times and things by this prophecy, as if God designed to make them prophets. . . . The design of God was much otherwise. He gave this, and the prophecies of the Old Testament, not to gratify men's curiosities by enabling them to foreknow things; but that, after they were fulfilled, they might be interpreted by the event, and his own Providence, not the interpreter, be then manifested thereby to the world."

theological positions, was particularly enraged by this image of God as an imperfect watchmaker, the mechanism that he created being in constant need of repair.

3. The Theological Meaning of the Triple Burden

Every aspect of Newton's space had, then, a physical reality to it; in every one of them, another ideal of science is stressed. Physical concerns induced him to assert its homogeneity, absoluteness, and infinity. But what is the subject in which time and space as absolute predicates inhere? What reality would they have (as they must) if all bodies were annihilated (as God can do)?[16] From early on Newton maintained that space and time are explicatory predicates to God's omnipresence and eternity, since these attributes should be understood literally and unequivocally. The presence of God in space allowed him not only to act in space—if pressed, Newton may have agreed that God could even act at a distance—but to be the actual carrier, or subject, of forces between bodies. And finally, space is indeed a *sensorium Dei*, a "sense organ" of God.

Note that Newton uses this enigmatic expression first and foremost by way of negation or remotion, that is, in a critical sense. We ought not say that all things *are* in God or participate in him, since God has no body:

> And yet we are not to consider the world as the body of God, or the several parts thereof, as the parts of God. He is an uniform being, void of Organs, Members or Parts, and they are his creatures subordinate to him, and subservient to his will; and he is no more the soul of them than the Soul of Men is the Soul of the Species of Things carried through the organs of Senses into the Place of its Sensation, where it perceives them by means of an immediate presence.[17]

In other words, the relationship between God and entities in space (creatures) is analogous to that between the sensing subject and his sensations. A sensation may be said to be "within" the "place of sensation"—the sensory—and yet its object is not part of the sensing subject. Newton disregards the asymmetrical aspect of his analogy. Whereas our sensations—the "sensible species" of the

---

[16] Grant, *Much Ado*, p. 242; on the principle of annihilation see below III.B.3, C.3, D.3; V.C.3.

[17] Newton, *Opticks* q. 31, p. 403. On the further, epistemological status of space see below II.H.5.

Scholastic idiom that Newton employs here—enter the sensory, the place of sensation, from outside it; this is not the case at all with God's sensory. In respect to God, "outside" and "inside" the *sensorium* cannot be viewed as *two* places: they are the very same space. Space is the place of objects and at the same time the place of God's intuition of these objects. Put differently, both objects and sensible species are the same for God. This being so, how could Newton avoid the conclusion that all beings are literally in God? At times he comes close to admitting it.[18] In his own way, then, Newton used the different properties of space to show how God is in things "by essence, power, and knowledge." Like his medieval predecessors, he was concerned not to forget God's transcendence over his immanence. Again like them, he could hardly succeed. But unlike all of them, his notion of ubiquity embodied the new ideals of unequivocation and homogeneity.

H. LEIBNIZ

1. Approaching Leibniz

"There was a time," Leibniz tells us, "in which I believed that all phenomena of motion could be explained with purely geometrical principles without assuming any metaphysical principles."[1] Even prior to the discovery that the "true measure of force" conserved is $m \cdot v^2$ (rather than Descartes's $m \cdot v$) Leibniz held, against Descartes, that motion, not extension, defines physical bodies; and that the existence of empty space can therefore be proven. A body at rest would be identical with the absolute space itself.[2] In his mature system, Leibniz insisted, against More and Newton, that space and time are just relations, while forces are intrinsic properties of bodies, prior to both extension and succession. On the face of it, this sounds like a contradiction in terms; the very mathematical expres-

---

[18] Clarke's fifth answer, GP, 7:426: "who [the omnipresent God] is not a mere *intelligentia supramundana*, [semota a nostris rebus seiunctaque longe], *is not far from everyone of us* [Acts 17:27–28]; *for in him we* (and all things) *live and move and have our Being*."

[1] Leibniz, GP, 7:280. On Leibniz's early philosophical views see Mazat, "Die Gedankenwelt des jungen Leibniz."

[2] GP, 1:71 (letter to Arnauld, n.d.): "corpus quiescens nullum esse, nec a spatio vacuo differe." Even in 1678 he writes: "Ego nihil agnosco in rebus quam corpus et mentes," and the souls he conceives of as points. Cf. B. Russell, *A Critical Exposition of the Philosophy of Leibniz*, pp. 77–78.

sion he offers for the true estimation of force involves both space and time. Evidently, true force differs from its appearance. It appears as a relational phenomenon, it expresses itself in relational magnitudes, but it is founded on an intrinsic property, a "principle of action" within things. It is hard to see what Leibniz had in mind without recourse to what he called "his system."[3] The question "What are physical bodies?" relates to the question "What are actual things (monads) and what are possible things (substances)?" or "How do the predicates of a true proposition inhere necessarily in its subject?"

That all these questions relate to each other need not mean that their answers are derivable from each other. Attempts to derive Leibniz's system "from his Logic" (Russell), or from his ontological position, or from epistemological or physical concerns are equally right and wrong. The *predicatum inest subiecto* principle did not commit Leibniz to believe that genuinely possible things (substances) must generate all their properties of themselves. The doctrine that "monads have no windows" and produce all of their states spontaneously does not commit him to attribute to all of them perceptions. Some philosophical edifices are indeed guided by one basic "philosophical intuition,"[4] but Leibniz's is not. The most promising approach to the reading of Leibniz is, I believe, the recognition that his problems and key terms are different and yet analogous on various levels of discourse. One such dominant problem is his endeavor to mediate between the absolute independence and absolute interdependence of things, to capture the unity-within-the-multiplicity of every single thing and of everything as a whole.

2. Predicatum Inest Subiecto: Four Reductions

"Always, therefore, the predicate or consequent inheres in the subject or antecedent, and in this very fact consists the nature of truth in general."[5] We know a proposition to be true if we can identify the

---

[3] E.g., GP, 6:563, 571 ("mon système"). Cf. *Essais de théodicée*, GP, 6:136 ("ce Système," referring to the Occasionalists). Cf. above I.A.n.9.

[4] Bergson, "L'Intuition philosophique," pp. 809-827, esp. pp. 809-811.

[5] Leibniz, *Opuscules et fragments inédits de Leibniz*, ed. Couturat, p. 518 (henceforth *OF*); Russell, *Leibniz*, p. v; cf. GP, 7:309: "In omni veritati universali affirmativa praedicatum inest subjecto, expresse quidem in veritatibus primitivis sive identicis . . . implicite autem in caeteris omnibus, quod analysi terminorum ostenditur . . ."; and in many other places. One possible source of Leibniz's ontological interpretation of the predicate-

reasons why its predicate merely explicates that which the subject implicitly includes. Leibniz, of course, says much more than that all propositions ought to be rewritten into the logical form SɛP, nor did he deny a truth-value to relational propositions.[6] He was, after all, a mathematician. But what he means exactly is obscured by at least four different meanings of the reduction of truth to analyticity. I shall call them, in turn, the qualitative, the quantitative, the modal, and the relational reduction. (i) Leibniz distinguishes, at times, between two *kinds* of "reasons" in the derivation of a predicate from the notion of the subject: in some propositions, SɛP is provable by virtue of the principle of noncontradiction. In most propositions, some or all steps of the derivation must rely on the principle of sufficient reason (PoSR), that "grand principle" which pervades every corner of Leibniz's thought; we shall postpone its discussion to the next chapter. It is a principle that "inclines without necessity,"[7] that is, without the logical necessity of the principle of noncontradiction (PoC). (ii) Leibniz also distinguishes, at other times, between finite

---

in-subject principle may be the controversy among the interpreters of Thomas in the sixteenth century over how, exactly, to understand the perseity of separate forms, e.g., angels, whose form involves *esse* "as a circle involves rotundity" (*Summa theol.* 1 q.50 a.5; above II.C.3 and n.33), and yet have their being depend on God's will. Cajetan distinguished between *perseitas simpliciter* and *perseitas physica*; in the summary of Franciscus de Silvestris, "Prima perseitas 'super habitudinem terminorum absolute sumptorum fundatur': secunda vero 'super habitudinem terminorum in esse naturali positorum.' " Franciscus objected that the latter implied the former: "Ut enim inquit S. Thomas Post. I lectio 14, 'si aliquod accidens de necessitate et semper inest subjecto, oportet quod habeat causa in subjecto, qua posita non possit accidens non in esse' ": *Comment. in Summam c. gentiles* 2 c.55 (Leonine ed. 13:396). See K. Werner, *Die Scholastik des späteren Mittelalters*, 4.1: *Der Endausgang der mittelalterlichen Scholastik*, pp. 325–26. Cf. below n.7.

[6] Ishiguro, *Leibniz's Philosophy of Logic and Language*, pp. 71–93. In his attempt to correct Russell, Ishiguro exaggerates the symmetry between relational and subject-predicate propositions. It is true that Leibniz held all or most of our concepts to be relational. But he believed in the existence of pure, nonrelational concepts, and hence also in ideal, nonrelational propositions that are primary to all others. A much deeper defense of the position that Leibniz "needs relation" is that of Hintikka, "Leibniz on Plenitude, Relations and the 'Reign of Law,' " in *Leibniz, A Collection of Critical Essays*, ed. Frankfurt, pp. 155–90. The following attempt to classify four meanings of reduction to analyticity is mine.

[7] GP, 7:301: ". . . sed semper aliqua eius ratio (inclinans tamen, non vero necessitans) reddi potest, quae ipsa ex rationum analysi . . . deduci posset." Cf. letter to Arnauld, GP 2:46. Cf. Thomas, *Summa contra gentiles* 2.30, *Opera* (Leonine), 13:338: "licet omnia ex divina voluntate dependeant, quae necessitatem non habent nisi ex sui proposti suppositione, non tamen propter hoc tollitur necessitas absoluta a rebus, quasi oporteat nos fateri omnia contingenter esse." Cf. also Franciscus de Silvestris *ad locum* and III.A. 1–2.

and infinite reductions.[8] What he calls *vérités de fait* against *vérités de raison* require, even in God's mind, infinitely many steps to prove the analyticity of S&P; the true concept of an *individuum* contains infinitely many well-integrated predicates; a possible world contains infinitely many compossible substances; a relational concept mirrors infinitely many relations. (iii) Again, it is one matter to reduce a proposition about possible subjects or states of affairs, and another to prove the analyticity of existential propositions. That Leibniz wanted even existential propositions to be analytic was conceded by Russell in the preface to the second edition of his seminal book.[9] Leibniz, after all, maintained that, in the absence of reasons to the contrary, being—existence—is in itself sufficient reason because it is "better than" nothingness (which, I am told, is not at all self-evident to Buddhists). (iv) And finally: relations, Leibniz believed, "sont le plus minces en réalité";[10] but truth, or right judgment, consists in the "conformity with the reality of things."[11] He admitted that most of our concepts of things contain relational aspects and tried to translate relational predicates into terms of properties. At least he must have believed that the necessity of relational propositions, whether by virtue of the PoC or of the PoSR, reflects the necessity of things. It is seldom clear in which of these four senses Leibniz "reduces" the predicate of a true proposition into the notion of the "subject." The ambiguity is due to Leibniz's heavy ontological commitments.

Truth reflects reality, and the tension that marks his notion of truth afflicts also his notion of reality: a proposition is true or false independently of any other proposition. But every proposition is connected to all other propositions. The reality of substance is likewise independent of other substances and likewise interconnected with all of them. Leibniz tried, as we shall see, to mediate between

---

[8] GP, 7:200: "Discrimen inter *veritates necessarias et contingentes* idem est, quod inter numeros commensurabiles et incommensurabiles . . . in surdis rationibus resolutio procedit in infinitum."

[9] *OF*, pp. 360, 376: "Ajo igitur Existens esse Ens quod cum pluribus compatibile est, seu Ens maxime possibile, itaque omnia coexistentia aeque possibilia sunt." Cf. Russell, *Leibniz*, pp. vi–vii. The consequences, though, are not any more "strange" than in his previously known doctrines.

[10] Leibniz, *Nouveaux essais sur l'entendement humain* (henceforth *NE*) 2.25; GP 5:210.

[11] *NE* 4.14; GP, 5:439.

these and similar opposites by means of postulating a continuum; but vices do not easily turn into virtues, and often not at all.

3. Reality and Individuation

"Reality" in Leibniz's terminology means first and foremost genuine possibility. The possibility of a substance is measured with two seemingly independent criteria. A genuine thing, a substance, must be individuated through and through. It must be capable of generating all of its accidents of itself. Leibniz was obsessed with individuality from his youth. He kept faithful to the inclination—manifest already in his dissertation—to seek the principle of individuation in the form of properties of things rather than, as Aristotle did, in matter. This was so even at a time in which he did not dismiss the reality of matter, and all the more so when he did. Aristotle assumed that all differences between singulars below the level of the *species specialissima* come from deficiencies in matter, because they cannot, by definition, be captured by a common form. The fact that this cow has a scar, while its twin does not, can be derived neither from the definition of cows nor from the specifications of a particular breed of them.[12] Ultimately, individual differences are not even intelligible since intelligibility means the cognition of forms. Aristotle must ultimately admit also the possibility of two or more singulars that are identical in all their properties, except that the bulk of (prime) matter in one does not coincide with a bulk of a similar amount of matter in the other. Leibniz—in this sense always a Scotist—insisted on the intelligibility of every genuine thing down to its individuality (*haecceitas*).[13] But only genuine, nonrelational properties individuate; things that differ only in number, or place, or time, or matter must be seen as one if they have all other properties in common. Leibniz deduced this "principle of the identity of indiscernibles" from the PoSR. Because parsimony is always better than waste, *nulla in rebus est indifferentia*.[14] It became his chief argument against Atomism or absolute space.[15] Only genuine individuals are truly

---

[12] Cf. below III.B.3.

[13] Cf. GP, 4:433 ("hecceité d'Alexandre"). Cf. already the *Disputatio metaphysica de principio individui* (1664) §18, GP, 4:23.

[14] *Leibniz-Clarke Corresp.*, GP, 7:373, 393, 395, 407; *Théodicée*, GP, 6:128.

[15] Leibniz, *Demonstratio contra Atomos sumpta ex Atomorum contactu* (1690) GP, 7:284–88.

possible, because their possibility neither includes nor precludes the possibility of another substance.

As a "principle of the better" (*principe de meilleur*) the PoSR, we recall, ensures that—*ceteris paribus*—being is always better than nothingness. A genuine possibility has, as such, a drive to exist (*exigentia existendi*).[16] But not all genuine "possibles" are compossible, not all substances can coexist in the same "possible world."[17] Leibniz had to postulate infinitely many possible worlds if he wished to avoid Spinoza's determinism or even monism. Spinoza's notion of reality, like Leibniz's, equates reality with independence, possibility, and perfection. If "no cause or reason can be given" for the non-existence of a being, it will necessarily exist, according to Spinoza.[18] This is certainly true of that which can be conceived as absolutely independent, *causa sui*. Every single reason for the independence of a substance is its attribute, or perfection, and increases its "reality."[19] Leibniz agreed to both ways of grading reality; his God is the *"ens realissimum"* on both counts. Spinoza also maintained that every genuine possible modification of the one substance possessing infinitely many attributes must be actualized.[20] Leibniz stressed even more strongly the parallel gradation of possibility and reality. But unlike Spinoza, he had to separate sharply reality and existence. The latter knows no gradation, nor is there room, as in Spinoza's system, for relative existence. Leibniz's substances are as real—as genuinely possible—as Spinoza's one and only substance. Not even God can make them impossible by thought or deed. They are also independent, that is, *causa sui*, at least as a concept. All have the same *exigentia existendi*. If all of them were to exist, there would be no room whatsoever for contingency. If only a few of them—though

---

[16] GP, 7:194: "Omne possibile exigit existere." *Principe de meilleur*, Leibniz, *Monadology* §46, GP, 6:614; *Théodicée* pref., GP, 6:44.

[17] Below III.B.4 (Scholasticism), III.E.2–3 (Leibniz).

[18] Spinoza, *Ethica* I prop. 11, schol., *Opera*, 1:44: "id necessario existere, cuius nulla ratio nec causa datur, quae impedit quominus existat."

[19] Ibid. I prop.9, p. 42.

[20] Ibid. I prop. 35, p. 66: "Quicquid concipimus in Dei postestate esse, id necessario est." Earlier, in the *Cogitata*, Spinoza distinguished between God's absolute and ordained power (above II.F.1; below III.A.1, E.3) and hence between actualized and unactualized possibilities; in the *Ethica* both powers became one. In a way, then, Leibniz had to return to a distinction like that between *esse essentiae* and *esse existentiae* (first articulated by Henry of Ghent). On the other hand, he did not distinguish between an essential and existential PoSR; below III.E.3.

still infinitely many—do exist, then this is only due to an external criterion. God endows some of them with the existence they strive for. But he does not do so arbitrarily, since he is guided by the principle of the better that "inclines him without necessitation." God's criterion is the maximum of compossibility. Compossible substances form infinitely many clusters of "possible worlds." Of these, God chose (as he ought to) the world inhabiting the maximal number of compossible substances. The degree of reality of a substance is, then, determined not only by its internal possibility, but also by its compossibility. A being compatible with the largest number of other beings is, we recall, *maxime possibile*;[21] Leibniz almost admitted there that such a being is more truly an existent than others—an admission as natural as it is detrimental to his intent *contingentiam salvare*. Leibniz never resolved the tension between these two criteria of reality, the *absolute* and the *comparative* reality; the former guaranteeing true individuals, the latter barring them from overcrowding the world and from infringing on the free will of God and man.

4. The Intrusion of Relations

At least, one may argue, he kept both criteria fairly separate. But it is hard to see how he could. I mean to say that, hard as he may have tried, there is no way in which he could, even in principle, avoid letting a possible world—the context of a substance—participate in its internal individuation. This is the same as claiming that there is no way in which Leibniz can avoid having some kinds of relations participate in the internal constitution of genuine things. In discussing substances, possible or existent (monads), Leibniz employs two sets of terms. Sometimes he speaks of their *properties* (or qualities), sometimes of their *states*. In one of his proofs for God's existence, of which he says that Spinoza agreed to it when he showed it to him in 1776,[22] Leibniz stipulated that genuine, simple qualities are such that they neither include nor preclude each other. They are, by def-

---

[21] Above n. 9.

[22] "Quod Ens Perfectissimum existit": GP, 7:261-62; cf. Russell, *Leibniz*, pp. 19–20. Leibniz refined his proof later, e.g., GP, 7:310, in that he omitted the recourse to existence as a perfection: the very *possibility* of such a being secures its existence. God is the *radix possibilitatis*. For a discussion of the same difficulty see also Rescher, *The Philosophy of Leibniz*, pp. 78–79.

inition, compatible, and therefore a being that has all of them is possible. If each such quality is named a perfection, an *ens perfectissimum*
is possible. Since existence is such a perfection, God exists. Let us
now assume that any other substance has fewer perfections. It is still
unclear why any of them should be incompatible with any other. If
two qualities, say P_1 and P_2, are always compatible, in what sense
could $S_1(P_1)$ and $S_2(P_2)$ be incompossible in the same possible
world? One avenue that Leibniz could have taken—and there are
some indications that he did—is to rule out, within a given possible
world, substances that have no single attribute in common. Or he
could have demanded that the distribution of a properties between
all substances of a possible world be somehow balanced. Incom-
possibility so conceived does not refer to logical impossibility; it
would mean that possible worlds are not arbitrary aggregates. But
surely, arbitrary aggregates must be in some sense possible, because
they contain no contradiction. Perhaps, then, they are meaningless
because they have the lowest degree of possibility. But if the organization of a possible world entails more than merely logical compossibility, and if the predicate "belonging to a certain possible
world" also inheres in a subject, then there is something in the relation between compossibles that determines the individual substance in its very individuality.

A further difficulty arises from the consideration of properties,
even more detrimental to Leibniz's intents. The individuality of a
substance is not a mere tautology. Some internal "principle of sequence" governs the coalescence of these rather than other predicates in this individual substance. But it would not be an *internal*
principle if the only rationale for a substance to have this rather than
another combination of predicates would be the circumstance that
this combination is the only one left for *this* substance to wear, because all other possible combinations are already occupied by its sis-
ter monads. Again Leibniz needs, without admitting it, a relational
criterion to explain the inclusion and exclusion of the properties of
a substance even though (simple) properties neither include nor exclude each other per se. Leibniz hints at a solution by endowing each
substance with all properties. Only God has all simple qualities simply, that is, to the highest possible degree. Other substances share
them in various degrees. "Every substance has something of the infinite, inasmuch as it involves its cause, God; [it has] even some trace

of his omniscience and omnipotence; now in the perfect notion of any substance of all of its predicates are included, the necessary and the contingent, past, present and future; indeed any substance expresses the whole world according to its position [*situm*] and point of view [*aspectum*]. . . ."[23] Leibniz may have argued as follows. That a certain property i having k degrees (P_i^k) cannot be followed or accompanied by another property P_j^l is not because the properties are incompatible—all simple properties are—but because of an imbalance of degrees. A person may possess somewhat more goodness than a sense of justice, but cannot be very good with very little sense of justice. Leibniz could even maintain that the consecutive order of properties follows from their degrees, or that a substance can acquire, at different states, various degrees of the same quality. The internal principle that determines the coalescence of properties can, however, not itself be a simple property; it must be an internal, relational structure of properties.

Leibniz preferred to speak of the "states" a monad occupies consecutively, though the term implies a temporal or at least nonreversible relation. He did not make serious attempts to translate states into atemporal properties—perhaps because he doubted that we can identify any simple property except existence. Nor, of course, do we have a complete notion of any individual. The terminology of states had many advantages. It permitted Leibniz to rephrase the internal, organizing principle of a monad as a principle of *change* and hence as a principle of *action*. At times he comes close to identifying a genuine substance with its active principle or its spontaneity.[24] A monad generates all of its states of itself, spontaneously, without interacting with any other monad. Monads "have no windows." The states of a monad follow not only each other, but from each other: the full knowledge of one state will include the knowledge of all of its states, past and future.[25] Now, the "states" of any monad must coincide with the states of all other monads in the same

[23] GP, 7:311. Note the marginal addition. Cf. also *OF*, p. 10.

[24] Leibniz, *Principes de la nature et de la grace* §1, GP, 6:598: "La substance est une Etre capable d'Action." Spontaneity: *Théodicée* §65, GP, 6:135, 562. Cf. also the so-called *Monadology* §11, GP, 6:608 ("principe interne").

[25] Leibniz, *Monadology* §7, GP, 6:607 ("Les Monades n'ont point de fenêtres"); §36, p. 613: monads cannot act on other monads nor be acted upon; their principle of action is "in themselves." To have a principle of motion "in itself" is, we recall, the sign of a body moving naturally in Aristotelian physics (below III.C.1).

world, or else they would have nothing in common and hence not belong to the same world.[26] This, I believe, is the minimal sense of the notorious "preestablished harmony" of monads: without interacting, all of them are nonetheless interconnected, and each of them mirrors in its own individual way all the others. According to the degree of their interconnectedness they can be seen as complex entities, subsets of monads with a dominating one which form together a *vinculum substantiale*.[27]

5. The Awareness of Monads

The states of a monad correspond, in Leibniz's theory of knowledge, to its various *perceptions* and apperceptions; the capacity to change states corresponds to its *appetitions*. All real things perceive, whether they appear to us animate or inanimate. The minimal degree or mode of perception are the *petits perceptions*—unconscious, instantaneous sensations that replace each other as they change.[28] In higher intelligences, perceptions do not drive each other out but combine and recombine with each other. But Leibniz does not distinguish too sharply between perceptions and apperceptions, or between sensation and conceptual knowledge; he rather envisages a continuum with varying degrees of clear and confused notions. He

[26] Leibniz, *Principes de la nature* §12, GP, 6:603–604; *Monadology* §78, GP, 6:620; *Théodicée* §59–63, 91, GP, 6:135–37, 289–90; Leibniz-Arnauld corresp., GP, 2:86 (concomitance or harmony between substances as a consequence of the individual substances's containing all its accidents and expressing the whole universe); Leibniz to the Landgraf of Hesse-Rheinfels (1686), GP, 2:12: "(9) Que chaque substance singuliere exprime tout l'universe à sa maniere, et que dans sa notion tous ses evenemens sont compris avec toutes leurs circonstances et toute la suite des choses exterieures. . . . (13) Comme la notion individuelle de chaque personne enferme une fois pour toutes ce qui lui arrivera à jamais. . . ." Cf. ibid., p. 17; GP, 1:382 (states, preestablished harmony); GP, 2:136; GP, 3:144 (three possible hypotheses: *influxus physicus*, Occasionalism, and the preestablished harmony). Cf. also Cassirer, *Leibniz' System in seinen wissenschaftlichen Grundlagen*, pp. 393f.

[27] The references were first collected by Russell, *Leibniz*, pp. 273–74. See, in particular, letter to des Bosses (1710), GP, 2:399, in which Leibniz suggests that the bond of monads (which Leibniz does not yet call *vinculum*) may solve the problem of transubstantiation without assuming the annihilation of monads. The term *vinculum* first appears in a letter to des Bosses, GP, 2:435. Of particular importance is GP, 2:481 (the *vinculum* now also ingenerable and incorruptible). Because the doctrine developed slowly, and for a while tentatively, the impression could be gained that Leibniz never really favored it: R. Latta, *Leibniz: The Monadology and Other Philosophical Writings*, p. 119. Cf. Boehm, *Le "vinculum substantiale" chez Leibniz: Ses origines historiques*, pp. 5–32.

[28] *NE* 2.9, esp. §1, 4, GP, 5:121–27; *Monadology* §14, 19–30, GP, 6:608–12.

speaks, indeed, of higher animals as possessing empirical knowledge; and he speaks of sense-notions and sense-truths.[29] Knowledge, it seems, is always conceptual knowledge of various modes and degrees of clarity and awareness. This ought to be kept in mind particularly in reference to the Clarke-Leibniz controversy.

Hindsight may be an effective way of elucidating the epistemological difference between Newton's and Leibniz's characterizations of space and time. Kant, who concurred with Leibniz in ascribing a phenomenal nature to both, tried nonetheless to preserve Newton's claim that they are not reducible to mere relations between things. Not only, says Kant, are they absolute in the sense that they precede the notion of things outside us and are presupposed by those, they are not *concepts* either. Concepts are won "discursively," intuitions are immediate. Concepts are imposed on experience by a linking, "synthetic" act of the understanding; intuitions are "given" (*gegeben*), and so also are space and time as the "pure forms of intuition." They pertain to the passive inventory of the mind.[30] This distinction between concepts and intuitions fits squarely into the center of the Clarke-Leibniz controversy. When Leibniz said that there is nothing more to space and time than the notion of the difference between things he probably meant the following. Assume a possible world (U) with three "things" (a, b, c) and three properties (F_1, F_2, F_3) in it. Assume further that a "thing" must have at least one property, and that (a) has all of them, (b) all but the last, and (c) only the first.

---

[29] *NE* preface, GP, 5:44: "Les consecutions des bestes sont purement comme celles des simples empiriques" etc. *NE* 1.1 §25, GP, 5:72 (*notions de sens*; Gerhardt misplaced the paragraph). *NE* 1.1 §18 (the senses provide propositions). On clear and confused notions 2.1. Appetition and force: Belaval, *Leibniz*, p. 402.

[30] Kant, *KdRV* B38–40 (esp. 3 and 4); B48 (esp. 4 and 5). "Immediacy" and "givenness" do not entail each other either in the history of the notion of intuition or in its subsequent employment (e.g., the immediacy of Being and Nothingness in Hegel's logic). Recently, Hintikka suggested a "conceptual" reduction of Kant's intuitions to concepts of singulars, "Kantian Intuitions," pp. 341–45; id., *Knowledge and the Known: Historical Perspectives in Epistemology*, pp. 126–34. Against this interpretation cf. Parsons, *Mathematics in Philosophy: Selected Essays*, pp. 110–49, esp. pp. 142–49. Of time, Kant will develop, later in the *Critique*, a theory that makes it into a mediating instance between pure concepts and sensations because it has an affinity to both (below VI.A. n. 6). Moreover, against Hintikka stands Kant's explicit opinion that singularity is not "given," but construed by a category (*Allheit*). Systematic consistency may be on Hintikka's side; Kant's text and the historical impulses imbedded in it favor the irreducibility of *Anschaulichkeit* to conceptuality. Cf., e.g., Cohen, *Kommentar zur Kritik der reinen Vernunft*, pp. 26–38.

This is the only content of the spatial expression that places (*a*) and (*b*) closer to each other than to (*c*) and also equidistant from (*c*). In a world of more properties and more things the mapping could turn three-dimensional. Space and time are merely relational concepts. Clarke, by contrast, protested that "*different Spaces are really different* or distinct from another, though they be *perfectly alike.*"[31] There may be things that share all properties yet are different in space or time; or three things in which the intermediate is equidistant from the others although it shares more properties with the one to its left than with the one to its right. The presence of things to the mind demands something else beyond their conceptual determination in themselves and against each other; while Kant believed that intuition (with its forms) need not be constitutive of a mind superior to ours, Newton made space into a divine sensorium that permits all things to be present to God and allows God to be present in all things. Kant, however, stated explicitly what Newton and Clarke assumed only implicitly: that the *Anschaulichkeit* of things is irreducible to conceptualization. To Leibniz, by contrast, all perception is conceptualization.

Since different monads have coinciding perceptions with each other, each monad can be said to *see* the whole world in which it is embedded from its own "point of view." The more consciousness a monad possesses, the more aware it will be of the interconnectedness of all phenomena and all things. Knowing itself it knows the world. God, who knows himself completely, knows *eo ipso* all the singulars in all possible worlds. Again, Leibniz sees a correspondence between the degree of awareness and the degree of reality, or possibility. An absolute awareness is the awareness of all the monads and all the ways they are interconnected so that if the slightest detail in any monad would be different, the whole world would have been different from its very beginning[32]—and therefore all other monads would have been different! Monads, their states and their perceptions, are as independent of each other as they are interdependent.

[31] Clarke's third letter, GP, 6:367. It is worth noting that a similar dispute arose in the epistemology of the fourteenth century: Aureoli postulated an *esse apparens*; Ockham and others refuted its necessity. Cf. below v.A.n.18 (lit.).

[32] Leibniz-Arnauld correspondence, GP, 2:40–43, 50–52 (the possible and actual Adam). Cf. esp. p. 40: "Car comme il y a une infinité des mondes possibles, il y a aussi une infinité des loix, les unes propres à l'un, les autres à l'autre, et chaque individu possible de quelque monde enferme dans sa notion les loix de son monde." Cf. ibid., p. 62.

6. Forces and Relations

The very same dialectic of interrelatedness and independence plagued—or, if you wish, enriched—Leibniz's concept of *force*. Any reader of Leibniz finds himself, at first, in a complex terminological maze. Leibniz speaks of primitive and derivative forces, of forces active and passive, dead and living, internal and external. And he claims—this was our starting point—that motion is relative, while force is, in some sense, absolute; wherefore it does not depend on spatial relations. Some interpreters translated his intentions into a simple two-worlds model—the metaphysical (real) and the phenomenal; the "spontaneity" or "action" of a monad is its "primitive active force": it is only represented, or has its "image," in velocities.[33] But what could a "passive force" mean in a monad? A monad neither acts on other monads nor need it resist action. The sum of its (apparent) interrelations coincides perfectly with its spontaneously generated events. The difference between "action" (force) and "resistance" (mass, impenetrability) can arise only in a lower level of reality, on the level of semi-substances.

Leibniz developed such a doctrine of semi-substances in order to lend precise meaning to Christ's real presence in the Host. It was not an occasional aberration into the domain of theology proper; Leibniz prepared a lengthy treatise on "catholic proofs."[34] The doctrine of the *vinculum* had the advantage that it could accommodate both a dogma of transubstantiation and (perhaps even better) a doctrine of consubstantiation; it pleased Leibniz's irenic temper. Aggregates of monads, if governed by one of them, are also substances of sorts. The unity of this aggregate has two aspects, or meanings, formal and material. As true unity, it is an *individuum*, all the "actions" or "events" of which inhere in the subject. The law governing the series of its actions is a *vis activa primitiva*. As an aggregate nonetheless, this unity must express their common ground, or homogeneity: this is matter, or the *vis passiva primitiva*, the resistance to any

[33] Buchdahl, *Metaphysics and the Philosophy of Science: The Classical Origins: Descartes to Kant*, pp. 394–405, 419–25 (p. 425: dynamics as "image of activity"), 461–69. Similarly already, Cassirer, *Leibniz*, esp. pp. 297–302.

[34] Cf. also Leibniz, *SB* 1.6, pp. 489–559, esp. 515–17 (Eucharist), 507–508 (transubstantiation); there are fragments of a lengthy work on proofs for Catholic dogmas. But note that the doctrine of the *vinculum* fits as much transubstantiation as it does consubstantiation. Cf. also *Théodicée*, discours préliminaire §18, GP, 6:60–61. An earlier, different solution: letter to Arnauld (n.d.), GP, 1:75.

disturbance of the common arrangement; in short, mass or impenetrability.

But absolute impenetrability is as impossible as the vacuum. Material bodies are neither infinitely penetrable nor infinitely impenetrable (rigid). Against all dynamic traditions from Descartes to Newton, Leibniz formulated his laws of motion and force upon perfectly elastic bodies. Indeed, we saw that the failure of Descartes to explain the phenomenon of elasticity guided many corrections of his laws (Spinoza, Hobbes, Malebranche); even Newton needed an additional unspecified force to account for it. "Force," for Leibniz, meant neither momentum nor impulse, but the common denominator of the sum of all causes and of the (a priori equivalent) sum of all effects in a given state—say, at impact. It is closer to the notion of the amount of work done. In a perfectly elastic body, force at impact would be *visibly* conserved: bodies would bounce back after impact to the height of their descent or to the measure of the distance traversed. The sum total of direction would likewise be conserved. Perfectly rigid bodies, if equal in mass and distance traversed, would come to a standstill—the absolute loss of motion or force.[35] Loss of force would also result if it were to be measured by $(m \cdot v)$ rather than, as Leibniz recognized with the help of Huygens, $(m \cdot v^2)$. In a perfectly elastic body, the motion or change can therefore be attributed as much to the body's capacity for change (*vis activa primitiva*) as it can be attributed to the impact of the other body and the transfer of forces. Both are perfectly equivalent descriptions; the former is only metaphysically preferable. At impact, two elastic bodies can be also perfectly interpreted as one body in which

[35] On this, Newton and Leibniz agreed against Descartes and Huygens. (*Principia* 2.46, AT, 8:1, p. 68 [first law of motion]; Huygens, *De motu corporum ex percussione*, *Oeuvres* 16:31.) Cf. Newton, *Opticks* 3 q. 31, p. 398: "For Bodies which are either absolutely hard, or so soft as to be void of Elasticity, will not rebound from one another." Leibniz, letter to Huygens (1692), *Mathematische Schriften*, ed. Gerhardt (henceforth GM), 2:145: colliding bodies would come to rest; only elastic bodies rebound. Cf. GM, 6:103; *NE* 2.13 §21, GP, 5:138; even inelastic bodies can be treated as if they were: Leibniz to Clarke (date?), GP, 7:414 (here reference to the above quoted passage from Newton). And, with a metaphysical lesson for the motion of a substance and its changes, Leibniz-Arnauld corresp., GP, 2:78: "L'Homme qui ne contient qu'une masse figurée d'une dureté infinie . . . ne sçavoit envelopper en luy tous ses estats passés et futurs, et encore moins ceux de tout l'universe." The famous proof for the measure of force from the impossibility of *perpetuum mobile* follows. Cf. Russell, *Leibniz*, pp. 89–90; Westfall, *Force*, pp. 291–92, 294–95, 302–303; Freudenthal, *Atom und Individuum*, pp. 71–77.

change is "occasioned." The "occasion" in which force is expressed
is the *vis activa derivativa*: Leibniz employs the word "derived" in
analogy to the sense in which, given a certain geometrical area, infinitely many—but not all—shapes can be derived from it. In this
sense, force is relatively independent of the question "Which body
really moved?"

Perfectly elastic bodies, as Leibniz knew well, are as much an
idealization as perfectly rigid ones; except that the former are a
right, the latter a wrong or misleading *ens rationis* (Leibniz never argued against the existence of atoms with the PoC, only on the
grounds of the PoSR).[36] On the other hand, all bodies are elastic
or—if you wish—systems of bodies. To the measure that they *seem*
to lose force at impact, they are more or less elastic; but they never
really lose force: the loss of $(m \cdot v^2)$ when translated into internal motion "is not lost to the whole of the universe." Only on this level of
incompletely elastic bodies, that is, of incompletely closed systems
(or genuine material substances), is it possible to mistake activity
within two bodies for their *interaction*. This is why $(m \cdot v^2)$ is a closer
"image" of reality (the activity of a monad) than mere motion. But
(physical) force is by no means an *ens*. Again Leibniz tried to mediate between the interconnectedness and independence of things,
here, of physical bodies. I should hasten to add that he did not prove
the priority and independence of force over and from relative motion, and thus he did not really refute the necessity of absolute
space. Physical forces, in his account, are also relative (to their system). And since even that which Newton called inertial motion
needs force, motion itself is both relative and absolute. The ultimate
derivation of force from action is only an "analogy of nature."
What, then, is the ontic status of forces? of relations in general?

7. Physical Homogeneity versus Metaphysical Unequivocation

Upon the status of interrelations between things or phenomena depends the validity of our knowledge of the natural world. Relations
are, in Leibniz's terminology, "a well-founded" phenomenon.

---

[36] Leibniz to de Volder, GP, 2:169: "Nempe hypothesi mea, quatenus corpora perfecte
elastica non sunt, vis intestinis partibus quae et ipsae Elasticae sunt recipitur neque adeo
perit, sed tantum sensibus subducitur, quae quidem consuetudini naturae et ordini id est
experientiae et rationi consentanea esse non negabis."

They represent reality without themselves being real; and so do phenomena, that is, the ways of perceiving the world. But how well are phenomena founded in reality? If relations possess the "least of reality," does Leibnız want his law of conservation of power to be less real (and therefore less true) than the least conscious substance in the least of all possible worlds? I believe that Leibniz could not afford an absolute dichotomy between the real (substances, monads) and the phenomenal (relations, perceptions) without continuous gradation between them; Gottfried Martin suggested a similar solution.[37] Some relations (or phenomena) are as real as the possible world that they order. "Indeed, God sees not only single monads and the modifications of each monad, but also their relations, and in this [fact] consists the reality of relations or truths."[38] The validity of God's knowledge stems from the circumstance that it is perspective-free, "iconography" rather than merely a perspective-bound "scenography."[39] Some of the relations so seen by God are ideal: they stem from the way monads see each other in a spatial or temporal order of coexistence. Other relations are real: the connection by which they move each other. These relations, both ideal and real, are valid not in the sense of being things themselves, but in the sense of being valid even *sub specie Dei*. An example of ideal relations—besides space, time, or matter—is undoubtedly the law of the lever. It assumes two perfectly balanced bodies, disregarding all other properties. There is no sufficient reason to assume that the one rather than the other body will descend, and, therefore, neither will. But it is the same principle of sufficient reason that precludes identical states or entities to begin with: *nulla in rebus est indifferentia*. Does it mean that this law is ontologically meaningless because it does not deal with things? But it does not deal with phenomena either. It is an abstraction, an idealization.[40] The mark of an idealization is that it disregards the context and assumes homogeneous entities or conditions. The *homogeneity* of matter is another such

---

[37] G. Martin, *Leibniz, Logic, and Metaphysics*, trans. Northcott and Lucas, pp. 158–72.

[38] Leibniz to des Bosses (1712), GP, 2:438.

[39] Ibid.: "Sunt enim scenographiae diversae pro spectatoris situ, ichonographia seu geometrica representatio unica est; nempe Deus exacte res videt quales sunt secundum Geometricam vertitatem, quamquam idem etiam scit quomodo quaeque res cuique alteri appareat, et ita omnes alias apparentias in se continet eminenter."

[40] Leibniz to Hartsoecker, GP, 3:519.

useful abstraction.[41] Such abstractions are not in themselves useful; they are useful when recombined into a contextual explanation. When so combining, they yield other relational concepts like force, and force is a *real relation*. The law of conservation of *vis viva* reflects, one might say, the coincidence of two or more world-lines of action of individual substances. It represents the force, or principle of action, which is the essence of substances. But even the law of conservation is only a (real) representation. Among the relation concepts, that of homogeneous space and time refer to a much lesser degree to properties of things than the concept of *force*. Force—or better, the conservation of *vis viva*—occupies, in Leibniz's thought, a position similar to covariance or to constants in modern physics: it is independent of a particular frame of reference or "point of view," though it can only be expressed in terms of this or that frame of reference. This interpretation commits me to attribute to Leibniz the view that complex notions (such as force) may be more "real" than their simpler constituents (space, motion).

There is yet a "more perfect" relation between substances, a real union into a "substantial chain." The union of mind and body is such a quasi-new substance made out of infinitely many monads of which one, the self, is dominating. There are, then, grades of reality or perfection to relations that lead straight into substantiality; of such a substance we may say that its component monads are its predicates, or even properties, whether or not it knows it. The lowest grade of reality is, perhaps, that of sense perceptions not yet subsumed under a cognitive pattern, a law. Even in those, Leibniz detected "quelque ressemblance avec la raison"; and memory, too, "imitates reason."[42] I suppose that in this, too, Leibniz could sense an *analogon rationis* (to use the language of his followers) that no two

---

[41] Leibniz to de Volder (n.d.), GP 2:190: "Etsi autem revera nulla actio in natura sit sine obstaculo, abstractione tamen animi separatur quod in re per se est, ab eo quod accidentibus miscetur. . . ." Leibniz to de Volder (1705), GP, 2:276: "universum corporeum compositum esse ex una substantia infinitis diversis modis affecta non dixerim, etsi materiam in se spectatam [!] (seu quoad passiva) ubique esse sibi similem dici possit. [Dici enim potest hactenus Realem esse Materiam quatenus in substantiis simplicibus ratio est quod in phaenomenis observatur passivi]. Cf. *NE* 3.10 §15, GP, 5:325 (uniformity of matter); mass, we learn from GP, 2:252–53, is an abstract and incomplete concept—contrary to substance.

[42] Leibniz, *Principes de la nature* §5, GP, 6:600; *Monadology* §26, GP, 6:611; cf. Baeumler, *Das Irrationalitätsproblem in der Aesthetik und Logik des 18. Jahrhundert*, p. 189.

drops of water are alike when observed closely, that every phenomenon, *like* every real thing, is as unique as it is, in the light of reason,
totally interconnected with others.

Absolute *homogeneity* is thus an abstraction useful in the construction of laws of nature. It has, of course, no meaning at all on the
level of real things. Absolute *unequivocation* would be possible only
with a complete knowledge of all individuals, a knowledge only
asymptotically approachable. The "being in things" of God, his omnipresence, has, of course, no spatial connotations:[43] it means his
operation in things, which again means his giving possibles existence; even more, God generates all possibilities in his thought spontaneously, all existents and all points of view. Having gone as far as
to interpret the mind–body relations in terms of the *vinculum substantiale*, nothing but the fear of being attacked could bar him from
seeing in the whole world, by analogy, a *suppositum* of God. God,
too, is the Monad-in-Chief. The possible and existing substances
are partial aspects of him, though independent of sorts. They surely
could be said to be almost as much the body of God as the monads
that constitute the nonmental portions of the self are its body.[44]

This sketchy interpretation of Leibniz's system—to which I shall
add a few more observations in the next chapter—may be wrong on
several counts. Yet if I could not do justice to its incredible ingenuity, I hope at least to have conveyed some impression of its elusive,

---

[43] *NE* 2.23 § 21, GP, 5:205: "Les Ecoles ont trois sortes d'*Ubieté* . . . *circomscriptive* . . .
definitive . . . La troisième Ubieté est la *repletive*, qu'on attribue à Dieu, qui remplit tout
l'Universe encor plus eminement que les esprits ne sont dans les corps, car il opere im-
mediatement sur toutes les creatures en les produisant continuellement, au lieu que les
esprits finis n'y scauroient exercer aucune influence ou operation immediate. Je ne scay,
si cette doctrine des écoles merite d'estre tournée en ridicule, comme il semble qu'on s'ef-
force de faire." It is closest to the opinion of Thomas of Strassburg, above II.D.2.

[44] Leibniz does distinguish between mere aggregates (from "The Chorus of Angels"
to "Cadaver") and organisms (complex substances, *vinculum*), e.g., Appendix to the letter to des Bosses, GP, 2:506. The former have no governing, or dominating, monads. But
the difference can be called into question in view of our world as a whole being more than
an aggregate; it is also governed by one monad, as indeed is also the "City of God," i.e.,
assemblage des esprits (Monadology §§85–89; *Principes de la nature* §15; GP, 6:621–22, 605;
Théodicée §§146, 247–48, GP, 6:196, 264–65). In what should "the chorus of the angels,"
"the city of God" and the monads under the dominion of God differ? In order, then, to
maintain the distinction between organism and aggregate and still see our universe as
governed by one monad Leibniz would have to distinguish between two meanings of
"dominion." He could not simply equate it with the distinction between mechanical and
teleological union, because every union, from stone to angel, is both.

almost slippery character. In one sense, Leibniz did claim that God contains all other things; in another sense, monads are conceptually absolutely non-identical with each other and with God. In one sense, God and monads are absolutely free; in another, they and their states are absolutely predetermined. It was not by mere malice that Leibniz was accused of hidden Spinozism. He could have just as easily been accused of Pelagianism. His doctrines could be accommodated to any desired political, ethical, or religious disposition. I am not even convinced by Russell's separation between Leibniz's esoteric and exoteric doctrine. True, Leibniz was too dependent on "the smiles of princes."[45] Yet if he had an esoteric doctrine, it was no less ambiguous than his exoteric one. He really wished to reconcile opposites and thought he could, to a measure, do so. His irenic temper knew no bounds.

I have argued that, for Leibniz, the absolute distinctness, unequivocation, or individuation of entities is absolutely meaningful only on the metaphysical level; while absolute homogeneity is meaningful only on the phenomenal level, and false if applied to real entities. If this interpretation is correct, it may help us to understand the historical root of Kant's theory of regulative ideals with which we started. Leibniz helped Kant to understand the tension between these ideals. But Kant did not accept Leibniz's distinction between their domains of validity to solve the tension. Both are, to Kant, metatheoretical postulates on the same level: the demand for homogeneity depends on its opposite, the demand for precise particularization. Against the principle *entia praeter necessitatem non sunt multiplicanda* stands the equally valid principle *entium varietate non temere est minuendo*.[46] Science must want to generalize, but it should never lose the sense of reality for the sake of generalization. The mathematization of physics between Galileo and Newton offers a vivid lesson of the difficult maneuvering between Schylla and Charybdis, between conflicting ideals of reason, none of which can be abandoned for the sake of the other.

---

[45] Russell, *Leibniz*, pp. 1–7, esp. p. 3; cf. p. 1: "To please a prince, to refute a rival philosopher, or to escape the censure of a theologian, he would take any pains." He was also timid, and denied having had any contact with Spinoza except for a letter about optical matters, even though they met and he read carefully the first part of the *Ethics*. And yet the difference between his published and unpublished positions are merely in matters of style; the monads are not a metaphor.

[46] Above, I.D.

Let me sum up this chapter. The medieval sense of God's symbolic presence in his creation, and the sense of a universe replete with transcendent meanings and hints, had to recede if not to give way totally to the postulates of univocation and homogeneity in the seventeenth century. God's relation to the world had to be given a concrete physical meaning. Descartes did so by maintaining the medieval sense of God's utter transcendence; the only relation of God to the world that could thus be rescued was that of causality, a relation that Descartes exploits to the extreme. More, on the other hand, rather translated the panpsychism, or even pantheism, of philosophies of nature in the Renaissance into a "clear and distinct" language. God thus acquired a body of sorts, or at least a *sensorium*. It may be of some significance that Descartes, a Catholic, avoided even the semblance of endowing his God with a body while More, the liberal Protestant, did not. Leibniz avoided both positions by denying bodies and places an absolute ontological status. All of them and most of the others believed that the subjects of theology and science alike can be absolutely de-metaphorized and de-symbolized.

It is clear why a God describable in unequivocal terms, or even given physical features and functions, eventually became all the easier to discard. As a scientific hypothesis, he was later shown to be superfluous; as a being, he was shown to be a mere hypostatization of rational, social, or psychological ideals and images. Our story thus comes to a halt. We have seen how and why God lost his body in Christian theology, how and why he regained it in the seventeenth century. Once God regained transparency or even a body, he was all the easier to identify and to kill. The study of his slow philosophical death—from Kant through Feuerbach to Nietzsche—is as fascinating as the story we told of his lost and found body. But that is another story.

III

DIVINE OMNIPOTENCE AND LAWS
OF NATURE

A. OMNIPOTENCE AND NATURE

1. Omnipotence and Necessary Truths

Some of Descartes's most enigmatic remarks concern the range of
the divine omnipotence vis-à-vis eternal truths. Eternal truths are
created in a radical sense of the word; even mathematical theorems
are contingent upon God's will. What the most radical defenders of
divine omnipotence in the Middle Ages hardly ever asserted, Descartes did without hesitation: that God could invalidate the most
basic mathematical operations, for example, $2 + 1 = 3$, or the laws
of mathematical physics, or create matter without extension.[1] He
said so not once or twice, but consistently and without much variation. How one can reconcile these extreme voluntaristic pronouncements with his ideals of rationality has long been an exegetical
problem; I shall address it later. Descartes may have wished to come
to the aid of post-Tridentine theology. Theologians and philosophers were at best reserved, at the worst horrified about the implications of what seemed to be a relativization of the principle of noncontradiction.[2]

Spinoza, who composed a guide, *more geometrico*, to Descartes's
system in the beginning of his philosophical career, added to it an

[1] Most of the references are assembled in Gilson, *Index scholastico-Cartésien*, p. 235 (s.v.
possibile). It is clear that Descartes develops his stand in opposition to some Scholastic
versions of "eternal ideas" as necessary even to God; see below III.D.n.26 (Suarez).

[2] Descartes himself seldom uses the Scholastic terms *potentia ordinata et absoluta*, but cf.
Meditationes, AT, 7:435; 8:2, 167 (*puissance ordinaire et extraordinaire*). Gassendi uses the
terms *ordinata-absoluta* in the same innocuous (though "voluntaristic") sense, namely as
the trite distinction between miracles and order: Osler, "Providence and Divine Will in
Gassendi's Views on Scientific Knowledge," pp. 549-60, esp. p. 554 n. 23. If Descartes
wanted to avoid the terms because they would enable an ill-wisher to identify his position
with a controversial theological position, it was done anyway; cf. Caterus, AT, 7:25:
"Deum negas posse mentiri aut decipere, cum tamen non desinit Scholastici qui illud af-
firment, ut Gabriel [!] Arminensis et alii, qui putant Deum absoluta potestate mentiri,
hoc sit contra suam mentem, et contra quod creavit, aliquid hominibus significare." Later
in the century the question became a much-debated one: Bayle, *Dictionnaire historique et
critique* (1740) 4:56 (s.v. Rimini).

appendix with his own deviating positions. Divine omnipotence
and necessity of nature, he argued, are one and the same, since all
that is really possible in the world is also as necessary as any mathematical truth. Only because this is beyond our immediate comprehension do we distinguish between the possible and the necessary.[3]
God's will (by which he loves himself, i.e., asserts himself) and
knowledge of himself as *causa sui* are likewise the same.[4] Already
Spinoza employs an exegetical technique he was to use later in deal-
ing with the traditional theological vocabulary, a technique that deserves the name neutralization through amplification: he expands
the connotation of a term to the point that it means everything and
nothing.[5] Later, in the *Ethics*, he became even more explicit: "Whatever we conceive to be in the power of God, is necessary."[6]

As always, Leibniz sought to mediate between extremes—Descartes's radical voluntarism and Spinoza's determinism; and found
a mediating formula in his famous distinction between logical and
physical necessities.[7] Logical necessity (*nécessité logique*) is grounded
on the principle of noncontradiction only, under which God's will
and even his thought are subsumed. Not only can God not create

---

[3] Spinoza, *Cogitata* 2.9 §2; note that in §4 the reference to the Scholastic *potentia absoluta–ordinata* distinction. But Spinoza's position was ambiguous in the *Cogitata* (cf. above
II.F.1). A remarkable attempt to reconstruct Spinoza's system out of the medieval philosophical heritage is Wolfson, *The Philosophy of Spinoza*. Even if one were to endorse this
approach without reservations, it should be noted that Wolfson neglects the strong influence of later Scholasticism on Spinoza, as, e.g., in this case.

[4] Spinoza, *Cogitata* 2.8, Van Vloten-Land, 4:217–19, esp. p. 219 ("Dividimus itaque
potentiam Dei in ordinatam et *absolutam*" etc.). The following distinction between *potentia
ordinaria* and *extraordinaria* falls within the *potentia ordinata*: it distinguishes between the
immutable and mutable order. Whether it is any more than a theoretical distinction,
"Verum hoc decernere Theologis relinquimus."

[5] Cf. below IV.B.3.

[6] Spinoza, *Ethica* 1 prop. 35, Van Vloten-Land, 1:66.

[7] Leibniz, *Théodicée*, GP, 6:50 (*nécessité géometrique* against *nécessité physique* or *morale*);
GP, 6:32 (*absolue–morale*); *Testamen anagogicum*, GP, 7:278 (*déterminations géométriques–d. ar-
chitectoniques*); *De rerum originatione*, GP, 7:303 (physical necessity as hypothetical necessity); *Principes de la nature*, GP, 7:603 (cf. 6:44; *impossibile* against *inconveniens*); GP, 2:62
(two kinds of a priori); *Noveaux essais*, GP, 5:387 (certitude . . . physique–nécessité, ou
certitude metaphysique). It has been argued that Locke's distinction between "verbal, trifling" and "serious, substantial" necessity (or certainty) reflects the distinction between
analytic and synthetic a priori: Woolhouse, *Locke's Philosophy of Science and Knowledge*,
pp. 25–32, esp. p. 27. I see at best a premonition of such a distinction: Leibniz, by contrast, argues for it by force of "principles." Locke's very choice of words shows his dis-
interest in the "trifling" necessity. See also above II.H.1 nn.5, 7 (necessity; "simple" and
"physical" perseity in later Scholasticism).

logical contradictions; he cannot even conceive of that which is logically possible (*possibile logicum* is one of the terms Leibniz inherited from the Scotistic tradition) as impossible.[8] More than that, God must think all possibilities and compossibilities by necessity of his nature. Any other necessity—"physical," "metaphysical," "moral," "architectonical"—is grounded on the principle of sufficient reason, a principle that "inclines God without necessity" (*inclinat sine necessitate*).[9]

2. Logical and Physical Necessity

The merits of Leibniz's distinction cannot be overstated. He articulated a central problem for all philosophies of science from the seventeenth century to our own days. Why should we assume that nature is well structured, and hence intelligible? In what sense are laws of nature "universal and necessary" (Kant)? Their necessity is evidently less than logical. The negation, say, of the universal law of gravitation does not entail a logical contradiction. A universe is indeed conceivable in which bodies repel (rather than attract) each other in direct proportion to their masses and in an inverse proportion to their squared distances. It may be an uncomfortable universe to live in, but it is not a self-contradictory one. Our intuition—or be it common sense—tells us nonetheless that there is a sense in which lawlike sentences are more necessary than statements about past contingents—say, that my house is colored brown. Leibniz derived extralogical necessities from the "great principle" of sufficient reason,[10] which rest ultimately on God's wisdom; but the principle leads him to opposite conclusions in different contexts. Kant detached his canon of "synthetic *a-priori* judgments" from theological considerations altogether. He wished to derive them from the immutable structure of any possible understanding.[11]

---

   [8] See below III.B.4.

   [9] Leibniz, *Théodicée*, GP, 6:127: "que celle raison incline, sans nécessiter." It is the mark of will in general, ibid., p. 126: "Astra inclinant, non necessitant." Cf. ibid., p. 414; Letters to Arnauld, GP, 2:12, 14 (in conjunction with the predicate-in-subject principle), 56. Cf. *Textes inédits*, ed. Grua, p. 479 (necessity to inclination as exact equation to approximation) and above II.H.2.

   [10] Leibniz, *Monadology* § 31, GP, 6:612: "Nos raisonnemens sont fondés sur deux grand Principes" etc. Cf. below III.E.3.

   [11] Kant himself was aware that his distinction between analytic–synthetic a priori is indebted to the Leibnizian two principles of contradiction and sufficient reason; against Wolff and Baumgarten he insisted, at least since the first critique, on the underivability

Should we wish to base less-than-logical necessities on the somewhat weaker ground of induction, we are led squarely into logical
paradoxes. Some of them, like Nelson Goodman's, obtain even if
we admit that laws of nature are never verifiable but at best falsifia-
ble. Every generalization of past observations in the form
$(x)[A(x) \supset B(x)]$ can be said to be "confirmed" (even in the modest sense
of "not falsified") regardless of whether we observe, in the future,
either $B(x)$ or $[\sim B(x)]$ to be the case, because the property held by us to
be $B(x)$ in the past may actually be the property $[B(x_{t<0}) \wedge \sim B(x_{t>0})]$.
Inductive inferences are a continuous exercise in begging the question:
they already assume the very constancy of certain properties that
they hope to establish.[12] A poor way out of the difficulty is the suggestion to treat all lawlike proportions as counterfactual conditional
statements. "All houses in Jerusalem are stone-built" would, then,
be true if, and only if, a house not built in Jerusalem would have
been stone-built if it were built in Jerusalem.[13] It may be objected
that the term "laws of nature" has all but disappeared from the active vocabulary of modern scientists—only philosophers of science,
or so it seems, are bothered by it. This is only partially true. Constants
of nature (such as the speed of light or Planck's constant), principles
(such as Heisenberg's uncertainty relation), and forces are burdened
with the same problems as previous laws—inasmuch as they imply
not only universally, but also some kind of necessity; if, that is, they
are not only descriptive, but in any sense prescriptive. Now, physicists
do not regard the speed of light as a mere empirical, temporary limit-
value: the question "What happens to a body moving twice the
speed of light?" they might say, is downright meaningless. Lawlike
statements did not altogether disappear from our scientific discourse.
Yet, the very meaning—let alone justification—of lawlike statements
is as dubious today as it was during the rise of early modern science.

There were good historical reasons why the status of laws of nature
became an oppressive problem. Many of its physical laws were counterintuitive. Whether or not counterfactual conditionals are indeed the

---

of the latter from the former. G. Martin, *Immanuel Kant: Ontologie und Wissenschaftstheorie*, pp. 83–88 (esp. 87f.) calls this self-interpretation "kühn," yet "wohlfundiert." Cf. already Latta, *Leibniz*, pp. 208–11.

[12] Goodman, *Fact, Fiction, and Forecast*, pp. 59–83; a good discussion of the paradox in
I. Scheffler, *The Anatomy of Inquiry*, pp. 295–326. "Entrenchment" is admittedly not a solution, but a practical guideline.

[13] Goodman, *Fact*, pp. 17–27, 119–22.

true form of lawlike statements, evidently some of the most fundamental laws of science were conceived and perceived as such. The
principle of inertia is a case in point. In the eyes of those who
first formulated it, the inertial principle embodied the emancipatory
achievements of the new mechanics: it freed physics from the childish
anthropomorphic notion that bodies have an *inclinatio ad quietem*, a
tendency toward rest.[14] Did it not prove, once again, that Aristotelian
science consisted of hasty generalizations from sense data? But that
every body tends to preserve a uniform, rectilinear motion in a given
direction is plainly a counterfactual conditional proposition. Cartesian
physics regarded all bodies as so many parts of one and the same
material continuum, differentiated only by the relative motions within
it.[15] The motion of any given body "in and of itself (*inquantum in se
est*) is only an "inclination."[16] Even in Newton's infinite space, a purely
inertial motion would be unobservable; in order to observe a body so
moving we have to approach it to a finite distance and exert some
attractive force that must alter the motion of that body ever so slightly.
The principle of inertia, like Newton's second and third laws of motion, does not describe nature, but rather a limiting case to natural
states, of which more will be said later. The logical—or metaphysical—status of such laws had indeed to be reexamined.

3. God's Absolute and Ordained Power

Leibniz had, then, good reasons to be proud of his distinction between logical and physical necessities. It abounded with consequences. It also allowed him to remain a good scientist while being
an obedient theologian, to believe both in the divine infinite free
will and in the preestablished order of the world with the largest
number of compossibles obeying the least number of simple principles. "This is," he says of his distinction, "the root of contingency, and I doubt that it has been seen by anyone else."[17] But he
should have known better. The distinction between logical and
physical necessity is but a reformulation of a distinction as old as the

---

[14] Below III.C.I, and n.2.

[15] Above II.E.I.

[16] Descartes, *Principia* 2.36–38, AT, 7:62–64. Cf. above II.E.n.3 (Lucretius).

[17] Leibniz, GP, 7:200: "Atque haec est radix contingentiae, nescio an hactenus explica-
tae a quoquam"; cf. GP, 6:127: "et c'est en cela que consiste la contingence" (here in the
sense of future contingents).

beginning of Scholasticism, namely the distinction between God's absolute and ordained power (*potentia Dei absoluta et ordinata*). The former considers God's power as such, and recognizes no limits to it, no confining law or order, except for the principle of noncontradiction; the latter considers God's power inasmuch as it is actualized or realizable in an order of things. It is a distinction well worth tracing; the classical world never had it. Aristotle did distinguish between absolute and hypothetical necessity; both, however, pertain to logics only *or* to physics only. Otherwise he always tries to prove the conceptual, even logical, necessity of that which is the established order of things.

Medieval theology introduced the distinction between the two aspects of God's power so as to enlarge as far as possible the horizon of that which is possible to God without violating reason. In the later Middle Ages, schoolmen were driven by an almost obsessive compulsion to actually devise orders of nature or orders of grace different from the one admittedly existing. Very few of them doubted the general correctness of the Aristotelian world-picture; and yet they insisted that, if God so wanted, the earth would cease to be the center of the universe or even the proper place of all heavy things.[18] God could move the whole system of the universe, spheres and all, in a straight line within an empty space indefinitely.[19] And had God wished, the saviour of the world could have been a stone or a donkey—*aut lapis, aut asinus.*[20]

Or again, God could deceive us at any time even in our most basic certitudes, or at least implant in us immediate notions of non-existing things—*notitia intuitiva rebus non existentibus*; whether or not such cognitions also generate erroneous judgments is a question often raised after Ockham's denial that this can be the case. Only our self-awareness is above error, since we cannot both exist and not exist.[21] Here is one of the sources of Descartes's malignant spirit,

---

[18] Below III.B.3.

[19] Above II.D.1, D.4 and n.20.

[20] Above II.D.1 and n.3.

[21] "Sum certus evidenter de objecto quinque sensuum et de actibus meis": Nicolaus of Autrecourt's letters to Bernard of Arezzo, in J. Lappe, *Nicolaus von Autrecourt, sein Leben, seine Philosophie, seine Schriften*, appendix *6; *Exigit ordo executionis*, prol. 2, *Medieval Studies* 1 (1939): 184; Weinberg, *Nicolaus of Autrecourt: A Study in 14th Century Thought*; A. Maier, "Das Problem der Evidenz," in *Ausgehendes Mittelalter* 2.367–418; id., *Metaphysische Hintergründe*, pp. 390–98.

that new broom with which Descartes hoped to clear the road for a totally new mode of philosophizing. At almost every turn of his argument, from the call for radical doubt to the discovery of the *sum res cogitans* as an Archimedian point from which to reconstruct the world, Descartes speaks the despised language of medieval theology. So does Leibniz.

"Laws of nature" may have had their finest hour in the seventeenth and eighteenth centuries. The term was of Scholastic, and perhaps of even older, origin.[22] In early modern science and philosophy it implied, to quote Kant again, "universality and necessity." "Universality" referred to the *homogeneity* of nature discussed in the last chapter. Seldom do we find this sense attached to the term "laws of nature" in the Middle Ages. "Necessity," on the other hand, referred to that peculiar notion of unconditional truths which nonetheless are not mere tautologies. This aspect of laws of nature constitutes the theme of this chapter. Late medieval Scholasticism was intoxicated with varieties of hypothetical reasoning; theologians and philosophers pursued the systematic development of imaginary orders and states so as to emphasize the utter contingency of the world. Where did this bent of mind come from, and in what way did it prepare for the scientific revolution of the seventeenth century? Was its significance philosophical only or did it also affect the practice of science? Indeed, we are also in danger of overstating our case. It is not my intention to blur the novelty of early modern sci-

---

[22] Reich, "Der historische Ursprung des Naturgesetzbegriff," in *Festschrift für Ernst Kapp zum 70. Geburtstag*, ed. Diller and Erbse, pp. 121ff., traced it back to the *foedera naturae* of Lucretius. Schramm argued for a medieval origin: "Roger Bacon's Begriff vom Naturgesetz," in *Die Renaissance der Wissenschaften im 12. Jahrhundert*, ed. Weimar, pp. 197–207 (*lex naturae universalis* that prevails, as an active force, over the *natura particularis* of bodies); above II.D.n.22. Milton, "The Origins and Development of the Concept of the 'Laws of Nature,' " pp. 173–95, argues against all classical or medieval origins of the term, but emphasizes the medieval roots of "voluntarism" and "Nominalism" as necessary background. Recently, F. Oakley, *Omnipotence, Covenant, and Order: An Excursion in the History of Ideas from Abelard to Leibniz*, has tried to set the covenantal *potentia absoluta–ordinata* dialectics against the principle of plenitude; in the seventeenth century, he argues, Boyle and Newton belong to the one, Leibniz to the other. My point of view, already delineated in "The Dialectical Preparation for Scientific Revolutions," in *The Copernican Achievement*, ed. Westman, pp. 177ff. and elsewhere, is different; Leibniz and Newton are both heirs to the medieval distinction, each with a different emphasis. Cf. below III.E.1., E.2. The counterpart to the principle of plenitude is not the *potentia absoluta–ordinata* dialectics, but the principle of economy (above I.D., Kant). Of great importance in Oakley's book is the treatment of political discourse.

ence. To the contrary, the better we comprehend the contribution of the medieval hypothetical reasoning to the scientific enterprise of the centuries that followed, the more precisely can we define that which was new in the reasoning of Galileo and of his contemporaries. Using Scholastic forms of reasoning, they discovered a new land. I wish to show, first, why Scholasticism saw this land from afar but could not enter it.

B. POTENTIA DEI ABSOLUTA ET ORDINATA

1. Patristic Sources

Viewing Judaism and Christianity as the only monotheistic island in a sea of polytheism is a historical mistake which originated in the Middle Ages. The *theologia naturalis* of the intellectuals of the pagan world often asserted the unity of the divine.[1] One of the strongest trends in Greek philosophy labored toward an ever more rarified vision of *one* God, from Xenophanes' attack against the anrthropomorphic pluralism of myths through Aristotle's unmoved mover and the Stoic *logos* down to the "one" of Plotinus. Neoplatonism crystallized negative theology into the form it was to maintain. Plotinus denied even that the "one" possesses self-knowledge. A reflexive act would impair its transcendental unity, because it could then be aspecticized into the one as knowing object and the one as known subject. At best, it may be attributed with a kind of nonreflexive intuition.[2] More often than not, the educated Graeco-Roman intellectual believed that, beyond the due political respect to the gods of the commonwealth (on par with our veneration of flags), religion is one in the variety of cults. Augustine puts similar words into the mouth of Porphyrius.[3] But not only Church fathers, even Jewish sages in later Antiquity were aware of a changing climate of opinion. One tradition lets a pagan philosopher by the name of Zeno tell Rabbi Akiba: "You and I both know in our heart that there is no substance to idolatry"; only Rashi, the celebrated medieval commentator to the Talmud, added: "And this Zeno was a Jew"—he could not imagine a pagan monotheist anymore.[4]

[1] Jaeger, *Die Theologie*, pp. 1ff., 50ff. (Xenophanes).
[2] Rist, *Plotinus: The Road to Reality*, pp. 38–52.
[3] Augustine, *DcD* 10.9ff., pp. 281ff.
[4] *Bab. Talmud, Avoda Zara* 55a; Rashi *ad loc.* More enigmatic is *Hulin* 13a: "Aliens

The classical tradition of anti-Jewish and anti-Christian propaganda is notable for the lack of serious attacks against the monotheistic idea.[5] The clash between Judeo-Christian and pagan theologies was not over the number of gods, but over the nature of the divine—be it one or many. To the Greek mind, God embodied the principle of universal, immutable order, self-contained and without any desire: he is in need of nothing. The image of God as a moral personality, choosing some people over others and active in history—an all-powerful busybody—insulted the Greek sense of harmony. That God should neglect the care for the immutable universal order in order to concentrate his attention on the affairs of a small, dirty nation in the provinces—is it not a "frog and worm perspective?" Celsus thought so.[6] Celsus, Porphyrius, and others directed their polemical ingenuity at the root of such presumptions. Their philosophical critique of the Old and New Testament laid the foundations for all subsequent critiques.[7] They asked, if God is immutable, how could he change his mind about the order of salvation? And if he wanted a given order of nature, what does it mean that he can act against it? And if he is omnipotent, can he reverse the past, or make the truth be false, or annihilate himself?[8] The invo-

---

abroad are not idolaters; they merely follow the customs of their fathers." On the meaning and knowledge of "idolatry," see Lieberman, *Yevanim ve yavnut be'erets yisrae* [Greeks and Hellenism in Jewish Palestine], pp. 236–52. It should be noted that the passage in *Hulin* was *not* employed by the *tossafists* of the twelfth century to buttress the decision that Christianity does not fall under the category of idolatry.

[5] To the contrary, because of its monotheistic creed, Judaism appeared to the early Greek authors as a "philosophical" religion: Stern, *Greek and Latin Authors on Jews and Judaism*, I: *From Herodotus to Plutarch*, pp. 10 (Theophrastus), 26–27 (Hecataeus); Levy, *Olamot nifgashim* [Studies in Jewish Hellenism], pp. 15–59.

[6] Origines, *Contra Celsum* 6.23, ed. Kötschau, p. 281; trans. Chadwick, p. 199. Cf. Andresen, *Logos und Nomos: Die Polemik des Kelsos wider das Christentum*, pp. 226–28. Cf. below n. 8.

[7] Harnack, "Porphyrius gegen die Christen," in *SB der königlichen Akademie der Wissenschaften, Phil. hist. Klasse I*. Literature: below IV.B. n.15.

[8] Galenus ridiculed the notions of Moses, for whom "it seems enough to say that God simply willed the arrangement of matter and it was presently arranged in due order; for he believes everything is possible with God, even should He wish to make a bull or a horse out of ashes. We however do not hold this; we say that certain things are impossible by nature and that God does not even attempt such things at all but that he chooses the best out of the possibilities of becoming (ἐκ τῶν δυνατῶν γενέσθαι τὸ βέλτιστον αἱρεῖσθαι). . . . We say thus that God is the cause both of the choice of the best in the products of creation themselves and of the selection of the matter"—an almost Leibnizian formulation: Galenus, *De usus partium* 11.14, trans. Walzer, *Galen on Jews and Christians,*

cation of the paradoxes of omnipotence was unfair: the ancients knew paradoxes of self-reference in various domains.

Patristic theology was much more troubled by concrete arguments against the divine providence than by abstract paradoxes. It usually answered them by denying that God wants to act *contra naturam* or that he wants to do that which does not fit or behoove (*decet*) his nature.[9] Already the Jewish-Hellenistic wisdom of Solomon contains a peculiar defense of God's power vis-à-vis the order of nature (*subest enim tibi cum volueris posse*) together with the emphasis on the reliability of the order of nature.[10] Origenes seems to have initiated the formula that would serve future discussions down to the thirteenth century: he distinguished in God between *agere per potentiam* and *agere per iustitiam*.[11] The latter means, in fact, that God does everything "in measure" and abhors the infinite. That the later distinction between *potentia absoluta et ordinata* corresponds to this older one was recognized already by Gregory of Rimini.[12]

Only once, in the most unexpected circumstances, do we meet a systematic attempt to interpret all miracles naturally. An Irish (?) monk of the seventh century named Augustinus wrote a little-known treatise under the name "De mirabilibus sacrae scripturae," in which the miracles in the books of Moses were systematically reduced to natural phenomena in the manner of a liberal Protestant or

---

pp. 26–27; on Celsus see below n. 11. The only classical school that had room for a notion of omnipotence was the Stoic: Cicero, *De natura deorum* 3.92 (Poseidonius). On the importance of force in their cosmology above II.B.3.

[9] Augustine, *Contra Faustum Manichaeum* 26, p. 480: "Deus autem, creator et conditor omnium naturarum, nihil contra naturam facit."

[10] *Sapientia Salomonis* 12.18 (quoted, e.g., by Petrus Damiani, *De divina omnipotentia*, pp. 599–600).

[11] Origenes, *Comm. ser. in Matt.* 95, Migne, *PG* 13: 1716: "quoniam in quantum ad potentiam quidem Dei omnia possibilia sunt sive iusta sive iniusta, quantum autem ad iustitiam eius . . . non sunt omnia possibilia" Cf. *Contra Cel.* 3.70; 5.25; *De principiis* 2.9.1; 4.4.8; and Rufinus, *De fide* 17, *PL* 21: 1131: ". . . impie Origenes ac nefarie fassus est, qui sic dixit: Non omnia quae voluit Deus fecit, sed ea tantum quae potuit continere et comprehendere" (quoted in Origenes, *De principiis*, ed. Görgemanns and Karpp, p. 400). Cf. also R. M. Grant, *Miracle and Natural Law in Greco-Roman and Early Christian Thought*, pp. 127–34.

[12] Gregory of Rimini, *Lectura super primum et secundum sententiarum* 1. d.42–44 q.1 a.2, ed. Trapp and Marcolino, 3:368: "Huic distinctioni satis concordat alia antiqua, qua dictum est quod quaedam deus non potest de iustitia, quae potest de potentia." The formula was used in the beginning of the thirteenth century by Praepositinus of Cremona: Grondziel, "Die Entwicklung der Unterscheidung zwischen der potentia Dei absoluta und der potentia Dei ordinata von Augustin bis Alexander von Hales," p. 31 n. 2.

Jewish theologian of the nineteenth century.[13] Had we no certain evidence of time and place, we would be tempted to doubt both since they are so much at odds with their putative *genius loci* and *Zeitgeist*.

At any rate, even if questions concerning the extension of God's will relative to the order of nature were not treated systematically in Antiquity—very few theological questions were—questions of that sort harbored an explosive potential.

2. Early Scholastic Discussions

It came to the fore in the eleventh century, when Petrus Damiani, one of Gregory the VII's most faithful supporters in the investiture struggle, attacked the recent fashion of logical argumentation in theology. To those who persisted in denying the power of God to restore virginity, Damiani's "De divina omnipotentia" answers in kind.[14] Aristotle once stated that even the gods cannot change the past *post factum*; here and there the matter was mentioned in Patristic literature.[15] Damiani argued that, if he wished, God could indeed reverse past events and have Rome not be founded, let alone perform lesser miracles such as the restitution of a virgin *post ruinam*, since the true propositions in God's mind do not have a temporal index.[16] Even moderate theologians like Anselm of Canterbury recognized immediately the danger in Damiani's argument. An omnipotence without bounds, Anselm retorted, is actually a weak-

---

[13] Anderson, "Divine Governance, Miracles, and Laws of Nature in the Early Middle Ages: The De Mirabilibus Sacrae Scripturae," esp. pp. 80–107; MacGinty, "The Treatise De Mirabilibus Sacrae Scripturae," has prepared an edition that soon will be published in the *Corpus Christianorum* and replace the unusable edition of Migne.

[14] Damiani, *De divina omnipotentia in reparatione corruptae, et factis infectis reddendis*, in *Lettre sur la tout-puissance divine*, ed. Cantin, pp. 384–489; 410–18 (necessitarian consequences of the opposite view); cf. Courtenay, "The Dialectic of Divine Omnipotence," in *Covenant and Causality in Medieval Thought*, pp. 2–3; Oakley, *Omnipotence*, pp. 42–44; Enders, *Petrus Damiani und die weltliche Wissenschaft*, pp. 16ff., esp. p. 17 n. 1.

[15] Aristotle, *Ethica Nic.* z2.1139b. 7–11 (referring to Plato, *Nomoi* 934ab; *Illias* 24.550–51, 522–24). It should be remembered that elsewhere Aristotle insists on the "necessity" of past contingents (the discussion of tomorrow's sea battle). On Porphyrius see Grant, *Miracle*, p. 131.

[16] Damiani, *De divina omnipotentia*, ed. Cantin, pp. 428ff., 474–78. Apart from this, Damiani also refers to a special version of a *non decet* argument: it behooves God to create something out of nothing rather than to turn something into nothing; and that which both is and is not is both bad and nothing (ibid., pp. 434, 436ff.).

ness.[17] A God that can create contradictions could also annihilate himself together with his omnipotence; a God that may be thought of as non-existing cannot be an *ens necessarium* either. God's will is at least bound by the principle of noncontradiction. As to the order God has established, "It does not behoove God to permit anything inordinate in his kingdom" (*Deum . . . non decet aliquid inordinatum in suo regno dimittere*),[18] for otherwise God could be interpreted both to want and not to want such an order. The object of God's will must remain consistent. From here onwards, almost everyone agrees to the subsumption of God's will at least under the condition of logical consistency; though not all agreed about what it is that constitutes a contradiction.

Nascent Scholasticism came also in contact with Maimonides' *Guide to the Perplexed,* and through it with a systematic exposition of the extreme voluntarism of the Moslem *'Ashari'a* which denied any kind of necessity of nature, including causality, and interpreted nature as a spatial and temporal sequence of discrete, atomic events absolutely independent of each other. Against their methodical anarchy, Maimonides developed a theory of contingent orders. We shall discuss it in the next chapter:[19] its most interesting feature, in view of subsequent treatments of the question, was the strict parallelism between laws of nature and the decreed laws of the covenant. Both are necessary and contingent structures at one and the same time; both are rational, yet must leave room for contingency; both are instances of divine accommodation to a resilient "matter" at hand. In places where Maimonides considered directly changes in the order of nature, his answer resembles Origen's seminal distinction between *posse de potentia* and *posse de iustitia.* We believe, he says, "that reality is eternal by nature, as He, may he be exalted, wanted; that nothing in it will change except in [accidental] details, even

---

[17] Anselm of Canterbury, *Proslogion* 7, in *Opera omnia,* ed. Schmitt, 1:105–106: "Sed omnipotens quomodo es, si omnia non potes? Aut si non potest corrumpi nec mentiri nec facere verum esse falsum, ut quod factum est non esse factum . . . ? An haec posse non est potentia, sed impotentia?" Cf. Petrus Lombardus, *Sent.* 1 d.42 c.2, 1:260) and Funkenstein, "Changes in the Patterns of Christian Anti-Jewish Polemics in the 12th Century," pp. 129–31; Courtenay, "Necessity and Freedom in Anselm's Conception of God," pp. 39–64.

[18] Anselm, *Cur Deus Homo* 1.12; cf. 2.17; ed. Schmitt, 2:80, 123.

[19] Below IV.C.2. On the influence of the *Kalam* see also Courtenay, "The Critique of Natural Causality in the Mutakallimun and Nominalism," pp. 77–94.

though He, may he be exalted, has the power to change it completely or to add or to subtract the one or the other nature from
among its natures . . . but his wisdom decreed to actualize every
creature as it is actualized . . . and not to change its nature."[20]

Patristic and other influence coalesced, then, in the Scholastic discussions about the horizon of God's omnipotence. Most alternative
positions were articulated in the thirteenth and fourteenth centuries;
they were held by fifteenth- and sixteenth-century Scholasticism.
The characterization of the main positions, to which I now turn,
leaves out many of the links and transitions in the ongoing, dramatic debate. My purpose is only to compare them to views in the
seventeenth century concerning the status of the laws of nature.
Both Scholastic and seventeenth-century philosophers of nature
asked for the meaning of modalities: what the contents of "necessity," "possibility," and "contingency" are in the view of God, of
the world, and of our understanding of both. The *reification of modal
categories* was a central problem to medieval Scholasticism and remained so in the seventeenth century—though with crucial changes
in emphasis.

The medieval debate focused, in turn, on the precise meaning of
(i) *contingency* (order), (ii) *possibility* (or the difference between logical and physical possibility), and (iii) *formal-logical necessity* (if applied to matters divine). This distinction, though somewhat schematic, will help me to link different, though interdependent,
problems.

3. Contingency and Singularity

The meaning of *contingency* was at the heart of the attempts to clarify
the relation between God's absolute and ordained power—the
terms first appear as a pair in Alexander of Hales. He understood
absolute power to refer to everything that comes to mind—whether
contradictory or not—while ordained power refers to the logically,
morally, or physically nonrepugnant.[21] Others employed the new

---

[20] Maimonides, *More Nebuchim* 2.29; cf. *Guide of the Perplexed*, trans. Pines, p. 346. Cf.
A. Ravitsky, "Keifi koach ha'adam—yemot hamashiach bemishnat harambam," in *Me-
shichiyut ve'eschatologia*, p. 217 and n. 67; Lasker, *Jewish Philosophical Polemics against
Christianity in the Middle Ages*, pp. 28–35.

[21] Alexander of Hales, *Summa theol.* p. 1 inq. 1 tr.4 q. 111 c.4, 1:236: "Ad hoc, nihil
temere asserendo . . . quod, si potentia Dei concipiatur ab anima absolute, non poterit
anima determinare nec capere infinitum pelagus suae potestatis. Sed cum anima specu-

terminology in the same sense of the distinction, inherited from Origenes, between *posse de potentia* and *posse de iustitia*—to separate the actual order of nature and grace from all other possible acts of God. The construction of the dichotomy was, then, either too large or too narrow. It was, I believe, Thomas Aquinas who raised this distinction to a higher level of reflection, who gave it a new dimension. The traces of Maimonides' theory are well recognized in his.

Potentia Dei absoluta meant, to Thomas, anything that does not violate the principle of noncontradiction, anything that can claim the status of a "thing"—whether within some order or outside any order.[22] By "noncontradiction" Thomas meant explicitly the *non repugnantia terminorum*, that is, a logical-formal property that needs no further clarification. God cannot create contradictory states of affairs because a subject that contradicts itself is not a "thing" (*res*). When asking what God can or cannot do, the terms "can" and "do" apply only to that which is *doable (factibile)*, things and not states (suffering); this new meaning of the "possible" was alien to the Greek philosophical tradition. *Potentia Dei ordinata*, on the other hand, means not only the actual order of nature, our universe, but also any other possible order of things inasmuch as it is an *order (ordo ad invicem)*. There are many such orders or, in Leibniz's later phrase, "possible worlds."[23] But Leibniz's "possible worlds" are, by defi-

---

latur divinam potentiam ut ordinatam secundum conditionem potestatis, veritatis, bonitatis, dico quod possibile Deo est quod posse potentiae est et non potest quod est impotentiae [this was Anselm's formula: above n. 17]. Secundum hoc dixerunt . . . quod quia contradicit suae potentiae vel veritati vel bonitati, potentiae maiestatis non conveniret ut faceret maiorem se; item, veritati eius contradiceret facere de eodem simul esse et non esse [later generations will subsume this impossibility under *potentia absoluta* rather than *ordinata*]; item, bonitati eius contradiceret damnare Petrum et salvare Iudam et peccare."

[22] Thomas Aquinas, *Summa theol.* 1 q.25 a.5 ad primum: "quod attribuitur potentiae secundum se consideratae, dicitur Deus posse *secundum potentiam absolutam*. Et huiusmodi est omne illud in quo potest salvari ratio entis, ut supra [a.3, where the condition of the *factibile* is shown to be the *non repugnantia terminorum*] dictum est. Quod autem attribuitur potentiae divinae secundum quod exequitur imperium voluntatis iustae, hoc dicitur Deus posse facere *de potentia ordinata*." *De potentia* q.1 a.5: "Ad quintum dicendum, quod absolutum et regulatum non attribuuntur . . . nisi ex nostra consideratione: quae potentiae Dei in se consideratae, quae absoluta dicitur, aliquid attribuit quod non attribuit ei secundum quod ad sapientiam comparatur, prout dicitur ordinata." In both texts, the distinction does not play a prominent role in elucidating the many meanings of that which is possible to God; it appears, as it were, not in the responsio, but in an answer to a particular objection.

[23] Thomas Aquinas, *Summa theol.* ibid. resp.: "Unde divina sapientia non determinatur ad aliquem certum ordinem rerum, ut non possit alius cursus rerum ab ipsa effluere

nition, nonrealizable inasmuch as they are concurrent with ours, a "logical possibility" in a sense which Scotus, as we shall see, was first to explicate. What Thomas had in mind was the creation of more or altogether different *species* than there are in our universe.

To the question of Petrus Lombardus: "Could God have created a better world than ours?" Thomas's answer differs from that of later schoolmen only in nuance. God could have created any number of totally different orders, there can be no extrinsic limit to the *ars divina*. To any given order, actual or possible, God could have created a better one if we take "better" as an adjective of that which is created rather than as a qualifying adverb of God's actions. Can any such possible world be called "best"? It is but a matter of semantics whether we say, with Thomas, that whatever God creates is best (*optimum*) or that, therefore, none is.[24] Since God wished to create, and to any things and orders he could have chosen there is a better one conceivable, God had to choose arbitrarily even if wisely:

> When we speak of bringing into being the whole universe, we cannot find anything beyond that which is created from which a reason could be elicited why it is such and such; since one cannot elicit a reason for the disposition of the universe either considering divine power, which is infinite, or considering the divine goodness, which is not in need of things, it is necessary to elicit its reason from the *simple will of the producer*; so that it is asked why the quantity of the heavens be such and not greater: one cannot give a reason except the will of the producer. And therefore also, as Rabbi Moyses [Maimonides] said, the Holy Scripture leads man toward the consideration of the celestial bodies, the disposition of which shows that everything is subject to the will

---

. . ."; ibid., ad tertium: "Unde, licet istis rebus quae nunc sunt, nullus alius cursus esset bonus et conveniens, tamen Deus posset alias res facere, et alium eis imponere ordinem." *Quaestiones disputatae I, De potentia* q. 1 a. 3 ad 8: "ars Dei non solum se extendit ad ea quae facta sunt, sed ad multa alia. Unde quando in aliquo mutat cursum naturae non propter hoc contra artem suam facit." Ibid. q. 1 a. 5: ". . . [divina bonitas] potest manifestari per alias creaturas et alio modo *ordinatas*"; cf. *De potentia* q. 6 a. 1 ad 12: "ars divina non totam se ipsam explicat in creaturarum productione." See also Courtenay, "Dialectic," p. 9.

[24] *Summa theol.* 1 q. 25 a. 6 ad 3; *De potentia* q. 1 a. 5 ad 15: "illud quod facit, est optimum per ordinem ad Dei bonitatem; et ideo quiquid aliud est ordinabile . . . est optimum." Robert Holcot, *In quatuor libros Sent.* II q. 2, s: "Tertio dico quod deus non posset facere optimum quod potest facere: quia quocumque bono dato deus potest facere melius." Holcot draws much of what he says here about the plurality of worlds from Ockham. On the differences between Ockham's and Thomas's views, vid. infra; but in respect to the question whether any possible world is best, Holcot draws the right consequence from both positions. Thomas also does not use the term "infinite" as willingly.

and providence of the creator. One cannot assign a reason why this star has such-and-such a distance from that star, or other such [unexplainable] matters which occur in the disposition of the heaven under consideration, except out of the order of God's wisdom.[25]

From these and similar references we obtain the following structure of the *potentia ordinata–absoluta* relation: whatever is not self-contradictory (*per se impossibile*) falls under *potentia absoluta* even if it were not well ordered; the conceptual clarification of "power" does not entail "wisdom." Under *potentia ordinata* falls not only our world, but also any other conceivable order of things, so that it is futile to ask why this universe was created—the question could be repeated infinitely regarding any order. It was a *necessarily* arbitrary act.

The distance between this position and Duns Scotus's proof that God must be the source of contingency (and that his will has primacy over his wisdom) is very small indeed.[26] We also find, in the above quotation as elsewhere, a premonition of the manner in which Duns Scotus was to modify the relation between both "powers" of God. Duns Scotus was first and foremost disturbed by the possibility, implicit in the notion of both powers even in Thomas, of calling any possible act of God, however imaginary, "inordinate." The terms "absolute" and "ordained" power, he insisted, do not stand for two *kinds* of divine power, but rather for two aspects of the same power.[27] It resembles the distinction of the lawyers between that which one is capable of doing de jure as against that which one can do de facto. The extension of the former is larger than the extension of the latter only where the law itself is not "in the power of the [free] agent" (*in potestate agentis*). If I steal a horse, my act would be inordinate though indeed possible. But in the case of an agent who has power over the law, the possible and the orderly are coextensive. Emperors, by definition, do not steal horses. *Quod*

---

[25] *De potentia* q.3 a.17 resp. Cf. below IV.C.2. (contingency in Maimonides).

[26] Scotus, *Ordinatio* 1 d.39 q.u. n.14, in *Opera omnia*, ed. Balič, p. 6: "Nulla causatio alicuius causae potest salvare contingentiam, nisi prima causa ponatur immediate contingenter causare, et hoc ponendo in prima causa perfectam causalitatem, sicut catholici ponunt." *That* there is contingency in the world can only be proven a posteriori: *Report. Paris.* 1 d.40 n.6 (*Opera omnia*, ed. Wadding, 11:220–22); cf. Gilson, *Scot*, pp. 327–28.

[27] Scotus, *Ordinatio* 1 d.44 q.u., 6:363ff.; Miethke, *Ockhams Weg*, pp. 145–49; Pannenberg, *Die Prädestinationslehre des Duns Scotus im Zusammenhang der scholastischen Lehrentwicklung*, pp. 68ff.; Bannach, *Die Lehre von der doppelten Macht Gottes bei Wilhelm von Ockham*, pp. 13–17.

principi placuit, legis habet vigorem.[28] The act of an emperor may contradict an existing statutory law, in which case either the act establishes a new law, or remains a special case which, by definition, cannot be generalized (and therefore be a law). In both cases we cannot
say of the emperor that he acted against the law. He is the law. But
in the second case—and herein lies Scotus's emphasis—we may say
that, although God acted *ordinate* (he cannot act *inordinate*), he nonetheless did not act *de potentia ordinata. Potentia ordinata* and *potentia
absoluta* name the very same range of divine acts. In the case of an
act that neither conforms to existing orders nor can be reasonably
assumed to establish a new one we may speak of God's absolute
power in the narrow sense, yet we must be careful not to call such
an act "inordinate." But this, I believe, was also the less articulated
sense of Thomas's referene to God's wisdom in the passage quoted
above. Yet the shift from the physical to the legal terminology is
significant. It paved the way toward the covenantal understanding
of the orders of nature and grace.[29]

[28] "Sed et quod principi placuit, legis habet vigorem, cum lege regia, quae de imperio
eius lata est, populus ei et in eum onme suum imperium et potestatem concessit": Inst. 1.2.6;
G.I.5; Dig. 1.4.31; Ulpians's original formulation probably was: "Quod principi placuit
legis habet vicem, utpote cum lege quae de imperio eius lata est populus et hanc potestatem conferat." Cf. F. Schulz, "Bracton on Kingship," p. 145. It also seems that the
name *lex regia* (rather than *imperatoris*) is no earlier than the third century. Nor was there
originally a link between this doctrine and the conception, eastern in origin, of the king
as νόμος ἔμψυχος: Wirszubski, *Libertas as a Political Idea at Rome during the Late Republic
and Early Principate*, pp. 130–36. On the medieval career see also Wilks, *The Problem of
Sovereignty in the Later Middle Ages*, p. 154 (and n. 1); Tierney, "The Prince Is Not Bound
by the Laws," pp. 388ff.; Miethke, *Ockhams Weg*, p. 146 n. 33, notes that Scotus accepts
the doctrine without restrictions. The most famous (and earliest) medieval restriction is
that of Manegold of Lauterbach, who viewed it as revocable by the people who conferred
it: MG, LdL, 1:365, 391 (cc. 30, 67); on the employment of *potentia absoluta—ordinata* in
the later political discussion see Oakley, *Omnipotence; id.,* "Medieval Theories of Natural
Law: William of Ockham and the Significance of the Voluntarist Tradition," pp. 65–83;
Courtenay, "Dialectics," pp. 10–13. An interesting imprint of this political interpretation
is Abarbanel (late fifteenth century), who develops his theory of kingship, with the use
of these same terms—*mukhlat* (absolute) as against *mugbal umesudar* (ordained); *Perush hatora* to I Sam. 8:4ff. Against Paulus of Burgos (from whom he took the terminology) he
argues that, since the social contract is absolutely binding and leaves no room for resistance, kingship may begin as constitutional, but is bound to turn into an absolute one. It
is better to have no kings at all, and the "laws concerning kingship" in Deuteronomy are
hypothetical only.

[29] Oberman, *The Harvest of Medieval Theology: Gabriel Biel and Late Medieval Nominalism, passim*; Hägglund, *Theologie und Philosophie bei Luther und in der ockhamistischen Tradition.*

Ockham's position was even simpler than that.[30] Every divine act can be analyzed in view of what could have been otherwise—*de potentia Dei absoluta*—and in view of that which was in the past. Only in the past do they form different aspects of God's capacity; in the future, they not only denote, but also connote the same act. God is in no one's debt—not even his own. "Orders" of all kinds are connotative, not denotative notions: God creates only *things*, and real things can always exist without each other; hence statements about aggregates of things, about structures and natural sequences, can never be much more than protocol-statements without any intrinsic necessity.

Franciscus de Mayronis summed up these positions.[31] For some, it is a distinction between all that God can do against what wisdom dictates (probably Thomas). Others compare moral order to every other possibility (possibly Bonaventure). Others confront God's possibilities before and after he acted (de Mayronis himself, later Ockham). Still a fourth way is the distinction de jure–de facto (Scotus). Again we note that the differences among these (and other) positions is not so much a difference of principles as it is a difference in emphasis. The principle that "everything is possible to God that does not entail a contradiction" is, of course, common to all of them. All of them agree that God can and does break the *communis cursus naturae*—or, as Thomas used to say, that God "frequently acts against the usual course of nature."[32] Even in the respect to causality, Thomas holds to the principle canonized later by Étienne Tempier's list that "everything which God does with the mediation of secondary causes he can also do immediately and without them."[33] Nor can

[30] William of Ockham, *Quodlibeta septem* 6.q.1; *Opus nonaginta dierum* c.95 (*Opera politica*, 2:719–24); Miethke, *Ockhams Weg*, pp. 150–56; Leff, *Ockham*, pp. 15–17, 455–68. On the influence of both Scotus's and Ockham's formulations on Gabriel Biel see Oberman, *The Harvest*, pp. 30–56 and *passim*. Bannach, *Die Lehre*, pp. 17–25.

[31] Franciscus de Mayronis, *In quatuor libros sententiarum* 1 d.43–44 q.6, p. 126v.c–f. He speaks of different modes of speech (*modi dicendi*) in respect of the distinction. Bonaventura, *Breviloquium* 1.7 (*Opera omnia*, ed. Quaracchi, 5:216a) distinguished only three senses ("secundum actum"; "secundum aptitudinem ex parte creaturae"; "secundum aptitudinem ex parte solius virtutis increatae"), of which the third, as Miethke (*Ockhams Weg*, p. 143 n. 24) remarks, subsumes that which was later subsumed under *potentia ordinata*. The second sense is identical to Mayronis's even in the formulation. From the conclusion it follows that the latter favors the third sense, wherefore one can take his different senses as actually different opinions.

[32] Thomas Aquinas, *De potentia* q.1 a.3 ad 1.

[33] Thomas Aquinas, *De potentia* q.3 a.7 ad 16; cf. Hochstetter, *Studien*, pp. 12–26, esp.

one argue that Ockham misused such principles to become an "Alleszermalmer," a radical skeptic: Ockham does hold the physical world-picture of Aristotle to be, on the whole, "correct." The difference between them is *not* that Ockham's world is more contingent than that of his predecessors. It is rather a difference in the *meaning* of "contingent orders" and of "things."

The radical change in the perception of the world that occurred between the generation of Thomas and that of Ockham is embodied in the latter's principle of annihilation. We quote it again: "Every absolute thing, distinct in subject and place from another absolute thing, can exist by divine power even while [any] other absolute thing is destroyed." Thomas admitted that God could have created other worlds, but each of the worlds that God could have created, much as ours, is such that the singular things that inhabit it are necessarily bound by some mutual reference-structure. For Ockham all things are immediate of God. For Thomas it is meaningless if not logically impossible to conceive of a good number of things out of any context. Ockham forces us to perform such ideal experiments with a critical intent: that which cannot pass the test of being conceived *toto mundo destructo* is not a "thing" (*res*).

What distinguishes one singular thing from another? In Scholastic terms, such a distinction would constitute the "principle of individuation" of singulars. It is one of the more complicated chapters in the history of Scholastic thought, complicated because of the incessant entanglement and disentanglement of theological, ontological, and epistemological issues: the question of what constitutes singularity is not easily separable from the question of how singulars come to be known. Thomas distinguished sharply between physical and nonphysical entities. Physical things, from stones and crocodiles to celestial spheres, are individualized through "matter signed with quantity."[34] Form is only a qualifying principle, it ac-

---

pp. 17f. (Ockham); Boehner, *Ockham: Philosophical Writings*, pp. xix–xxi; Denifle, *Chartularium*, nr. 473 §43 (p. 546): "Quod primum principium non potest esse causa diversorum factorum hic inferius, nisi mediantibus aliis causis"; cf. §60; Miethke, *Ockhams Weg*, p. 157; Blumenberg, *Die kopernikanische Wende*, pp. 37–38 ("Postulat der Unmittelbarkeit").

[34] Thomas Aquinas, *De ente et essentia* c.2, ed. Roland-Gosselin, pp. 10–11: "Materia non quolibet modo accepta est individuationis principium, sed solum materia signata. Et dico materiam signatam, que sub determinatis dimensionibus considerata." Cf. ibid. c.5 (separate intelligences) and *Summa theol.* 1 q.3 a.2 ad 3, as well as below n. 37.

counts for *what* a thing is; but there can be, and are, many things that belong to the same *infima species* and share the same "essence." Slight, accidental differences in material constellations account for slight differences in the instantiation of inferior, sublunar species, for example, for the fact that an infant is born with a pimple whereas his twin is not. They may even be identical and yet "two in number"—only in the sense that the bulk of matter of the one is not identical (though in all other respects similar) to the bulk of matter of another.[35] Neither Aristotle nor Thomas could endorse Leibniz's *principium identitatis indiscernibilium*, which forbids us to assume any two identical real entities, because that principle is based on the assumption that form individuates, not matter. For those who take matter to be the principle of individuation, it follows that separate forms, for example, angels or separate intelligences, constitute a species each for itself; each is one of its kind by definition.

Thomas's universe constitutes a hierarchy of substantial forms of various levels of perfection. They are interdependent: the world is called "one" because things in it are structured in a mutually supporting order (*ordo ad invicem*), they are "ordained toward each other" (*ad alia ordinantur*). A plurality of worlds like ours is impossible; it contradicts the notion of a center of gravity and other proper places. *Our* universe, supposing all the things inhabiting it (*suppositis istis rebus*), could not be "better."[36] A single substantial form cannot be made better even *de potentia Dei absoluta*, just as the number four cannot be made to be greater than it is. God could add essential properties to existing things; but this would disturb the harmonious interdependence of things, their *ordo ad invicem*; such a universe, albeit possible, would be worse than ours and not reflect God's wis-

---

[35] This was understood in the Middle Ages, as it is still by most interpreters of Aristotle, to be Aristotle's doctrine. A more cautious note was struck by Anscombe, "The Principle of Individuation," in *Articles on Aristotle*, 3: *Metaphysics*, ed. Barnes et al., pp. 88–95. True, "the statement that matter is the principle of individuation does not mean that the identity of an individual consists in the identity of its matter." But regarding matter as a principle of individuation is based on more than just *Metaphysics* z7. 1033b24. Aristotle insists that matter is the source of contingency, matter is to blame for deviations from the normative form (monsters)—e.g., *De gen. anim.* Δ 3.778b16–18. Matter, then, is the source of particularization *below* the *infima species*; by extrapolation we may therefore say that it is the source of individuation. Again it is true, however, that "principle of individuation" is an expression to which there is "no counterpart . . . in Aristotle" (Anscombe, "Principle"). Cf. above II.H.3 and below IV.C.2.

[36] Above n. 24, cf. below n. 75.

dom. It is presumably impossible *de potentia Dei ordinata*, which "pertains to God's wisdom." Lastly, God could create other things altogether, or add things (i.e., species) to ours, but this would be another universe.

A singular thing taken out of its context would be meaningless if not impossible, and hardly conceivable. All the more so since singular things are not even immediately recognizable, at least not to the *viator*. We have no immediate cognition of material objects; they are sensed through the mediation of the sensible species and known through the mediation of intelligible species. In both cases, only qualities are transmitted. Matter as such can never be conceived, since as such it is formless and cognition means assimilation of the same by the same, that is of forms by the mind. But matter, the sole individualizing principle of material objects, cannot be assimilated by the immaterial mind. We would have to eat a table in order to assimilate it as a singular thing, matter and all, and still we would not be performing a cognitive act. God, of course, knows singulars immediately and does not need the mediation of the senses. Nor do angels or we once *in patria*. But God's knowledge of singulars is not a passive knowledge generated by a confrontation with objects. It is an *active* knowledge generated by doing, by creating those objects. And he imparts his knowledge—his ideas—of singulars to spiritual entities.[37]

Even before Thomas, not everyone saw matter as the sole principle of individuation—more precisely, of things being multiple in

[37] Thomas Aquinas, *Quaest, disp. IX: De veritate* q.2 a.5 resp.: "Et ideo simpliciter concedendum est quod Deus singularia cognoscat. . . . Ad cuius evidentiam sciendum, quod scientia divina, quam de rebus habet, comparatur scientiae artificis. . . . Artifex autem secundum hoc cognoscit artificiatum per formam artis quam habet apud se secundum quod ipsam producit. . . . Omnis autem forma de se universalis est; et ideo aedificator per artem suam cognoscit quidem domum in universali. . . . Sed si forma artis esset productiva materiae, sicut est formae, per eam cognosceret artificiatum et ratione formae et ratione materiae. Et ideo, cum individuationis principium sit materia, non solum cognosceret ipsam secundum naturam universalem, sed etiam in quantum est singulare quoddam. Unde, cum ars divina sit productiva non solum formae, sed materiae, in arte sua non solum existit ratio formae, sed etiam materiei." Cf. *De veritate* q.8 a.11 resp. (angels); *Summa theol.* 1 q. 14 a.11: "unde, cum virtus activa Dei se extendat non solum ad formas . . . sed etiam usque ad materiam . . . necesse est quod scientia Dei usque ad singularia se extendat, quae per materiam individuantur." The divine intellect as *ars factiva*, God's comprehension of that which he produces *sicut artifex intelligit artificium* is commonplace; e.g., Graiff, ed., *Siger de Brabant: Questions sur la Métaphysique* 2.16, p. 71.

number but of the same *species*.[38] After Thomas, this doctrine certainly was most fiercely attacked. It had not only philosophical difficulties. It also seemed to infringe on God's power, to make it impossible for him to create more than one sample of immaterial substances. At least five items in Tempier's condemnation list brand it in various guises—without, however, tying it to epistemological issues.[39] None of the alternative theories were, I believe, successful until Scotus. He recognized clearly that to use negation as a principle of individuation (as did Henry of Ghent)[40] will hardly do: if it is neither matter nor form, it must be a positive principle added to all other "formalities," the "thisness" of a singular into which all other forms constituting it "contract." Formal principles individualize everything, including prime matter.[41] This was a revolutionary turn: a thoroughly individualized thing—this chair, this crocodile—could be said to be possible even without existing—as a *possibile logicum* marked by consistency only (*non repugnantia terminorum*).

Thomas's question, how does God (or another purely spiritual being) know singulars, now lost its ground. Singulars constitute formal knowledge. God cannot but know them, they are imprinted

[38] E.g., William of Auvergne: Moody, *Studies in Medieval Philosophy, Science, and Logic: Collected Papers 1933–1969*, pp. 26–27, 78–80. It may be said to be a Neoplatonic tradition; certainly Ibn-Gebirol's *Fons vitae* sees the form, rather than matter, as the cause of particularization. Finally, Plotinus himself sometimes posits a reason-principle as the core of singularity: Rist, *Plotinus*, pp. 109–11; id., "Forms of Individuals in Plotinus," pp. 223–31.

[39] Denifle, *Chartularium*, 1.473, pp. 543–55 nn. 27, 79, 81, 96, 97, 191.

[40] E.g., Henry of Ghent, *Quodlibeta* 5 q.8, 1:246b: ". . . in formis creatis specificis, ut specificae sunt, ratio individuationis ipsarum . . . est negatio, qua forma . . . ut est terminus factionis, facta est indivisa omnino in suppositis et individualis et singularis privatione omnis divisibilitatis per se et per accidens et a qualibet alia divisa. Quae negatio . . . [est] duplex, quia est removens ab intra omnem plurificabilitatem et diversitatem, et ab extra omnem identitatem." Fourteenth-century schoolmen used to distinguish four answers concerning individuation: by matter, by form, by negation, by quantity; e.g., Johannes de Bassolis, *In secundum sententiarum questiones* d.12 q.4, ff. 76rb–84vb—one of the clearest expositions (from a Scotistic point of view).

[41] Against Thomas, Scotus argues that having matter a principle of individuation amounts to the discarded doctrine that accidents individuate (*Ordinatio* 2 d.3 q.4 n.111, 7:446). Against Henry of Ghent he insists that it must be a positive principle complementing the quiddity of a thing (*Ordinatio* 2 d.3 q.6 n.15). The term *hecceitas* was used sparsely by Scotus, and became widespread among his followers. Of an individual form he does not speak at all, because individuality to him is *not* a common nature or quiddity. For an excellent summary see M. M. Adams, "Universals in the Early Fourteenth Century," in *CHM*, pp. 412–17.

in his nature, since even he cannot make the possible impossible. We, at least in this life, do not know their specific difference but must—as assumed hitherto—infer their singularity from accidental impressions. But if (even material) singulars can be fully conceived without existing, how does a spiritual being, or any intelligence, know singulars qua existents? The shift in the doctrine of individuation is the deeper reason why Thomas asked only about God's knowledge of *singulars*, while Scotus had to postulate a special mode of cognizing *existents*, both sensitive and intellective. We have a non-mediated sensory knowledge of existents (*notitia intuitiva sensitiva*); and at least with the beatific vision, those of us who deserve it will also have a nonmediated cognition of God, of angels, and, of course, of material singulars. There is no cogent reason why, in moments of grace, we may not possess such intellective intuitive cognitions even *in via*.[42]

The revolution deepened in the fourteenth century, not only due to Ockham's influence. Time and again the Venerable Inceptor, as Scotus before him, applies his criterion of annihilation for identifying possibly real things: real things are only those capable of being created independently of any other thing. "Forms" and "natures" cannot be attributed with any ontological status; a "form," even if it existed only as a divine idea, could be annihilated while its presence in that which it informs is conserved.[43] Ockham's universe of

---

[42] The difference between knowledge of singulars and knowledge of singulars qua existents has not been sufficiently stressed in the literature concerning the origins and career of intuitive cognition. Scotus is unambiguous: *Report. Paris.* 4 d.45 q.3, 14:575: "cognitio intuitiva non est tantum singularis, inquantum est cognitio intuitiva, sed essentialiter est ipsius naturae existentis ut existens est." Cf. *Opus Ox.* 3 d.14 q.3; *Ordinatio* 2 d.3 p.2 q.2, 7:553; and Day, *Intuitive Cognition: A Key to the Significance of the Later Scholastics*, pp. 65f.; Tachau, "Vision and Certitude," pp. 22–23; Boler, "Intuitive and Abstractive Cognition," in *CHM*, pp. 465–66. Older, mainly Franciscan traditions did assert immediate knowledge of singulars: Matthew of Aquasparta, Robert of Marston, Richard of Middleton (Überweg and Geyer, *Grundriss*, 3:482, 488–89); Matthew even speaks of *species singulares: Quaest. disput.*, ed Quaracchi, p. 309. Another tradition, likewise Franciscan, wished to abolish the *species intelligibiles* (Tachau, "Vision and Certitude," pp. 10–26; Godfrey of Fontaines, Henry of Ghent, Petrus Olivi). Only after Scotus could both traditions meet (William of Ockham)—because he shifted the question toward the knowledge "not only of singulars, but of singulars as existing"; and Tachau, "The Problem of *Species in Medio* at Oxford in the Generation after Ockham," pp. 394–443, has shown that the abolition of intelligible species also continued afterwards to be separable from the postulate of intuitive cognition.

[43] William of Ockham, *Sent.* 1 d.29 q.4D; 2 d.2 q.4 Q, p. 115. This, of course, is only one of his arguments against the reification of universals; others are semantical, logical,

things consists of substances and their absolute qualities alone. Any absolute quality can be subtracted from a subject, or added to it if it is compossible with other qualifications of that subject. No principle of individuation is necessary to account for signulars; on the contrary, any reference structure between singulars needs justification.[44] Matter, even prime matter, is always actual: "alia est prima materia mea, et alia est prima materia tua";[45] it is always the matter of this or that single existent. Correspondingly, no mediation between our cognition of singulars as existing and these singulars themselves is necessary, either sensible or intellective. Existential judgments are caused in part by the intuitive cognition caused by the object, in part by the object's existence.[46] Our concept, if well-constructed, should refer to singulars either directly or obliquely. Treating a connotative notion as a denotative one leads to false hypostatization; and superfluous connotative notions may lead to more distinction than warranted by the phenomena—as we shall soon see in the case of violent motion.[47] This utter primacy of the empirically given singulars protects Ockham's epistemology from the imposition of any logical, let alone physical, necessity of relations, structures, "natures."

The shift is clearly noticeable in the handling of the question whether God could have multiplied worlds. The schoolmen answering it had always in mind Aristotle's proof that our world is

---

and epistemological: Leff, *Ockham*, pp. 104–23; Miethke, *Ockhams Weg*, pp. 160–61; Adams, "Universals," pp. 417–22 (critique of Scotus), 434–39; Blumenberg, *Die Legitimität der Neuzeit*, pp. 109–10. The epistemological and ontological priority of singulars: Vignaux, *Nominalisme au XIV* siècle*, pp. 11–45.

[44] *Sent.* 1 d.2 q.6 Q: "Quaelibet res extra animam seipsa est singularis. . . . Nec est quaerenda aliqua causa individuationis . . . sed magis esset quaerenda causa quomodo possibile est aliquid esse commune et universale." It seems to me that Ockham's formulation is close to Aureoli's: "Quaerere aliquid, per quid res . . . singularis sit, nihil est quaerere. Hoc enim quaerere est . . . quod faciat eam particularem" (Aureoli opts for quantity), Petrus Aureoli, *Sent.* 2 d.9 a.3, p.112bD; cf. P. R. Dreiling, *Der Konzeptualismus in der Universalienlehre des Franziskanerbischofs Petrus Aureoli*, pp. 159–70, esp. p. 160 n. 1.

[45] *Summulae in libros Physicorum* 1, 14, p. 18b; Moser, *Grundbegriffe der Naturphilosophie bei Wilhelm von Ockham*, p. 44. Moser's (as later Shapiro's) analysis of Ockham's concept of matter, as well as other aspects of his natural philosophy, relied mainly on the *Summulae*; for a deeper and more thorough analysis cf. now Goddu, *The Physics of William of Ockham* (matter: pp. 95–111).

[46] Hochstetter, *Studien*; Day, *Intuitive Cognition*; Miethke, *Ockhams Weg*, pp. 163–92; M. M. Adams, "Intuitive Cognition, Certainty, and Scepticism in William of Ockham," pp. 389–98; Boler (above n. 42); Goddu, *Physics*, pp. 23–51.

[47] Below III.C.2.

unique. The possibility of a *pluralitas mundium*, raised by the Atomists, was kept alive throughout the medieval discussions on the horizon of God's power.[48] It then assumed a new career in the cosmological speculations awakened by the Copernican revolution. Aristotle first proves (*De caelo* A 9.276a18–277b5) the logical contradictions involved in the Atomistic assumption. Another universe means another system of natural places. A body moving "naturally" in that universe to its "earth" would be moving away from our earth, and therefore moving both naturally and by constraint, which is impossible. Until now, Aristotle has only proven that if other worlds are organized as ours is, determined by the same forces, then there could not be any matter outside our world. But what if we chose to imagine a universe totally different from that which we happen to inhabit? It seems as if this is the question that led Aristotle to add, probably as a later insertion, another argument (*De caelo* A 9. 277b27–279b4) in which he shows (i) that the world included within the outermost sphere of the heavens contains all the matter that can be and (ii) that, therefore, there can be no additional possible forms, for they would exist without a possible substrate.[49] The *actual* forms are also the only *possible* ones. Note that Aristotle is not attacking here the theory of ideas as such. He stresses that his argument must be compelling even to those who (unlike him) disengage the existence of ideas from their instantiation in a substrate. Even they have to concede that the number of ideas can not exceed their possible instantiations; there can be no absolutely vacant forms even if we assume that sensible things only "participate in" and "imitate" ideas. "World" and "this world" are coextensive notions. There is, as we know from other passages, only one additional form beyond the universe that is as unique and necessary as the form of the universe itself. Being outside the universe, it does not *share* anything with the universe, yet it is a necessary precondition for the world and its motions. The prime mover permits no *metexis; mimesis* is the only way in which it is said to "cause" in a manner analogous to "desire" awakened toward it. No wonder that later interpreters saw in it the "form of the universe."

[48] For a short history cf. Blumenberg, *Die Legitimität*, p. 113–25 (pp. 120ff.: Ockham); Dick, *Plurality of Worlds*, pp. 23–43 (31–35: Ockham).

[49] Elders, *Aristotle's Cosmology: A Commentary on De Caelo*, pp. 137–49; Solmsen, *Aristotle's System of the Physical World*, pp. 222–49 (unmoved mover and world-soul).

Thomas does not deal with the problem in the context of the question, inherited from Lombard's sentences, whether God could create a world better than ours. While discussing the rationale for the proliferation in creation, he concludes with the question whether the world is altogether one.[50] The very order of things created by God manifests the unity of the world. Because things are structured in a mutual order (*ordo ad invicem*), ordained toward each other (*quaedam ad alia ordinantur*), we speak of this world as one; order entails unity. A plurality of worlds means a plurality of coexisting orders without mutual experience; it is assumed by those who deny the existence of an ordering wisdom (*sapientiam ordinantem*) and attribute everything to chance, such as Democritus. If, however, many orders like ours would exist, particularized only by matter (so that the many universes would be exact replicas of each other), then Aristotle's argument, that if there were pieces of "earth" outside our universe they would have to fall to our earth, holds. There can be no many earths (*non enim est possibile esse aliam terram quam istam*). In other words, God could multiply worlds absolutely speaking, but not *ordinate*, if by "worlds" we mean partial, different orders. If we mean our order, it seems that Thomas denies the possibility of its numerical multiplication even *de potentia Dei absoluta* on logical grounds.

For Ockham the question of plurality does bear relevance to the question of better worlds. "Better" has three meanings—essential (qualitative), substantial (quantitative), and accidental.[51] In the first sense, God cannot make a *species specialissima* better without changing it, but he could add or subtract species—Ockham is not concerned with the *ordo ad invicem*. God can also replicate our universe, as it is, infinitely. Against Thomas, Ockham counts it as a possible amelioration and does not even invoke the principle of economy (later called Ockham's razor) against it. Aristotle's argument from the amount of matter does not hold in view of God's power to increase the amount of matter *ex nihilo*. Matter, it should be remem-

[50] *Summa theol.* 1 q.47 a.3. Petrus Lombardus's question whether God could have made a better world is in *Sent.* 1 d.44 c.1; cf. Thomas, *Summa theol.* 1 q.25 a.6. The comparison of the world to a musical instrument, "sicut, si una chorda plus debito intenderetur, corrumperetur proportio ordinis," is also in Hervaeus Natalis, *In quatuor libros sententiarum commentaria* d.41 a.1, p. 170a.

[51] *Sent.* 1 d.44; Goddu, *Physics*, pp. 60–75 (in conjunction with possible-worlds semantics); Funkenstein, "The Dialectical Preparation," pp. 193–98.

bered, is (against Aristotle or Thomas) always actual even as prime matter, though not necessarily quantified (this, too, against Thomas); and it is always the matter of this or that singular: "my prime matter is not your prime matter." Aristotle's (and Thomas's) argument from structure, that is, from the absolute nature of simple motions, does not hold either—it rests on a treatment of connotative notions as denotative ones. Proper places do not denote either an absolute subject or an absolute quality in it. Aristotle contended that, given two or more separate "worlds," a body inhabiting any one of them would have two or more proper places toward which to move. If it moves toward one center of gravity l_1, it moves ipso facto away from its twin l_2 in the other world, that is, it moves both naturally and by coercion. Ockham meets this argument by pointing out that one could argue in the same way against Aristotle: namely that, even within our unique world, a fiery body moving "naturally" upwards toward point l, within its natural place can also be said to move away from the opposite point on the periphery. If one argues that the natural places in this world form at least one contiguous body, which they do not if they are distributed between many worlds with as many earths, waters, etc., Ockham retorts that, once we do not mean "proper place" as a point but rather as a generic concept, it need not be a continuum. "Natural places" are relative (connotative) notions to begin with, referring to singular bodies and the nearest mass of their predominant element. There could be many earths, separate from each other, and all of them called by the generic name "earth." Gravity could be understood as an *actio in distans* (a finite distance, of course), even though Ockham did not think of it in analogy to magnetic attraction. In short, for Thomas, "the world" meant, first and foremost, the unity and cohesiveness of its structure. For Ockham it was derived from the brute fact that it is one aggregate. That it is well ordered he does not deny, but does not assume any order as a necessary condition for "this world" to be one.

It may be doubted whether Ockham's version of the principle of immediacy—the immediacy of singulars—guarantees God's omnipotence any better than the Thomistic, and Scotistic, assumption that some *ordo ad invicem* is a constitutive element in every substance, that is, than the assumption that in some cases it would involve a logical contradiction if God wanted a certain thing without its "nature" or formal determination.[52] Ockham, however, believed it to be the case: namely, that the criterion of isolation through imaginary destruction of contexts was necessary to save the utter contingency of the world.[53] But his very insistence, in the name of saving the contingency, on the primacy of concrete singular things led him to postulate another kind of necessity. What guarantees our intuitive notions? Not their logical independence—it may serve only as a clue. Nor indeed any *adequatio rei ad intellectum* through the mediation of *species*: there is no similarity or identity between concepts and things.[54] Left is only a strictly causal dependence. But this is problematic on two counts. If an intuitive notion is caused by singular objects only, how can there be a negative intuitive notion? But there must be such notions to back the judgment "*p* is not there," since it is an existential judgment.[55] If intuitive notions depend casually on the existence of extramental objects, how can God cause a *notitia intuitiva de rebus non existentibus*? But he must be capable of doing so, since notion and object are two different things and therefore, by divine omnipotence, one could be destroyed while the other is conserved.[56] It seems that Ockham—and many of his generation—exchanged the physical necessity of orders and structures for the physical necessity of efficient causality, at least at times.

Anneliese Maier has shown that fourteenth-century Scholasti-

---

[52] Above n. 43.

[53] It is not altogether correct to say that the principle permits Ockham to define positively that and what a thing *is*; it rather enables him to identify such constructs that are *not* things because they cannot be thought of apart from other entities.

[54] Boehner, "The Realistic Conceptualism of William of Ockham," in *Collected Articles on Ockham*, pp. 156–74, esp. pp. 161–62.

[55] Hochstetter, *Studien*, pp. 55–56, recognized that both problems were related—the knowledge of non-existence of a singular and the *notitia intuitiva de rebus non existentibus*.

[56] Ockham, *Quodlibeta* 3 q.3, in *Philosophical Writings*, ed. Boehner, pp. 128–33; Boehner, "The Notitia Intuitiva of Non-Existents According to William Ockham," in *Collected Articles*, pp. 268–300; and the literature already quoted. Ockham put asunder that which Scotus had united: the notion of a singular and of an existent. So strong is his faith, however, in the causal link between the existent and the intuitive cognition that leads to a positive existential judgment, that he cannot conceive the latter without the former even *de potentia Dei absoluta*. In the natural course of events, only an existent can cause an intuitive cognition of it; *de potentia eius absoluta*, it can be caused by God, but *without* ever (if it is a genuine intuitive cognition, rather than an illusion) causing a wrong existential judgment. In such a case, God, rather than the thing, is the immediate cause of my intuitive cognition—just as rain can come down from the blue sky (the postulate of immediacy). Perhaps one can use Wittgenstein's idiom and say: in such cases of a *cognitio intuitiva de re non existente*, say the sun, we "see" God "as" the sun. Cf. below v.A.2.

cism paid much greater attention to mechanisms of efficient causality, and that efficient causality came to be the prime meaning of
causality.[57] In an earlier, equally seminal article she drew our attention to the shifts in the meaning of "necessity" and "contingency"
between the thirteenth and fourteenth centuries.[58] Employing the
Aristotelian distinction between "absolute" and "hypothetical" necessity to efficient causal connection (which was usually *not* identified with the distinction *potentia absoluta-ordinata*), "necessity"
meant primarily that which is always the case (*necessitas ut semper*);
that which is often (*ut saepe*) or sometimes (*ut raro*) the case was only
hypothetically necessary—the presence of the effect shows a conditional necessity (*quoad causa*).[59] Fourteenth-century authors reversed the terminology. Natural processes are now interpreted not
as contingent per se and necessary at best in consideration of their
causes, but rather as necessary per se (if nothing intervenes) and contingent *secundum quid*.[60] The real center of this "dynamical" understanding of causal processes is not the effect, but the cause: a cause
will always act in a given way unless impeded, and an event resulting from an impeded cause is "contingent" in only a conditional
sense. The sum total of all natural causes known only to God would
still determine that *this* event be their result. Only voluntary acts are
contingent in and of themselves. Human or divine intervention
alone may alter the course of nature. It is clear that this shift in perspective could have come about only because efficient causality became the sole causality in the proper sense; and that this again happened because singulars were, from now on, the backbone of the
universe of discourse.

4. Possibility, Real and Logical

A growing attention, from the end of the thirteenth century, to the
various meanings of possibility brought the distinction between
God's powers both closer to and farther from the Leibnizian two necessities. Leibniz, who sometimes praised himself for having dis-

[57] A. Maier, *Metaphysische Hintergründe*, pp. 273–99 ("Das Problem der Finalkausalität
um 1320"), 300–35 ("Die Zweckursachen bei Johannes Buridan"). Cf. also Bannach, *Die
Lehre*, pp. 276–314 and Crombie, *Robert Grosseteste*, pp. 167–77.

[58] A. Maier, *Die Vorläufer*, pp. 219–50.

[59] Below n. 63.

[60] In this way Ockham viewed the causal nexus between intuitive cognition and the
presence of the intuited object: above n. 56.

covered "the root of contingency" with his notion of contingent orders, refers at times to Aristotle's distinction between absolute and
conditional necessities as a precursor.[61] In this he was wrong: Aristotle's distinction—to which Moslem logicians added many in-be-
tween grades—was either purely logical (the principle of noncontradiction is absolutely necessary, a syllogism only conditionally so)
or purely physical (the movement of the spheres is conditionally
necessary since they are "capable of many states"; the prime mover
is absolutely necessary).[62] In other words, "absolute" and "hypothetical" stand for the necessity of the first link in a chain as against
its successive members; for Leibniz, it is the necessity of the whole
chain, including its first member, that is contingent. Moreover, Aristotle felt, intuitively, that logical and physical modalities coincide
in that they are mediated by temporal meanings. "Necessary" is
that which is always true, "possible" that which is sometimes true,
and there is also a "necessity" to contingent facts in that they are
true now.[63]

Almost from the outset, the Scholastic discussion about the extent of God's power abandoned implicitly the strictly temporal understanding of modalities. Peter the Lombard was certain that God
could do many things that are "neither good nor just, because they
never are or were" or will be.[64] Nor is God's power confined to
those *rationabilia* that he did in fact create. We have seen how
Thomas sharpened the latter point by exchanging *rationabilia* for or-

---

[61] Leibniz, *De rerum originatione radicali*, GP, 7:303. He is more careful in the fifth letter
to Clark (GP, 7:384–90) where he distinguishes both between *nécessité absolue / nécessité
hypothétique* and between *nécessité logique / nécessité morale*. That the ambiguity may be a
phase in his thought is argued by R. M. Adams, "Leibniz's Theories of Contingency,"
pp. 1–41, esp. pp. 6–9.

[62] Absolute and hypothetical necessity: *Metaphysics* E5.1015a20–1015b15; *Physics*
B9.199b34–200b8; *De gen.* B11.3377b14–29. Cf. Dühring, *Aristoteles, Darlegung und Interpretation seines Denkens*, pp. 243–44; Hintikka, *Time and Necessity: Studies in Aristotle's
Theory of Modality*, pp. 130–31. (The "possible" in one sense as both what is contingent
and what is necessary; in another reserved to the contingent only.) Aristotle seems to have
also a third "contingent" necessity, as in *De interpretatione* 9.18b5–19b4. See also Anscombe, "Aristotle and the Sea Battle," pp. 1–15; Rescher, *Studies in the History of Arabic
Logic*, pp. 43–54 (51ff.: a comprehensive bibliography). Anscombe also notes the lack of
a distinction, in Aristotle, between physical and logical necessity, a distinction whose origin I see in the medieval discussions.

[63] Cf. Hintikka, *Time and Necessity*, pp. 93–113; Mansion, *Le jugement d'existence chez
Aristotle*, pp. 68–74.

[64] Petrus Lombardus, *Sent.* 1 d.43 c.u., 1:264.

ders. Thomas stresses that there are in God's mind ideas of things he never did or will ever create.[65] In another instance, Thomas reversed the link between eternity—or even immutability—and necessity in a yet more concrete way.[66] Our world, he said, *pace* Maimonides, would be contingent even if it were eternal or immutable (which it is not); it could have existed from eternity and yet depended for its existence on God's resolve not to destroy it. We have also seen that Thomas defined possibility explicitly as logical non-contradiction (*non repugnantia, non incoherentia terminorum*).[67]

This atemporal meaning of possibility or even contingency became explicit with Scotus. He emphasized it not so much in view of God's power but in order to classify the structure of will. The free will of a *voluntas creata* does *not* mean the freedom to have chosen differently in the past or to do so in the future.[68] Imagine a will at an instant (a) choosing A; and imagine that it exists only at this instant; then the freedom to choose $\sim A$ in the future is meaningless. It is a category-mistake to confound modality with temporality. Even at the instant of choice (willing), the will has the purely logical power (*potentia logica*) to choose $\sim A$, albeit never realizable. "I do not call contingent that which is not necessary or not always, but the opposite of which could have happened at the very same time it actually did." Scotus thus distinguished between logical and real possibility. The former accompanies the latter and is marked only by *non repugnantia terminorum*. Both Thomas's "possibile" and Scotus's "possibile logicum" are characterized by *non repugnantia terminorum*. They differ profoundly in that the former, though realizable, can remain forever unrealized, while the latter may be *unrealizable*, as in the case in which it accompanies a decision once made to decide its opposite. There is no doubt that Leibniz assumed from here the distinction between the possible and the compossible—either directly

[65] Thomas Aquinas, *Summa theol.* 1 q.14 a.9 resp.; but the knowledge of such non-entia, says Thomas, is not "scientia visionis, sed simplicis intelligentiae."

[66] *Summa theol.* 1 q.46 a.1 resp.; 1 q.46 a.2 ad 2 (*ex nihilo* means only *non est factus de aliquo*, not necessary in time); ibid. ad 6. Maimonides, *Guide* 2.21 (Pines trans., p. 314).

[67] Above n. 22.

[68] Scotus, *Lectura* d.39 q. 1–5 n. 49, in *Opera Omnia*, ed. Balič, 16:494 (*potentia logica*). The definition of *possibile logicum* ("cuius termini non includunt repugnantia"), e.g., *Ordinatio* 1 d.2 q. 7 n. 10. Cf. Normore, "Future Contingents," *CHM*, pp. 368–69; Knuuttila, "Modal Logic," *CHM*, pp. 353–55; Deku, "Possibile logicum," pp. 1–21; Pape, *Tradition und Transformation der Modalität*, I: *Möglichkeit-Unmöglichkeit*, pp. 35–60.

or, perhaps, through Suarez—and with it the most important logical facet of his "possible worlds." I will return to his doctrine of contingency after a few detours.

There is no sign that Scotus wished to identify this *possibile logicum* altogether with God's absolute power. There are good reasons to assume that even the latter is subsumed under *potentia realis*, and that not even God could make me, at the time that I will A, also will $\sim A$ (to avoid the pitfalls of psychoanalytical ambivalence-theories, we may have to translate it today into terms of knowing that or believing that A; though I do not think that ambivalence really means willing opposites in Scotus's sense). The only way in which God could do so, is by changing the past. He cannot make me wish A and $\sim A$ simultaneously; but perhaps he could, after I wished A, annihilate the past up to that moment and cause me to wish $\sim A$. It seems that this was Ockham's one-sided, and probably mistaken, interpretation of Duns Scotus.[69] If there were no power, even infinite, which could reduce my willing A at instant (a) to will $\sim A$ at that instant, it is a redundant notion. Now it could be argued that such an (absolute) power exists: not in such a way as to be able to make the proposition "X willed A at (a) " be false by annihilating the past. Ockham argues that this is a logical fallacy; it implies the very same contradiction in terms it seeks to avoid. But if the past is necessary and cannot be changed—assuming that Duns Scotus joins the *consensus philosophorum et theologorum* on that issue—so is the present.

For those who wished to retain Scotus's impulse, only two ways were open. That even God cannot actualize a logical possibility by annihilating the past may be grounded on a less-than-logical necessity: perhaps it is impossible that an event of the past be made not to happen not because it is self-contradictory, but because it is incompossible with the rest of the world's history. And perhaps this less-than-logical necessity binds God *absolute*. This seems to have been John of Mirecourt's argument.[70] It should be emphasized that Ock-

---

[69] Ockham, *Tractatus de praedestinatione et de praescientia Dei respectu futurorum contingentium*, in *Opera theologica*, ed. Boehner and Brown, 2:534; Normore, "Future Contingents," pp. 370–73.

[70] To the accusation that he said that God could make it true that his father never was while he still exists, or that God could (*ex post*) make it true that the world endured for but a day, Mirecourt answers that he denied that this is possible, yet that it is not *evident* to him that it is either possible or impossible. Cf. Stegmüller, "Die Zwei Apologien des

ham recognized less-than-logical restrictions even on God's absolute power: "I say that omnipotence . . . does not pertain to anything that does not include a [logical] contradiction; that is to say,
the omnipotent cannot make everything which does not include a
contradiction, because he cannot make [another] God. The omnipotent can nonetheless do everything doable that does not include a
contradiction."[71] Remember that, according to Ockham, it cannot
be demonstratively proven that God is one. Indeed, his objection to
Duns Scotus's definition of contingency was not that it imposes on
God an extralogical impossibility, but that it imposes a logical one.

Another answer to Ockham's challenge was the serious attempt
to defend the reversibility of time *de potentia Dei absoluta* by Bradwardine and Gregory of Rimini. Both of them were moved not so
much, as once Damiani was, by the desire to prove God's immutability (the symmetry of past and future contingents) as by the desire
to defend God's absolute freedom and the contingency of his will in
the Scotistic sense of the term. Gregory of Rimini had a comprehensive knowledge of the history of the problem.[72] Like Ockham,
he assumes that *potentia absoluta* excludes, except for instances of *repugnantia terminorum*, also instances of self-reference: God cannot
deceive (*mentiri*). While God cannot make a thing that is not-be, or
to make the truth that a thing was be false, he can nonetheless make
it (now) never to be, or never to have been true that this thing was,
without acting on that thing (changing it) and without having to re-
write history. Unfortunately, by force of the same argument, God
could also make *me* not to have existed now. And Rimini, who
mentions the argument, fails to respond to it specifically.[73] Because

---

Jean de Mirecourt," pp. 40–78, 192–204, esp. p. 48; Courtenay, "John of Mirecourt and
Gregory of Rimini on Whether God Can Undo the Past," pp. 224–56; pp. 147–73.

[71] Ockham, *Sent.* I d.20 q.u., *OT*, 4:36.

[72] Gregory of Rimini, *Lectura super primum et secundum sententiarum* I d.42–44 q.1,
3:362–84; cf. Courtenay, "John of Mirecourt," pp. 159–62. It is interesting that, with all
his knowledge of and reliance on the sources, Gregory does not mention Damiani even
once.

[73] Gregory of Rimini, *Lectura*, pp. 375.28–376.6. The answer, p. 382.3–7, is too general: "licet album non potest deus facere esse nigrum, postquam fuit album, sine illius
mutatione, posset tamen illud facere nigrum sine mutatione, quia posset facere illud numquam fuisse album et semper fuisse nigrum." Gregory overlooks Buckingham's main
point, namely, self-certitude. Gregory, as once Damiani, separates language from states
of affairs; but his theory of propositional objects (*complexe significabile*) is, of course,
much subtler than Damiani's, and permits him to say that since the *complexe significabilia*

he, like Ockham, could not acquiesce with empty logical possibilities, he could define the difference between *potentia absoluta* and *ordinata* in terms of possibility and compossibility. Scotus did not, and perhaps could not, do so. The medieval argument has thus come full circle: it started with the reaction to Damiani's contention that God could change the past. Now it rediscovered its merits.

5. Necessity in Divine and Human Logic

The ultimate difficulties in the application of modal categories went even deeper than the difficulties with the reification of the possible and much deeper than the meaning of the distinction between the two powers of God. In the fourteenth century, the distinctions of the persons in the Trinity led some—like Holcot—to the conclusion that Aristotelian logic is absolutely valid only in the realm of creation:[74] it needed a revision when applied to God, a revision of our concept of necessity *de dicto*. From the premises "The divine essence is the father" and "The divine essence is the son" one cannot draw the (syllogistic) conclusion "The father is the son"; that which is necessary to us is not necessary to God, and that which is contradictory to us is not always absolutely contradictory. Holcot does not suggest that his divine logic be exempt from the principle of contra-

---

are untensed in God's propositions, and since they are the only bearers of truth or falsehood, God can indeed "change the past."

[74] Robert Holcot, *In quatuor libros sententiarum questiones argutissime . . .* I q. 5(4) H: "Similiter, non est inconveniens quod logica naturalis deficiat in his quae fidei sunt . . . rationalis vel logica fidei alia debet esse a logica naturali. . . . Sunt enim in logica fidei tales regulae . . . 'quod unitas tenet suum consequens ubi non obviat relationis oppositio,' et ideo concessis praemissis dispositis in modo et in figura, negatur conclusio quia illa conclusioni obviat relationis oppositio, sicut si sic arguitur: Haec essentia est Pater, haec essentia est Filius, ergo Filius est Pater . . ."; the text after Gelber, *Exploring the Boundaries of Reason: Three Questions on the Nature of God by Robert Holcot OP*, pp. 26–27 n. 72; cf. Prantl, *Geschichte der Logik im Abendlande*, 4:6–7; on the similar views of the author of the *Centiloquium* see Boeher, "The Medieval Crisis of Logic and the Author of the Centiloquium attributed to Ockham," in *Collected Articles*, pp. 351–72; a thorough discussion in Gelber, "Logic and Trinity: A Clash of Values in Scholastic Thought 1330–1335." It should be stressed that the claim that syllogistic logic does not apply to matters divine and yet the principle of noncontradiction not be violated (under which even the *logica divina* is subsumed) presupposes that the *syllogismus expositorius* be a concatenation of independent propositions. This was so perceived in the Middle Ages; Aristotle, however, understood a syllogism—premises and conclusion—to be one proposition: Patzig, *Die aristotelische Syllogistik*, pp. 13–14. Medieval syllogism, he shows, was not a proposition, but a "rule of inference." For later echoes of the divine logic see Maierù, "Logica Aristotelica e Teologia Trinitaria: Enrico Toffing da Oyta," in *Studi sul XIV secolo in Memoria di Anneliese Maier*, ed. A. Maierù et al., pp. 481–512. Cf. also Leibniz, *Théodicée*, disc. prel. §22.

diction, nor does he really anticipate three-valued logic. True propositions cannot be contradictory even to God; but the passage from one proposition to another, our rules of inference, are sometimes invalid when transcending the domain of creation. Such were the fortunes of the attempts to *reify* modal categories—contingency, possibility, necessity.

In many ways the medieval attempts to reify modal categories or to invest them with meaning resembles or even approaches the seventeenth-century conception of laws of nature as contingent orders. In two respects, however, the developments described and their echo in sixteenth-century Scholastic thought differ radically from their early modern counterparts. On the one hand, we saw a shift from the understanding of order as inherent in things—almost organically—to an emphasis on singulars, whereby their order is comparable to a compact. Indeed, when discussing the order of salvation, it is a covenant; the seventeenth century, by contrast, was much more interested in the relation between things than in the *relata*. The very notion of things was made to fit the mathematical relations governing them, even while conceding that the latter are, in some sense, contingent.

Moreover, throughout the Middle Ages, the distinction between absolute and lawlike necessities was carried on with an emphasis on the contingency of our world or of any other possible order—contingency not only in the sense that things could have been otherwise. Ultimately, even those thinkers who stressed the perfection and order of our world believed that the choice of God to actualize *this* order was unaccountable and arbitrary. This is true even of Thomas. Nothing typifies the change from the Middle Ages to the seventeenth century more than the answer to the question that both ages asked: Can God create a better world? Thomas, as we saw, not ony affirmed it; he also believed that, since the number of "better worlds" is unlimited, there *can* be no objective rational criterion for why our world was created. If God wanted to create *a* world, he had to choose arbitrarily. God is the source of all contingency. Such was still the point of view of Suarez.[75] Leibniz, too, held that the number

---

[75] Suarez is even more emphatic than Thomas, who put strictures on the perfectibility of species within the universe. In reviewing all positions, he sharpens the difference between two camps: "Alterum extremum vitandum est quorumdam Theologorum, qui dixerunt divinam potentiam non posse semper facere plures aut meliores species rerum, sed posse ab ipso Deo cognosci aliquam speciem creabilem adeo perfectam, ut non possit

of possible worlds was infinite. But he also held that ours is the *best* of all possible worlds, the world with the largest number of compossibles subsumed under a minimum number of laws. The principle of sufficient reason, which "inclines God without necessitating him," guarantees both. From the source of all contingency that he was in the Middle Ages, God became the guarantee of the absolute rationality of the world.

These changes occurred, not in the least part, because of the new confidence in the advance of physics. Awareness of these differences should not obstruct our view of the role of Scholastic thought in establishing some of the conditions necessary for the emergence of early modern science. Before we return to describe the theological and philosophical foundations, we ought to assess the contributions of the patterns of hypothetical reasoning developed in the Middle Ages to the emergence of classical mechanics. I do not mean only its contribution to the discussion over the *status* of natural laws, that is, to the metalanguage of science, of which we saw some examples earlier. I rather claim that it had also a significant impact on the *modus operandi* of early modern science. Which is not to say that nothing new happened during the scientific revolution. To the contrary, only a careful examination of the ancient and medieval modes of hypothetical reasoning will allow us to determine, with precision, that which was new in the seventeenth century, even while expressed in an older idiom.

C. IDEAL EXPERIMENTS AND THE LAWS OF MOTION

1. Ideal Experiments in Aristotle: Reductio ad Impossibile

The excellence of modern physics has been, ever so often, ascribed to its courage to become counterintuitive. The failure of Aristote-

---

Deus perfectionem efficere. Quod sensit Durandus in 1 d.44 q.2 et 3; et Aureolus in eadem fuit sententia, ut ibidem refert Capreolus. Nec videtur repugnare Scotus in 3 dist. 13 q.1. Quod solum fundatur in hoc, quod non potest dari progressus in infinitum in perfectionibus specierum": *Disputationes metaphysicae* 30 d.17 a.19, p. 212. If, as he said, he sides with Thomas, he gave him again (as in the question of omnipresence, above II.D.1 and n. 21) a Nominalistic interpretation. Since the problem whether our world is the best was, as we saw, discussed in the Middle Ages in terms of the perfectibility of species, Suarez's review amounts to the distinction between those who think that our world is the best possible and those who do not. Leibniz may indeed have read it.

lian physics to give a clear account of motions was ascribed accordingly to its "method of reasoning dictated by intuition." Experience tells us that bodies come to a standstill when their motive force ceases. Only "the idealized experiment shows the clew which really forms the foundation of the mechanics of motion—namely that bodies would continue moving forever if not hindered by external obstacles. This discovery taught us that intuitive conclusions based on immediate observation are not always to be trusted."[1] Einstein's account does not differ much from that of seventeenth-century scientists. "From early on we judge those motions sustained by causes unknown to us to cease of their own . . . as we grow older we assume that what we often witness is always the case: that they cease of their own or have an inclination to rest."[2] Only an ideal experiment could establish the principle of inertia. The conditions under which a body will continue to move indefinitely and uniformly in a given direction are unobservable if not downright counterfactual.

Depending on our methodological predilections we may call these conditions "empty" (Hegel), "ideal" (Cassirer), "idealized" (Einstein), "fictional" (Vaihinger), "mythical" (Quine), or simply "counterfactual" (Goodman).[3] We may even argue that what is true of some laws (like Newton's first three laws) is true of all, that all

---

[1] Einstein and Infeld, *The Evolution of Physics*, pp. 6–9. "Ideal experiment" stands here for an experiment that cannot be carried out. It is the sense in which I will henceforth use the term. It can also stand for experiments that actually were not carried out, but could have been; in this, broader, sense it is used by Koyré, e.g., in *Metaphysics and Measurement*, pp. 44ff.

[2] Descartes, *Principia philosophiae* 2.37, AT, 8.1, pp. 62f. Cf. (Arnauld), *La Logique ou l'art de penser* 1.9, trans. J. Dickoff and P. James, *The Art of Thinking: Port Royal Logic*, p. 69; Hobbes, *Leviathan* 1.2, ed. Macpherson, p. 87: "But that when a thing is in motion, it will eternally be in motion, unless somewhat els stay it . . . is not so easily assented to . . . and because [men] find themselves subject after motion to pain, and lassitude, [they] think everything els growes weary of motion and seeks repose of its own accord." Hobbes, however, rightly observes that the assumption that "nothing can change itselfe" is shared by both the old and the new science.

[3] Hegel, *Vorlesungen über die Geschichte der Philosophie*, in *Werke*, ed. Moldenhauser and Michel, 19:193 ("Die Vorstellung, die himmlichen Körper würden sich für sich in gerade Linie fortbewegen, wenn sie nicht zufälligeweise in die Anziehungssphäre der Sonne kämen, ist ein leerer Gedanke"). Cassirer, *Substance and Function*, pp. 120–22 (ideal experiments); Vaihinger, *Die Philosophie des Als Ob*, pp. 28–36, 105–109, 417–25, 451–71 (ideal experiments as useful "fictions"; "abstraktive neglektive Fiktionen"; "schematische Fiktionen"); Quine, *Words and Objects*, pp. 51, 248–51 (the "utility" of "limit myths" and other "entia non grata" despite their inconvenience); Rescher, *Hypothetical Reasoning*, pp. 7–8 (the counterfactual status of conditionals in thought experiments; 89 (bibliography); Goodman, *Fact* (above III.A.n.12).

explanatory models or even controlled experiments have an ideal aspect to them.[4] Evaluations aside, it is clear that such *experimenta rationis* do not just assume *p* to be the case while, in fact, it is not, so as to select, from the body of all pertinent factual propositions, all those propositions that are compatible with *p* or cohabitable with it in a "possible world."[5] They function as limiting cases. An imaginary experiment isolates a phenomenon and allows one or more variables in it to assume different values; the counterfactual case serves as the limiting case when a variable assumes some unattainable limiting value—zero friction, for example, in the case of bodies rolling on a plane.[6] So much extrapolation from the factual into the imaginary is evidently worthwhile only if we abandon hope to win valid generalizations from so-called immediate sense data.[7]

Perhaps we ought to feel uncomfortable with this almost paradoxical mediation between the factual and the imaginary—I do not.[8] But seventeenth-century philosophers of nature were proud of

[4] An attempt to formalize the process of idealization involved was made by Nowak, "Laws of Science, Theories, Measurements (Comments on Ernest Nagel's *The Structure of Science*)," pp. 533–48.

[5] This technique (recommended by Rescher, *Hypothetical Reasoning*) applies better to counterfactual contingent statements—as, e.g., the question of what impact Napoleon's victory at Waterloo would have had. See also Rescher, "Counterfactual Hypotheses, Laws, and Dispositions," pp. 157–78, esp. pp. 164f.; Lewis, *Counterfactuals*, esp. pp. 84–117. On the use of counterfactuals in the medieval literature of obligations see Stump and Spade, "Obligations," in *CHM*, pp. 315–41.

[6] It is a twofold process of idealization. A set of counterfactual conditions p_c is construed in which L (a law) is manifestly valid; which is then projected on a set of quasi-factual similar conditions p_f in which L, albeit valid, is not the case: $\sim L(p_f) \rightarrow L(p_c)$. The sequence of instances of *p* is construed under the additional, and likewise counterfactual, assumption that *p*—a "phenomenon"—can be isolated from its context so as to consist of a limited number of variables of which at least one is taken to change gradually. Cf. n. 4.

[7] As an integral part of the inductive process, Whewell describes the "method of curves," which consists "in drawing a curve, of which the observed quantities are ordinates, the quantity on which the change of these quantities depends being the abscissa." This method enables us not only to order "good observations," but also to obtain laws "from observations which are very imperfect" or even to arrive at "data which are more true than the individual facts themselves." Whewell, *On the Philosophy of Discovery*, pp. 206–207. Whewell realized that such abstractions are complicated by the circumstance of interconnectedness of laws of nature. Ducasse, "William Whewell's Philosophy of Scientific Discovery," in *Theories of Scientific Method*, ed. Madden, p. 205.

[8] The insecure status of imaginary experiments led Popper, *The Logic of Scientific Discovery*, pp. 442–56, to interpret them as mere auxiliary measures, permissible only as a "concession favorable to the opponent." This characterizes, as I shall try to show, at best Aristotle's use of imaginary experiments, but not their use in physics since the seven-

this new mode of abstraction, and called it the method of resolution (and composition), a name taken from the tradition of their Scholastic adversaries.[9] Because, as they believed, Aristotle and "The Schools" were unable to rise above the level of descriptive generalizations, mechanics was freed only recently from infantile notions and crude inductions. Johann Clauberg, a so-called Cartesian Scholastic whose understanding of Aristotle was subtler and more intimate, offered a better interpretation. "The common philosophy does not consider a thing adequately as it is in itself and in its own nature, but rather as it behaves in relation to others; hence, its inner nature often remains obscure."[10] Considering a "thing in itself," isolating a phenomenon from its natural *context*, is a move that "common philosophy," as Clauberg rightly observed, forbade. But why?

Not because Aristotle or medieval physics neglected altogether the mathematical analysis of motion, nor because he and his followers failed to consider imaginary conditions, but rather because they saw no mediation—either in principle or in practice—between the factual and counterfactual conditions of the same "body" (or, as we would say, the same phenomenon). Aristotle, and with him medi-

---

teenth century. Popper may have reacted to Vahinger or Cassirer (though neither is mentioned). Kuhn, "A Function for Thought Experiments," in *The Essential Tension: Selected Studies in Scientific Tradition and Change*, pp. 240–65, emphasized their pedagogical function; his notion is somewhat narrower than mine—I doubt that he would, e.g., subsume the inertial principle under the category of thought experiments proper.

[9] Randall, *The School of Padua and the Emergence of Modern Science*, pp. 15–68, tried to show how already the pre-Galilean, Paduan theory of science succeeded in transforming "the demonstrative proofs of causes into a method of discovery" (p. 31), by its understanding of the resolutive-compositive method; Crombie, *Robert Grosseteste*, pp. 290–319, has dated this methodological shift even earlier. Yet none of these precursors used counterfactual conditionals constructively, let alone tried to justify their usage, as did Galileo (below III.C.4).

[10] "Vulgaris philosophia non tam accurate considerat rem, ut in se et sua natura est, sed potius prout se habet respectu aliorum, quo ipso tamen interna ejus natura plerumque occulta manet. Cartesiana scrutatur cujusque rei propriam ac internam naturam, ut constet, quaenam sit ejus propria forma, ex qua deinde facile definiri potest, quae similitudo vel dissimilitudo inter hanc rem et aliam quamvis intercedat, si modo et illius alterius rei interna proprietas simili ratione ante cognita sit": Johann Clauberg, *Differentia inter Cartesianam et in scholis vulgo usitatam philosophiam*, in *Opera omnia philosophica*, 2:1217–35. That "vulgar" philosophy is the Aristotelian-Scholastic tradition becomes even clearer by the allusion to "obscure" qualities. The "inner nature" of things regarded "in themselves" are Descartes's absolute, "simple natures" (*Regula ad directionem ingenii* 5.6, AT, 10:379, 381ff.).

eval physics, saw both as *incommensurable*. For Aristotle this also meant that they are impossible; not so for medieval physicists trained in consideration of possibilities *de potentia Dei absoluta*. While arguing the incommensurability of motion in the void and motion within the medium—and be it in order to reduce the former *ad impossibile*—Aristotle anticipated some arguments and even techniques of early modern physics. Scholasticism went even further, and turned many of Aristotle's impossibilities into well-argued, interconnected logical possibilities. New in early modern physics was certainly not the employment of imaginary, counterfactual states but the insistence on their commensurability. Limiting cases *explain* nature even while they do not describe it; and they can be actually *measured*. Seventeenth-century scientists may not have erred altogether in their judgment of previous traditions, but their views ought not be our sole guide for a historical retrospection.

Aristotle used ideal experiments on several occasions. Some were advanced for purely illustrative purposes, to demonstrate a conceptual necessity; and he could have exchanged them for any number of other similar hypothetical illustrations.[11] Some are genuine thought-experiments: in the context of the dialectical argument in which they occur they are unexchangeable. Most of them belong to a distinct group of *arguments from incommensurability*,[12] and they are of particular interest to us inasmuch as they are set forth with a basic technique which resembles Galileo's resolutive method. A finite body is imagined under a series of conditions in which one variable diminishes or increases gradually; the relation between the factors involved could be easily expressed as a function (though neither Aristotle nor Galileo did so). Yet unlike Galileo, the task of such ideal experiments is not to formulate a general law valid for factual and limiting cases alike, but to reduce a false universal characteristic *ad*

[11] In this way Aristotle establishes (*De anima* Γ 1.425b4–10) the conceptually necessary connection between the diversification of sense organs and the perception of the "common sensibles." In another argument (*De caelo* A9.278a23–b9) he establishes the necessary relation between the number of forms and amount of matter (above III.B.3). In such cases, the counterfactual examples are but illustrative to the general rule and could be replaced by others. On this type of argument see Patzig, *Die aristotelische Syllogistik*, pp. 158–59.

[12] Aristotle distinguishes, it seems, between (i) irrational (ἀσύμμετρον), e.g., $\sqrt{2}$, (ii) incomparable (yet still capable of proportionality), e.g., line and curve (ἀσύμβλητον), and (iii) having no proportion or comparison (ἄλογον), as between zero and magnitude (cf. below V.B.n.12). The Latin (or English) term "incommensurable" covers them all a fortiori: iii and ii are by definition also i. I use the term first and foremost in this sense. It leads, as I shall show, to a wider sense—that of theoretical incommensurability.

impossibile. No mediation is possible between factual statements or generalizations about our world and counterfactual assumptions with their implications. They are incompatible because they describe incommensurable conditions.

In this way Aristotle argues against "movement in the void" (*Physics* Δ 8.215a–216a26) and later against "weightless bodies" (*De caelo* Γ 2.301a20–b16). The logical argument is always clearly separable from the dialectical; the aim of both is to make the void unpalatable. Since "place" is inseparable from the body whose place it is, and the void is a place with nothing in it, it cannot exist. In the (dialectical) arguments *ad hominem* that follow, Aristotle makes a series of concessions to the enemy. He concedes the void and asks whether indeed it is, as claimed (by the Atomists), the condition of movement.

Suppose there were space: there could be no motion in it, since it could be neither forced nor natural. Suppose there were forced-like and natural-like motion in it: the forced-like motions would have to be infinite. Suppose they were finite: then we obtain a contradiction. Suppose there were natural-like motions in space: we obtain another contradiction. The argument, in detail, sums up as follows.

Aristotle first examines a supposed locomotion in the void from the vantage point of direction or goal (*Physics* Δ 8.214b13–215a14), which also bears on the duration of such movement. It could be neither natural nor forced. Not natural, "for there is no place to which things can move more or less than to another." Empty space has no natural places and bodies in it cannot have natural motions, they cannot be determined by "a cause of motion within themselves" to go toward a specific place. A body dropped in empty space would simply remain where it was. Forced movement in space is, therefore, a conceptual impossibility, since it presupposes natural motion. But even if conceded, it could not be like the movement of a projectile since there is no medium to carry the body on. If we suppose it nonetheless, such a forced movement would be indefinite: "For why should it stop here rather than there? So that a thing will either be at rest or must be moved *ad infinitum*, unless something more powerful gets in its way." Presented as the untenable consequence of a false assumption, this was nevertheless the clearest anticipation of the inertial principle before the seventeenth century.[13]

---

[13] *Physics* Δ 8.215a19–22; cf. *De caelo* Γ 2.301b1–4; Sambursky, *Chukot shamayim vaarets*

Once again Aristotle sets aside the conceptual absurdity of the notion of "natural" and "forced" motions in the void and imagines analogous motions in the void (but is careful now not to call them so), and proceeds with a systematic argument from incommensurability to prove their impossibility on purely physical-mathematical grounds. The parallel argument in *De caelo* teaches us what many commentators failed to see: again Aristotle distinguishes between a putative force movement sidewards in the void (*De caelo* 215a24–216a11) and a quasi-natural, up-and-downwards motion (216a11–21). Unlike the previous argument, he reduces them not *ad absurdum*, only *ad impossibile*. I mean to say that, although he never learned to distinguish between logical and physical impossibilities, his arguments from incommensurability prove the latter only. His ancient and medieval commentators were therefore led time and again to ask for the exact properties of these imagined motions in the vacuum. Are they consecutive or instantaneous? If consecutive, can they be assigned a definite value? Such questions would have been superfluous had he concluded, as hitherto, that motion in the void is a contradiction in terms. Instead, Aristotle shows that it is incommensurable, and hence incompatible, with any conceivable motion in the plenum. He examines two cases of quasi-violent motions: up and downwards (where the medium impedes motion because it goes contrary to the proper motion of the medium itself) and lateral motion (which is somewhat faster, since the medium, being "at rest," impedes the motion less). Other things (force or weight) being equal, the velocity of a body moving (in analogy to forced motion) in the void must always be greater than the velocity of an equal body moving in a medium, however rare, since velocity increases in an inverse proportion to resistance, that is, in direct proportion to the rarity of the medium. Nowhere does Aristotle suggest, as do many of his interpreters to this day, that because of this, motion in the void would be instantaneous or with infinite speed, only that it would be "beyond any ratio." The temptation is strong to render his intentions with the equation $\lim F/R_{R\text{-}0} = \infty$ $(v = F/R)$,

---

[Laws of Heaven and Earth], p. 97; Apostle, *Aristotle's Physics*, p. 254 n. 12. Aristotle, it seems, draws the utmost conclusions from the (Platonic) assumption of elements as geometric planes: *De caelo* Δ 2.308b36f; Plato, *Timaeus* 53c–55c. On the medieval treatment (or lack thereof) of these passages of Aristotle see E. Grant, "Motion in the Void and the Principle of Inertia in the Middle Ages," pp. 265–92.

but it would be wrong. He argues only that velocities in the plenum are commensurable in the proportion of their media, i.e., $v_1/v_2 = m_1/m_2$, and that this equation becomes meaningless when $m_2 = 0$ (void), since there is no proportion between zero and a finite magnitude. The movements of two equal bodies moved by equal forces in the void and in the plenum have no common measure.

Aristotle now concedes for a while a (forced-like) motion in the void which bears a common ratio with a similar motion in the plenum. Let z be a segment of the void with length l_z traversed by a body A in the (finite) time t_z. Let B and D be segments of the plenum of the same size so that $l_z = l_D = l_B$, traversed by A in the times t_D, t_B respectively. Evidently t_D is greater than t_z, so that in the time t_z, A will traverse only l'_D of l_D in D. We assumed t_z/t_D to have a definite ratio: if D will be "thinned out," the distance l'_D traversed in it by A at the time t_z will approach l_D. If D is thinned out in the proportion of l_D to l'_D, A will traverse in this thinned-out body (B?) at the time t_z the distance l_D. It will have traversed the same distance in the same time in the void and in the medium—an impossibility.

So much for quasi-forced motions in the void. As to motions in the void in analogy to natural motions, a simple argument suffices to exclude them. Differences of weight or lightness are caused by the fact that heavier or lighter bodies penetrate the medium faster than less heavy or light bodies in the direction of their proper place, that is, going, respectively, down or upwards. But if there is no medium, heavy or less heavy, light and less light bodies will move with equal velocities. "But this is impossible."

The bipartite structure of Aristotle's argument becomes still clearer when compared with *De caelo* Γ 2.301a20 ff., where he introduces the assumption of "weightless bodies" only to discard it in a similar way.[14] We need not follow Aristotle here in the excessive usage of letters for variables. Imagine, he says, a weightless body,[15] and com-

---

[14] Dühring, *Aristoteles*, pp. 320f., misunderstood this bipartite structure of *Physics* Δ 8.214b2ff and assumed that Aristotle is speaking throughout the argument about the same kind of (unspecified) motion in the void. But "dass alle Körper sich mit der gleichen Geschwindigkeit bewegen" is a consequence only of assuming a natural-like motion in the void. Otherwise, *Physics* Δ 8 would contradict *De caelo* Γ 2. A similar imprecision is in Ross, *Aristotle*, pp. 87–89 and Apostle, *Aristotle's Physics*, p. 254 n. 12.

[15] In the sublunar domain only; this I take to be meant by ἔνια ἔχειν (301a22). The supralunar bodies are, of course, neither heavy nor light. As to the whole argument cf. *De caelo* A6.273a21–29 (the refutation of infinite weight).

pare it to a heavy or light body of the same size. He then examines their behavior, as before, in analogy to natural and to forced motions: the weightless body will always traverse a smaller distance laterally (or in the direction opposite to the natural motion of the analogous body with weight). One could then cut the heavy or light body—or augment it—until it traverses the same distance as the body without weight. "But this is impossible," because the weightless body must be imagined as always moving a longer distance when moved by force and a shorter distance when moved in the direction of places when compared with a body having weight or lightness; it does not matter how big or small they are in comparison. Both in the case in which weight$_1$/weight$_2$ nearly equals distance$_1$/distance$_2$ (natural motion) and in the case that weight$_1$/weight$_2$ nearly equals distance$_2$/distance$_1$ (forced motion), the proportion becomes meaningless when weight (or lightness) = 0. "Nothing" has no proportion to any finite magnitude. Again he suggests that a weightless body moving (laterally) by constraint "will continue infinitely"—which clearly does not mean infinite speed, but infinite distance. Here Aristotle is saved from circularity in that he does not assume from the beginning that a "weightless" body could not initiate a motion downwards by itself. This is not even his conclusion; all he proved is that such a motion of a body, however large, would be *incommensurable* with motion over the same distance of a however small body with weight. In the whole passage Aristotle seems to draw the utmost conclusions from the Platonic assumption of elements as geometrical planes.

Note that Aristotle's proof rests on a further, tacit assumption that *some* bodies evidently move up and downwards without constraint. This we "see with our eyes."[16] He probably thinks of the elements earth and fire. Without this assumption, all he would have proven here is that either every (sublunar) body is weightless, or none is. Sense perception decides here between two alternative and equally exhausting theories: since some bodies can be seen to move by gravity or levity only—their motion is simple in every respect—all bodies must so move. The Aristotelian induction (ἐπαγωγὴ) does

---

[16] *De caelo* Δ 4.311b21–24: εἰ τοίνυν ἔστι τι ὃ πᾶσιν ἐπιπολάζει, καθάπερ φαίνεται τὸ πῦρ καὶ ἐν αὐτῷ ἀέρι ἄνω φερόμενον . . .; cf. *De caelo* Δ 1.308a24; *Ethica Nic.* 8.1.1145b2–6.

not consist of immediate generalizations from sense data; the generalization follows a complex theory—in this case, the examination of idealized cases.

Aristotle's confidence in the immediate and pure manifestation in nature of gravity and levity may help us understand the odd position he took in respect to projectile motions. Aristotle discusses it briefly immediately following the proof for the universality and necessity of gravity or levity.[17] The problem is well known. What keeps a thrown body in its constrained motion after it lost its immediate physical contact with the original mover? Not an inner principle, since such a principle could be attributed only to natural motions, nor the mover, which is now distant. Aristotle takes recourse to the medium in its capacity of being "both light and heavy at the same time." The original mover imparts on the air layer next to the moved body both movement and the capacity to act as a mover, and this air layer imparts them on the next air layers along the path of the projectile. In each layer of air, the "capacity to act as a mover" is actualized more slowly than motion itself; the body is carried a certain distance and has to rest for an imperceptibly short while, otherwise its motion would be instantaneous. This translational causality—the "capacity to act as a mover"—decreases from one air layer to the other; when it fades, the object is carried one more air layer and then drops down by its own "heaviness."

Aristotle's solution is complicated and clumsy. I find it odd that Koyré praised it as a "measure of his genius."[18] Aristotle seems to violate some of his most sacred hermeneutic principles. Not only was it easy to marshal a good many arguments "from experience" against the putative behavior of the air, as many adherents of other explanations soon did. Aristotle himself, it seems, for once abandoned his basic trust in sense data and took instead counsel from his enemies. His language suspiciously resembles the language of the Atomists.[19] Much as Leucippus and Democritus replaced the mis-

[17] I largely follow my analysis in "Some Remarks on the Concept of Impetus and the Determination of Simple Motion," pp. 329–48.

[18] Koyré, *Metaphysics and Measurement*, p. 27.

[19] The critique of Atomistic epistemology *Met.* Γ 5.100b7ff; *De gener. et corrup.* A2.315b7–15. For the authenticity of the first reference see Zeller, *Die Philosophie der Griechen*, 1:1132. Cf. also Owen, "Tithenai ta phenomena," in *Aristote et les problèmes de la méthode, Symposium Aristotelicum*, pp. 83–103.

leading appearance in nature of material continuity with the assumption of imperceptible spatial gaps and likewise imperceptible atoms in bodies, so also Aristotle resolves here the misleading appearance of continuity of projectile movement into a series of imperceptible shifts and pauses. The motion of a projectile, he says, is not continuous "but only seems so." Besides all that, it was already recognized by the ancients that Aristotle did not really avoid ascribing an intrinsic principle of sorts to projectile motions; he merely shifted it from the body moved to its medium.[20]

Why, then, was Aristotle never willing to concede an acquired, accidental capacity to move to the thrown bodies themselves? Why was he willing to assume it only for the medium and with such high methodological costs? It has, I believe, nothing to do with the principle that "whatever moves is moved by another."[21] The received accidental property "to act as a mover" or to aid motion could, in principle, be conceived as an invisible "something else" accompanying the body rather than in the medium. Such indeed was later the status of Philoponos's or de Marchia's *virtus derelicta* or the later medieval *impetus*. Yet the very context in which Aristotle offers his solution in *De caelo* indicates the source of his obstinacy. The medium is only relatively "heavy or light," while earth and fire are absolutely heavy or absolutely light. The proof for the universality of these properties hinges on their pure simplicity and immediate perceptibility. Had Aristotle conceded the possibility that an absolutely heavy body may at times move, and move by constraint of an accidental, accompanying other property, he would have rendered the distinction between forced and natural motion imperceptible. It would remain as a probable, theoretical distinction only, which can never be ascribed with absolute perceptual certainty to any body

---

[20] Simplicius, *In Aristotelis physicorum libros . . . commentaria*, ed. Diels, 1349.26; quoted by Samburksy, *Das physikalische Weltbild* p. 465. The argument was repeated, e.g., by Benedict Pereira, *De communibus* 4, 3, p. 781.

[21] *Physics* II.1.241b34: Ἅπαν τὸ κινούμενον ὑπό τινος ἀνάγκη κινεῖσθαι. There is no exception; even natural movements assume a previous removal of the object from its οἴκεος τόπος. A body moving by its nature has but "a cause of movement in itself" and is not a "self-mover." Cf. Wieland, *Die aristotelische Physik*, pp. 231ff.; and Weisheipl, "The Spector of *Motor Coniunctus* in Medieval Physics," in *Studi sul XIV secolo*, pp. 81–104 (Aristotle: pp. 83–91).

whatsoever.[22] But such absolute certainty was needed if he wished to prove that *all* bodies are heavy or light because *some* are.

To sum up: nowhere does Aristotle distinguish between logical-absolute and physical necessities, though he does distinguish between absolute (or simple) and hypothetical (or conditional) necessity both *de re* and *de dicto*; at times he seems to postulate even a contingent necessity of sorts to distinguish the past, "which even the gods cannot change," from future contingents. In fact, Aristotle tries to prove time and time again the conceptual absurdity of that which is physically impossible. He accumulates arguments, at times begging the question. Yet, his arguments from incommensurability could, in retrospect, be read with such a distinction in mind. In the course of such arguments, Aristotle engaged in imaginary experiments in order to teach his adversaries how to imagine even that which is properly impossible. The excellence of his reasoning is proven by the fact that he deduced many of the characteristics of Newtonian space: that it separates extension (dimensionality) from matter or material place; that bodies in it, once moved, will continue moving; if dropped, they may stay without motion. These assumptions are shown to be necessarily impossible not because of merely logical-conceptual considerations (at least not within the boundaries of particular argument), nor because they do not immediately correspond to our intuition, but because they imply states that are altogether incommensurable with any natural state. This group of arguments has a typical pattern: a body is subjected to a regular change of one variable; each stage of this change is commensurable with the others—until we abolish the variable altogether; then we would have taken the object out of its natural *context* and placed it under counterfactual conditions, or, as we might say, "in another world." This totally incommensurable state stands in opposition to imaginary, even counterfactual, but perfectly commensurable limiting cases in early modern physics. Alternative

---

[22] This would not, however, be the only case in which theoretical and factual discernibility would differ. "Nature would like to distinguish between the bodies of freemen and slaves . . . but the opposite often happens, that some have the souls and others have the bodies of freemen." "Conventional" slavery adds to the problem (*Politics* A5.1254b25–32). Actual slavery can also be "conventional" through the law of the victor (A6.1255a3ff.).

worlds are, in Aristotle's eyes, strictly disjunctive; and since ours exists, they do not. Our universe is unique, and nothing in it could profitably be taken out of its *context* and examined under ideal, non-existence conditions. These are the deeper reasons why Aristotle was not willing, as Clauberg rightly observed, to see things "as they are in themselves" but always insisted that we should see them "as they are in respect to each other."

2. Idealization and the Impetus Theory

Aristotle's theory of motion, like many of his doctrines, underwent serious transformations and modifications, in Antiquity, by the Moslem *falasifa*, and later in medieval western Europe. The driving force for such changes was either purely interpretative or theological or both; and the starting point of even adverse theories was more often than not Aristotle's arguments against his own positions. The wish to expand the horizon of God's omnipotence provided, throughout the Middle Ages, the initiative for taking up many of his "impossibilities" of nature so as to prove their possibility, if not reality; but Aristotle had taught medieval theologians and philosophers how to do it, how to construct an imaginary experiment properly.

This was already recognized in the fifth century by Johannes Philoponos, one of Aristotle's most critical commentators. In his commentary on the *Physics*—he wishes at this point to defend the notion of space as dimensionality independent from the material continuum although coextensive with it—Philoponos acknowledged that Aristotle taught him the method of thought-experiments and their value.

> We often assume the impossible, so as to understand the nature of things in and of themselves[!]. Aristotle indeed asks those who claim that the earth is motionless because of the fast rotation of the heaven this question: In what direction would the earth move if we let, by supposition, the heavens stand still? And in the following he imagines a body stripped of all qualities or form and considers it in itself. We, too, followed our imagination to separate all forms from matter and consider it by this method bare and in itself. . . . Plato, too, separated in thought the origins of the cosmic order from the cosmos itself and asked how the totality behaves in and of itself, separate from the God.

And even though it is not possible that one of these assumptions become real, reasoning separates what is together according to nature in thought, so as to manifest how everything behaves in itself according to its specific nature.[23]

I do not know whether Philoponos recognized, while writing this passage, the slight but decisive difference in the roles assigned to imaginary experiments between himself and the Stagirite. Aristotle, in *De caelo* B 13.295a10–15 and even more so in the other considerations to which Philoponos alludes, does use imaginary experiments as a *reductio ad impossibile*. Philoponos wants to learn the nature of a thing "in and of itself"—note how often this phrase is repeated. But whether or not he was aware of the difference, it permeates his interpretations throughout. He insisted that bodies moving in the void, either "by nature" or by constraint, move in a specific time, each according to its weight. The medium in both cases is only a hindering factor. This means, as Averroës rightly remarked against Avempace's similar theory, that bodies falling in the void would fall mostly according to their natural motion, while less so when they fall in the medium;[24] Philoponos, as we saw, actually admitted that much; to Averroës this seemed a conceptual absurdity to call most "natural" a counterfactual conditional. It is not likely that Philoponos's theory of the fall was influenced by Hipparchus. In fact, Philoponos does not offer the most appealing explanation given by Hipparchus to the acceleration of falling bodies: the original thrust given to them wears down when they move upwards un-

---

[23] Johannes Philoponos, *In Arist. Physicorum libros quinque posteriores commentaria*, ed. Vitelli, pp. 574.46–575.10; esp. ll. 21ff.: ὅταν γὰρ ὑποθέσει τινὶ ἔπηταί τι ὃ μὴ ἐνδέχεται γενέσται, τότε ἐκ τοῦ ἀδυνάτου εἶναι τὸ ἑπόμενον ἐλέγχομεν τὴν ὑπόθεσιν, ἐπείτοι γε τοῦ συνιδεῖν ἕνεκα τὴν πραγμάτων αὐτῶν καθ᾽ αὐτά φύσιν καὶ τὰ ἀδύνατα πολλάκις ὑποτιθέμεθα . . . pp. 575.8ff.: καὶ ὁ Πλάτων δὲ τὸν τῆς τάξεως τοῦ παντὸς αἴτιον καὶ ἐπίνοιαν τοῦ κόσμου χωρίσας, ζητεῖ πῶς ἂν ἔχοι τὸ πᾶν αὐτὸ κατ᾽ αὐτὸ θεοῦ χωρισθέν . . . Cf. Wasink and Jansen, eds., *Timaeus a Calcidio translatus commentarioque instructus*, in *Plato Latinus*, ed. Klibansky 4:301.14–18: "Idemque nudae silvae imaginem demonstrare et velut in luce destituere studens detractis omnibus singillatim corporibus, quae gremio eius formas inuicem mutantur et inuicem mutant, ipsum illud quod ex egestione uacuatum est animo considerari iubet." It seems that Chalcidius quotes Numenius; cf. 278.17–279, where the editor notes the Aristotelian origin of the phrases *sublatis, ademptis* (ἀφαίρεσις). Cf. also below v.B.1,7 (Proclus).

[24] Averroës, *Aristotelis opera cum Averrois commentariis* 4.71, 4, fol. 132v; quoted by Moody, "Galileo and Avempace," in *Studies*, p. 231.

til it becomes smaller than gravity; then the body falls, but the amount of the thrust still maintained slows its downfall, though less and less so.[25] Nor does Philoponos criticize the theory.

Philoponos explained acceleration as well as projectile motion, against Aristotle, by his own theory of imparted force: this imparted quality, which wears off by itself even in space, wears off more rapidly when a body moves in a medium. "Space" is neither a location outside the finite material continuum, nor is it an accident of bodies only: it is rather the immutable, dimensional container coextensive with all bodies. Does this theory refute Aristotle's contention that motion in the void is incommensurable with motion in the plenum? Not in the least, and again Philoponos emphasizes it. Whether we render his (and Avempace's) intention in the modern notation velocity (v) = force in space (f) − resistance (r) or whether we read it differently, say $v = p/(r + d)$ (as Anneliese Maier suggested for similar medieval theories), or whether (as I believe) its closest expression is $(df − f)/rd = v$,[26] it is clear that the velocity of a body in the void, according to Philoponos, cannot be assigned a concrete value that would function as a limiting case. Philoponos makes it clear that different media do *not* relate to each other in a corresponding proportion to the velocity of bodies. This means that the only way to assess the resistance of a medium of any degree of rarity is to let a body pass through it. We cannot predict what amount of resistance another medium thinner than this one would slow down, or hinder, the motion of a body. Which means that we would have to create a vacuum—which Philoponos believes to be

[25] Simplicius, *In Aristotelis de caelo libros commentaria*, ed. Heiberg, p. 264. Cf. below III.C.4 (Galileo's *De motu*).

[26] A. Maier, *Zwischen Philosophie und Mechanik*, pp. 239–85, esp. p. 278, against Moody (above n. 24). Both agree that Avempace permitted the treatment of *velocitas* and *tarditas* as extensive magnitudes. Maier's formula represents Thomas's interpretation of Avempace. As to Philoponos, it is unclear that his intentions can be captured with a formula. M. Wolff, *Fallgesetz und Massbegriff: Zwei wissenschaftshistorische Untersuchungen zur Kosmologie des Johannes Philoponos*, pp. 30–35, recognized that Aristotle does not combine the relation force–medium and force–weight: they stand unmediated as two expressions of velocity. He shows that Philoponos mediated between them, but refrains from abstracting a formula from his three tables (cf. also his n. 29). Philoponos claims that the total times of velocity of the same body in various media are not proportionate to the densities of these media (below n. 28), but the added *tarditas* is. If f/r be the velocity of a body in the void, then, in the plenum, $v = f/r − f/rd = (fd−f)/rd = f(d−1)/rd$. It fits Philoponos's remark that the *tarditas*-factor (f/rd) becomes ever smaller, and (d) is measured proportionally from a given density.

impossible—so as to assess the motion of a body in it. Since we can only approximate but never attain a medium with zero density, the motions of bodies in the vacuum, whether natural or constrained, are still *incommensurable* with their motion in the void.

This is all the more true in the case of those medieval commentators who followed Avicenna against Averroës, and denied that motion, even in the void, would be instantaneous.[27] In the thirteenth century this was the common opinion. It did not commit those who held it to adopt an impetus theory of projectile motion; most of them ascribed the successive nature of motion in an imaginary (and for most of them impossible) void to distance itself, the *distantia terminorum*. But with or without an impetus theory, motion in the void could remain utterly incommensurable; it could not be an actual measure for existing motions.[28] This was to remain true till Galileo.

Why, then, was the impetus theory revived—or perhaps reinvented? It seems that theological rather than physical concerns provided the initiative. This may be true of Thomas Aquinas and Johannes Olivi; it is certainly true of Franciscus de Marchia.[29] The

[27] Averroës, *Comm. in Phys.* 4.71, in *Opera Ar.* 4:130ff.; Thomas Aquinas, *Comm. in Phys.* IV lect. 11–13, *Opera* 18:351ff.; Scotus, *Ordinatio* 2 d.2 q.9, ed. Balič, 7:299ff; Lasswitz, *Geschichte der Atomistik vom Mittelalter bis Newton*, 1:207–208; Wolfson, *Crescas' Critique of Aristotle: Problems of Aristotle's Physics in Jewish and Arab Philosophy*, pp. 183, 205, 403–409; Maier (above n. 26); E. Grant, "Motion in the Void," *passim; id., A Source Book in Medieval Science*, pp. 334–50.

[28] "It is impossible to find the ratio which air bears to water . . . i.e. to find how much denser water is than air, or any one kind of air to another": Philoponos, ibid., p. 682. Which is to say: even if a *finite* time is assumed for motions, natural or coerced, in the void—Crescas later spoke of "rudimentary time" (*zeman shorshi*); *Or Adonai* 1.2.1, p. 16—it would still be not only a limiting value of motion in the medium, but also unknowable, since the exact ratio of motion through air and a putative thinner medium that is not void ("pure" fire perhaps) is unknown: the limit value is, even in principle, unknowable. Galileo, on the other hand, considered the resistance of the air as simply negligible, or at least accountable.

[29] Following A. Maier, *Zwei Grundprobleme der scholastischen Naturphilosophie*, pp. 142–200. Thomas (ibid., pp. 135–41) rejected the theory emphatically in his *Physics* commentary, but refers to it in passing in *De potentia* q.3 a.11 and *De anima* q.u. a.11 as if accepting it. Recently, M. Wolff, *Geschichte der Impetustheorie: Untersuchungen zum Ursprung der klassischen Mechanik*, though stressing the theological origins of Philoponos's δύναμις, argues for the origin of Olivi's *inclinatio* in his money theory (pp. 174–91). But cf. J. Naphtali, "Ha'yachas sheben avoda le'erech bate'oriot hakalkaliot shel ha scholastika bameot ha-13 veha-14" [The Correlation between Labor and Value in the Scholastic Economic Theories of the 13th and 14th Centuries], pp. 12–17. The text of de Marchia was translated by Clagett, *Mechanics*, pp. 526–31 (without the theological context).

office of a theologian is not to deny the existence of true miracles, but to make them plausible; to show that they are at least consistent with everything else we know about the world. This is certainly true of the Catholic Church, which is, as far as I know, the only religion to have institutionalized miracles—a sociological *hapax legomenon*. The sacrament of the Eucharist is the most predictable of them all. It has the capacity to move the believer toward God, to infuse him with the *gratia gratum faciens*; but God is not present in it directly, only indirectly or instrumentally. The Host has a *virtus*, a power, of its own, imparted by God but not identical with him. What kind of causality is this instrumental causality of a distant mover? De Marchia wished to show that this mode of causality is not unknown in nature. Projectiles move in an analogous way. A "force" is left behind by the original mover, which permits them, given the right circumstances, to continue moving.

Recent discussions of the later medieval career of impetus mechanics have generally suffered from the wish either to establish or to disprove its approximation to the law of inertia. The Scholastic term is, indeed, untranslatable into the vocabulary of early modern physics. "Impetus" is a quality somewhat analogous to heat. It is a motive *power* accounting for motion and thus still conceived on the basis of the assumption that "omne quod movetur ab alio movetur," and hence on the distinction between rest and movement.[30] Inertia, on the contrary, is not a force, but a state under which both rest and uniform motion are subsumed, and distinguished from change of either velocity or direction. In regard to them only the quest for immediate causes is meaningful.[31]

[30] In this respect, I follow Maier, *Zwei Grundprobleme*, pp. 113ff., esp. pp. 126; 217ff.; 223ff.; *Ausgehendes Mittelalter: Gesammelte Aufsätze zur Geistesgeschichte des 14. Jahrhunderts* 1:353ff., esp. pp. 376f.; 431ff.; Koyré, *Metaphysics and Measurement*, pp. 28–32; id., "Galileo and Plato," pp. 400ff. A different assessment of the impetus—in particular, Buridan's indefatigable impetus—as an approximation to either the notion of inertia or that of momentum in early modern physics. Duhem, *Etudes sur Léonard de Vinci*; Clagett, *The Science of Mechanics in the Middle Ages*, pp. 523f.; Moody, "Laws of Motion in Medieval Physics," in *Studies*, pp. 189–201; Dijksterhuis, *The Mechanization of the World-Picture*, pp. 111–15 (momentum).

[31] A clear insight into the difference between a state and a cause was attained by Thomas Aquinas—not, of course, apropos the inertial principle (which he lacked even if he may have held some form of *impetus*), but in the clarification of what is meant by "natural" motion. He polemizes (*In libros Arist. de caelo . . . expositio* 3.2 lect.7, in *Opera*, 3:252) against Averroës, who accepted the medium as necessary agent—i.e., as an effi-

Or was perhaps Ockham's radical simplification of the problem an anticipatory step toward the recognition of uniform motion as a state? Ockham simply rejects the theory of impetus as well as the Aristotelian account of projectile motion by denying the existence of the problem.[32] The term "movement," as extension or quantity, is a connotative term, denoting an object and connoting a series of places that it occupies consecutively. Both movement and conservation of movement are two expressions for one and the same phenomenon. We need only one cause to explain why a body left l_1 and reached l_n through $l_2 \ldots l_{n-1}$. If it left l_1, it necessarily occupies other places. Yet Ockham, because he was preoccupied with the reduction of our concepts to those singulars and absolute qualities that they stand for *in recto* or *in obliquo*, stopped exactly at the point where he might have hit upon the distinction underlying the inertial principle—namely the distinction between uniform motion (or rest) and change of motion. Ockham succeeded, however, in detaching the problem from its original theological context. It seldom appears again as an explanation for instrumental causality.

Buridan introduced the impetus hypothesis almost as a concept won by induction.[33] From his predecessors he inherited a list of "arguments from experience" against the Aristotelian interpretation of

---

cient, or active, cause—of natural motions (Averroës, *Physica* summa 4 text 82, in *Opera*, 4, fols. 195vb–196va; *De caelo* 3 summa 3.2 text 28, in *Opera*, 5, fols. 91vb–92va). Thomas argues that when we speak of "forms" as the "cause" of gravity or levity we do not refer to them as a *movens*, as an active source of motion; the form is but a *passive* cause in this case. In other words, when encountering natural motion, we may abandon the search for causes. On the necessity of an analogous principle for every explanation of motion see Koslow, "The Law of Inertia: Some Remarks on Its Structure and Significance," in *Philosophy, Science, and Method: Essays in Honor of Ernest Nagel*, ed. Morgenbesser et al., pp. 552–54 (condition of normalcy).

[32] Ockham, *Sent.* 2 q.26M; Boehner, *Ockham: Phil. Writings*, pp. 139–41; Clagett, *Mechanics*, pp. 520f. Cf. Moser, *Grundbegriffe*, pp. 91–111 (Ockham as interpreter of Aristotle); Shapiro, "Motion, Time, and Place according to William of Ockham," pp. 213–303, 319–72; Goddu, *Physics*, pp. 193–205. In his *Physics* commentary, Ockham tries more seriously to save the Aristotelian explanation. The air as such has slower or quicker *parts* that help carry the body; Ockham replaces, then, Aristotle's "air layers" with "air currents," divided, presumably, not vertically but horizontally. But this would render the motion of a projectile *cessante movente* unpredictable, and Ockham makes it clear that it is not his opinion, but rather the only way to rescue Aristotle's view on the matter. Ms. Berlin Lat. 2°41, fol. 202va–rb.

[33] *Questiones super octo physicorum libros Aristotelis* 8.12; Maier, *Zwei Grundprobleme*, pp. 207–14; *Questiones super libris quattuor de caelo et mundo* 2.12; 13; 3.2, ed. Moody, pp. 180–84, 240–43.

projectile movements; he adds to them some of his own; and after having introduced his own interpretation he returns to "experience" and points at a group of movements which, although they were explainable by the displaced theory of the active air layers, are better explained by the new one. From there he goes even further to consider the celestial movements as conserved by their impetus. Once introduced to explain one group of movements in which it seems manifest, "impetus" soon becomes a key concept to explain every movement as one of its aspects. The universality of impetus is a logical consequence of its postulation. There can be no sufficient reason why it should be confined to projectile bodies only. It has rather to be conceived as a universal factor in every movement.

Buridan is thus led to understand all movements as an interaction of impetus and the natural inclination of the body: (i) The gravity (or levity) of a body, together with the resistance of the air, is a retarding factor in the case of forced movements. Buridan gives no detailed analysis of these movements, though we have every reason to believe that they can be described as an inverted analogy to the natural movements. The impetus of forced movement decreases constantly owing to gravity (or levity), if left to itself. Only if force continued to be exerted on the body would it presumably continue its motion indefinitely, since its impetus is renewed. For a physicist operating without a clear notion of friction, this is the most logical interpretation of continuous movements. (Buridan, it is true, recognizes the resistance of the air as a factor in movements; but it is a constant, not a variable that can be abstractly reduced toward zero.) Note that, all of a sudden, projectile movements become the rule, continuously forced movements the exception, or rather a secondary phenomenon. (ii) The gravity (or levity) of a body is an accelerating factor in the case of natural terrestrial movements. This is not to say that impetus is not always the cause for the *preservation* of motion, but rather that gravity acts as an intrinsic force imparting to a body both movement and impetus. Since "movement" is steadily imparted to the body, its impetus (determined by force and *quantitas materiae*) must also increase. Movement as such never produces impetus, but rather the force causing movement is also the force causing impetus—in falling bodies as in reflexive motions. (iii) The natural inclination of celestial bodies neither retards nor accelerates their motion; it does not interfere with their impetus at all, since this inclination means nothing more than potential circular movement.

Given the initial impetus, the celestial bodies will continue to move *uniformiter* indefinitely. Buridan could thus discharge the "intelligences" as the efficient cause moving the spheres. That he intended a "mechanical" interpretation of the universe is, however, an exaggeration. His celestial impetus is entirely hypothetical; he did not want to link the celestial and terrestrial impetus to the point where the rejection of the former would mean the rejection of the latter.[34] Consequently, this aspect is not discussed in the *later* questions on Aristotle's book on heaven and earth.

In short, Buridan describes terrestrial motions as impetus plus or minus gravity (or levity), celestial motions as impetus + or − 0. The methodological shift underlying this analysis of motion is considerable. *All* terrestrial motions, including natural movement, are complex motions, a product of the interaction between the natural inclination of the body moving and the impetus it acquired either by external force or by its very inclination. Gravity and levity are but a single factor of all movement. Even natural movements are not "simple" in their totality.[35] Their simplicity is but one aspect of them, isolated not in the immediate observation, but as a mental concept. Buridan has introduced, as Philoponos before him, an important methodological change, and he seemed to have been aware of it. Yet even for him, the motion of a body by impetus only was incommensurable with its motion in the plenum.

3. Idealization and Calculationes

There existed a second group of imaginary considerations concerning the nature of movement, totally theological in nature. An infinite motion of the universe in straight line in the void was, we re-

---

[34] Above I.A.n.4. On the history of the problem of the intelligences as movers see Wolfson, "The Problem of the Souls of the Spheres from the Byzantine Commentaries on Aristotle through the Arabs and St. Thomas to Kepler," pp. 67–93.

[35] In the *Quest. super de caelo* 1.5, ed. Moody, pp. 20–57, Buridan defines simple motions as (i) "simplices respectu medii" and (ii) "qui est ab unico simplici motore." In the latter sense, free fall is not simple: "ad illum motum concurrit aliud movens praeter gravitatem naturalem quae a principio movebat et quae semper manet eadem"—and this *aliud movens* (impetus) is the cause of acceleration (ibid., pp. 179f.). The distinction between forced and natural motion pertains to *one* aspect only. Once this aspect is isolated, it is absolute: "potentiae . . . et virtutes cognoscuntur per motus et operationes" (ibid. 4.1, p. 245), and yet the motions of each of the four sublunar elements are absolute, not relative, *virtutes* (ibid. 4.6, pp. 261–64). Aristotle saw in water and air only relatively light or heavy elements. In short, simplicity is the result of a factoral analysis and not, as it was to Aristotle, immediately given.

member, asserted as a possibility *de potentia Dei absoluta*. The disciplined imagination of fourteenth-century theologians examined the preconditions for such a hypothesis. Such a void must be absolute, that is, immobile: otherwise the motion of the universe— or even of a single body if the rest of the world were destroyed— would be relative only to itself, that is, objectively indistinguishable from rest. We discussed it in the previous chapter. The advantage of the method of annihilation becomes clearer if we imagine not merely a single body *toto mundo destructo*, but a single spiritual substance, say an angel: how can God, if he so wishes, move it? If it moves as bodies do, then even an *ubi definitivum* requires the presupposition of an imaginary dimensional space.[36] We are then, in fact, examining the preconditions for motion as such, be it of a point.

Here, so it may seem, we found at last a real methodological anticipation of the sort of imaginary experiments underlying the early modern dynamics. Was not the Galilean analysis founded upon the very same principle of the extrapolation of a phenomenon from its context? Indeed, the career of the "method of annihilation" may have its roots in the Terministic analysis of our concepts; but the differences between the use made of this method in the fourteenth century, and the place which it occupies later in science (Galileo) or philosophy (Descartes, Hobbes) are considerable. The *Terministae* used the heuristic *topos* (isolation through annihilation) mainly with a negating, that is, critical, intent, much as Thomas earlier used the distinction between logical and contextual necessity. In a general sense, they wished to establish what the Aristotelian cosmos—which neither Ockham nor his followers wanted to destroy—is *not*: namely, in any sense "necessary." In particular, they wished to establish what *things* are not: connotative notions such as extension, motion, or time should not be hypostatized nor have any claim to an ontic status. It is significant that one of these abstractions (extension), rather than the individual *res extra animam*, became the material "substance" of Descartes, guaranteed by a *cognitio intuitiva* (as were Ockham's *res*).[37] The resolutive method demanded the extrapola-

---

[36] Johannes de Ripa, 1 *Sent.* d.37, ed. Combes, p. 232; Grant, *Much Ado*, p. 131; cf. above II.D.2 and n. 17.

[37] Below III.D.3.

tion of relations rather than of things, and with a constructive rather than critical intent.

Closest to the spirit of science in the seventeenth century are indeed a third group of imaginary experiments: those influenced by the mathematical techniques of the "*Calculatores.*" The study of proportions led them to distinguish between uniform change and uniform change in the rate of change (*motus uniformis, motus uniformis difformis*), to reduce uniform acceleration into terms of uniform motion by proving that $(v_o + v_1)/2 = v$.[38] Oresme later applied this "rule of mean speed" to every segment, however small, of the distance (or the amount of change) that an accelerated body acquires; and realized that it amounts to the arithmetical sum of the sequence of odd numbers with as many members as there were segments divided by two—or, as we would write, $S = at^2/2$.[39] These were simple cases of one variable; the Calculatores soon learned to handle proportions of variable proportions, or proportions of proportions to any desired exponential degree.

It is true that, except for a few cases (these are an important exception) all of these mathematical exercises were valid only *secundum imaginationem*, from the simplest down to the set-theoretical considerations mentioned earlier, or to considerations of minima and maxima. Even Oresme did not do that which to us seems obvious: apply the formulas of uniformly accelerated change to the doctrine of impetus. Indeed, Oresme did improve upon Buridan's theory of impetus in that he recognized the ambiguity of Buridan's concept. Impetus, for Buridan, served to explain both the continuation of projectile (coerced) motion and the acceleration of falling bodies; it was both the cause of motion and caused by motion. Oresme reserved it for the explanation of acceleration alone, and hence ceased to regard impetus as a *res natura permanens*. Still, Oresme refrained from employing mathematical terms to describe the motion of falling or projected bodies—perhaps because he believed

[38] C. Wilson, *William Heytesbury: Medieval Logic and the Rise of Mathematical Physics*, pp. 122–26.

[39] *Questiones super geometriam Euclidis per Magistrum Nicholaum Oresme* q.14, in *Nicole Oresme and the Medieval Geometry of Qualities and Motions*, ed. Clagett, pp. 562–64: "Secunda conclusio est quod subiecto taliter diviso, et vocetur semper pars remissior prima, proportio partialium qualitaum et habitudo earum ad invicem est sicut series imparium numerorum, ubi prima est 1, secunda 3, tertia 5 etc. . . ." Cf. introduction, pp. 72, 104; cf. ibid., pp. 158, 164.

that in both cases not only the rate of change of velocity, but also the rate of change of acceleration, changes.[40] Not until Dominicus de Soto did any schoolmen consider applying a mathematical formula to free fall, and even he did so only in general terms.[41]

And yet, whether real or imaginary, the Calculatores did—for the first time in the western tradition—mathematize the concept of change of motion. And more, they conceptualized the notion of processes in general. I shall discuss it later.

4. Galileo: Idealization as Limiting Cases

Only a meticulous historical examination could show how much of the accrued tradition of medieval imaginary experiments, or the work of the Calculatores, was available to Galileo; interest in their work seems to abate in the beginning of the sixteenth century. Nor is it clear what versions of impetus mechanics Galileo knew, and from which sources.[42] My concern, then, is not to explain the evolution of his mature thought out of medieval residues. I rather wish to compare his law of free fall, praised by physicists from Newton to Einstein as the dawn of modern physical thought, with the medieval employment of the scientific imagination. The latter was to a

[40] Oresme, *Le Livre du ciel* 2.13, ed. Menut and Denomy, p. 416: "Et pour se, en mouvement violent a iii estas ou iii parties. . . . Secondement, quant la chose meuve violemment est separée de tel instrument ou premier motif, encore va l'isneleté vient en cressant, . . . et lors l'isneleté ne crest plus ne cette qualité ou redeur." It seems as if the impetus causes acceleration: the original force, or *movens*, causes an increase in the rate of acceleration (first phase); when the impetus, which wears out of itself, takes over (second phase) the body still accelerates, but in a decreasing *rate*; in the third phase, the body decelerates until gravity overcomes its violent motion. Impetus, then, never accounts for motion—and motion as such (say, uniform violent motion) needs no explanation—it does not exist. Nor does uniform acceleration—therefore his formula was of no use for the examination of free fall. Its rate is actually the derivative of a derivative. Cf. Maier, *Zwei Grundprobleme*, pp. 236–58, esp. pp. 254ff.; Clagett, *Mechanics*, p. 552; Wolff, *Impetustheorie*, pp. 228–38.
[41] Clagett, *Mechanics*, pp. 555–56; see also Maier, *Zwei Grundprobleme*, pp. 299–302 (gravity analogous to impetus); Wallace, "The Enigma of Domingo de Soto: *Uniformiter Difformis* and Falling Bodies in Late Medieval Physics," pp. 384–401.
[42] Wallace, *Galileo's Early Notebooks: The Physical Questions*; Clagett, *Mechanics*, ch. 11, esp. pp. 653–71; Murdoch and Sylla, "The Science of Motion," in *Science in the Middle Ages*, ed. Lindberg, pp. 249–51; Maier, *Zwei Grundprobleme*, pp. 291–314; Dijksterhuis, *The Mechanization*, pp. 329–33, believes that Beeckman arrived at his formulation of the law of free fall inspired by the Calculatores, but his evidence is circumstantial only. Yet, even Leibniz still admired the use of *latitudo formarum* in ethics: *Initia et Specimina Scientiae Novae Generalis*, GP 7:115; GP 7:198.

measure also his own in his early attempts to solve the problem of free fall.

In his *De motu*, an earlier treatise that he wrote while he was still a young professor of mathematics in Pisa, Galileo recognized two factors in the free fall of bodies—gravity and impetus.[43] His explanation resembles that of Hipparchus in Antiquity: when a body is thrown upwards, it overcomes gravity by virtue of the force (impetus) given to it by whomever or whatever projected it. This force wears out gradually; when it equals the force of gravity the body changes its course, and even when it falls, it has still enough of the initial impetus left in it to slow down its motion downwards, which would otherwise be faster. The more the impetus continues to wear out, the more the body accelerates its downwards motion. If the impetus were to wear out before the body hits the ground—but it does not—the body would continue in its fall *with uniform velocity*. Hipparchus did not consider that case, which Galileo admits to be an imaginary condition. Galileo also ascribed, in a hydrostatic analogy, a capacity of bodies to receive impetus that is proportional to their specific gravity; he already insisted that bodies of different weight but of the same specific gravity fall with equal velocities.

It is noteworthy that Galileo, already in *De motu*, discarded levity altogether: the speculative reduction of forces to attraction and repulsion was a mark of many Renaissance philosophies of nature. Later, in formulating his mature law of free fall, he also got rid of the *impetus*. One force only remained, acting on all bodies homogeneously. Which means that Galileo not only recognized—as many impetus theoreticians in the later Middle Ages also did—that constant application of a force causes acceleration, but that he also

[43] Galileo Galilei, *De motu dialogus* (ca. 1590), in *Opere*, ed. Favaro 1:319–20, 404–408, esp. p. 407; cf. Galileo, *Discorsi . . . intorno a due nuove scienze*, in *Opere* 8:201 (presented as Sagredo's opinion, and dismissed). Clavelin, *The Natural Philosophy of Galileo: Essay on the Origins and Formation of Classical Mechanics*, trans. Pomerans, pp. 120ff., esp. pp. 132–33 (emphasizes the hydrostatic model and the universality of heaviness); Drake, *Galileo at Work: His Scientific Biography*, pp. 21–32, esp. pp. 28ff. (Pereira as source of the Hipparchian hypothesis). I lay particular stress on Galileo's imaginary condition of a falling body losing the retarding effect of the *impeto* in midway and continuing to move uniformly (admittedly this does not happen). In his mature theory of free fall, "gravity" took the place of "impetus," wherefore the body decelerates moving upwards, and accelerates (by gravity) downwards; if gravity were to cease in midway, again the body would continue to move downwards uniformly. It seems to me a much better premonition of inertia than Drake's "neutral" motion: Drake, *Galileo Studies*, pp. 240–56.

recognized that *only* the application of force causes acceleration or change of motion. Uniform motion needs neither causes nor forces. Gravity became an external, attractive force rather than an internal quality or "form" in bodies; only thus could it attract unequal masses so that they descend with equal speeds. Though we should be cautious not to credit Galileo with a precise notion of force, denying it altogether would be likewise mistaken.[44] Galileo construed an elaborate imaginary experiment to prove that there is no difference between natural and coerced motion, gravity and impetus. Imagine, he says, a hole through the center of the earth, and a body falling through it, passing the center due to the impetus it has accrued. Once having passed the center, the motion, hitherto natural (gravity), becomes of itself "coerced" or artificial, that is, "by force" only.[45] Finally, even in *De motu*, the resisting factors other than the specific weight (or capacity to absorb impetus) are negligible. That was neither the case for Philoponos or Avempace, nor for Buridan. For the first time, then, the descent of free-falling bodies could be determined empirically under all essential conditions.

There was nothing new in the mathematical formulation of the law—Oresme already possessed it—though Galileo did in all likelihood reinvent it, and even then it took him some time to recognize its precise physical meaning.[46] Indeed, much of the work of Tartaglia, Benedetti, and Galileo resembles the usage made by the Calculatores of the theory of proportions. In one decisive sense, however, Galileo differs altogether from his medieval counterparts. His kinematical definitions and theorems are not an exercise in the systematic imagination of a "rational physics." They are an instrument with which he hoped to reconstruct reality. They had to be experimentally testable, even if obliquely; and Galileo spent effort and ingenuity in the attempt to devise precise measuring instruments.[47]

---

[44] As does Westfall, *Forces*, ch. 1, esp. pp. 46–47 (acceleration as due to natural inclination, not force).

[45] *Dialogo . . . sopra i due massimi sistemi de mondo Tolemaico, e Copernicano* 2, *Opere*, 7:262–63. This is true because of the *impeto* (here: energy) accrued; but it is the same *kind* of motion before and after passing the center.

[46] The date on which Galileo decided to substitute time for space is uncertain; see Drake, *Galileo*, pp. 91–133; Clavelin, *The Natural Philosophy*, p. 287 (*Opere* 9:85); Wallace, *Galileo and His Sources*, pp. 272–76.

[47] Bendini, "The Instruments of Galilei," in *Galileo, Man of Science*, ed. McMullin, pp. 256–89; Settle, "Galileo's Use of Experiment as a Tool of Investigation," in ibid., pp. 315–

Benedetti and Galileo, Huygens and Descartes, Pascal and Newton used their imaginary experiments in a definite way which differs
toto caelo from their medieval predecessors not in discipline and
vigor, but in their physical interpretation. Counterfactual states
were imagined in the Middle Ages—sometimes even, we saw, as
limiting cases. But they were never conceived as commensurable to
any of the factual states from which they were extrapolated. No
number or magnitude *could* be assigned to them, even if schoolmen
were to give up their reluctance to measure due to their conviction
that no measurement is absolutely precise. For Galileo, the limiting
case, even where it did not describe reality, was the constitutive element in its explanation. The inertial motion of a rolling body, the
free fall of a body in a vacuum, and the path of a body projected had
to be assigned a definite, normative value. And Galileo was well
aware of the absurdity of his procedure in the eyes of orthodox Aristotelians. In the *Dialogo* he ridiculed Simplicio for his reluctance to
assume that to be the case, which in fact could not possibly be so.
Simplicio, after protesting once again against the exaggerated use of
mathematical modes to explain physical phenomena, is taught by
Salviati that the true *filosofo geometra*, when he wishes "to recognize
in the concrete the effects which he has proven in the abstract, must
overlook the material hindrances."[48] An even stronger argument to
that effect is put forward in *Two New Sciences*. The consideration of
the paths of the projectile are simplified even as a theoretical model
without the "material obstacles": the effects of gravity are represented by parallel lines along the path rather than by radii directed
to the center of earth; but this procedure in his statical consideration, he says, is no different from Archimedes' and others' who also
"imagined themselves, in their theorizing, to be situated at infinite
distance from the center."[49] Galileo, as Blumenberg rightly empha-

---

37. The famous discussion, initiated by Koyré, whether Galileo's mechanics rested on
actually performed experiments should perhaps be reformulated: even if one admits a
crucial role to experimentation, how much are they complemented by imaginary experiments in the narrow sense, i.e., such that cannot be performed?

[48] Galileo, *Dialogo*, in *Opere* 7:242. Cf. Koyré, "Galileo and Plato"; *id., Metaphysics and
Measurement*, p. 37; Cassirer, *Das Erkenntnisproblem*, 1:383; Wallace, *Galileo and His
Sources*, pp. 278–80, 286.

[49] Galileo, *Discorsi*, in *Opere* 8:274–75; cf. also Boyer, "Galileo's Place in the History of
Mathematics," in *Galileo, Man of Science*, ed. McMullin, p. 239 (commensurability of
curve and line).

sized,[50] does not compare an "ideal" state to a "deficient" reality; the very deviation of the real from the ideal can be measured and explained with an ever more complicated model. Rather than comparing reality to the ideal, he compares the complex to the simple. The scientific revolution of the seventeenth century learned to assert the impossible as a limiting case of reality.

But "reality" is a vague notion. In some way medieval Scholastics were more "realistic" than seventeenth-century science—as in their claim that we can never isolate phenomena from their context altogether or conduct precise measurements. In other ways their range of imagination was much broader. In a few instances they even conceived of ideal states as a limiting case. Conversely, Galileo was in some ways more attached to the real, if by real we mean daily experience and praxis. Rather than confronting "idealism" with "realism" in the Middle Ages as against the seventeenth century, I suggest we attend to the shift of function of imaginary experiments. In the Middle Ages their function was throughout critical—except in some of the work of the fourteenth century. In the seventeenth century's science and philosophy, they became a tool for the rational *construction* of the world, of the *machina mundi*. This is true of the principle of annihilation in Descartes's or Hobbes's philosophy. This is true of imaginary motions of simple bodies under simplified conditions in Galileo's physics. This is also true in the actual construction of experiments in chemical laboratories, under the useful fiction that one can isolate a system of chemical substances and construct them in such conditions as to study pure processes. The study of nature in the seventeenth century was neither predominantly idealistic nor empirical. It was first and foremost *constructive*, pragmatic in the radical sense. It would lead to the conviction that only the doable—at least in principle—is also understandable: *verum et factum convertuntur.*[51]

The medieval confrontation between the ideal (or imaginary) and the real was mainly *critical*; the seventeenth-century mediation between the imaginary and real was *constructive*. Perhaps I can fortify this distinction with yet another consideration.

Scientific instruments until the nineteenth century fall roughly into two classes. Some are observational, others are manipulatory.

---

[50] Blumenberg, *Die Genesis der kopernikanischen Wende*, pp. 470–88, esp. 482ff.
[51] See below ch. v for a detailed discussion of this theme.

Measuring instruments or magnifying instruments are observational; phials and burners and conductors are manipulatory. The former are not intended to alter the object; the latter manipulate it, and they do so by *isolating* an object or a group of things from their environment as best as possible so as to induce certain processes which can then be claimed to be well regulated. The Middle Ages knew mostly the former (the astrolabe, Jacob's staff)—except in the cases of optics and alchemy.[52] Alchemists wanted to manipulate nature, to extract pure substances, by imitating the conditions of creation. A good part of their procedures were symbolic or based on assumptions of nature—symbolism. Their manipulatory instruments were, however, inherited by early modern chemists, who were already committed to the ideal of unequivocation. Not all of them gave up the hope of transforming baser metals into gold. But inasmuch as they were chemists they desymbolized both nature and this scientific language. They used—we use—laboratories in order to isolate, to study phenomena under ideal conditions, which, of course, can only be done approximately. In one limited segment, the scientist reconstructs nature in order to understand it. I do not overlook the fact that the distinction between observation and manipulation is relative. But not until modern particle physics was there reason to suspect that any observation of certain objects must manipulate them. The method of isolation and reconstruction, whether in thought or in praxis, was the seventeenth century's mediation between the real and the ideal.

D. DESCARTES, ETERNAL TRUTHS, AND DIVINE OMNIPOTENCE

1. 2 + 1 ≠ 3: Possible Interpretations

Returning to the question with which I started,[1] what did Descartes mean when he claimed that God created and, therefore, could in-

---

[52] But see Schramm, "Roger Bacon" and E. Grant, "Medieval Explanations and Interpretations of the the Dictum that Nature Abhors the Vacuum," pp. 327–55, esp. pp. 332–47 (Clepsydra). See also Fisher and Unguru, "Experimental Science and Mathematics in Roger Bacon's Thought," pp. 353–78.

[1] Above III.A.1. The following after my article, "Descartes, Eternal Truths, and the Divine Omnipotence," pp. 185–99. In a recent interpretation, Curley, "Descartes on the Creation of Eternal Truths," pp. 569–97, suggests, following Geach, a structure of iterated modality as a solution. Also close to my position is M. D. Wilson, *Descartes*, pp. 120–31.

validate eternal truths such as $2 + 1 = 3$? Only three interpretations seem possible. Descartes either meant to exempt God from the principle of contradiction (which, for the seventeenth century as for the Middle Ages, usually included the principle of the excluded middle);[2] or he somehow distinguished between real analyticity and analyticity for us, that is, allowed for that which seems contradictory to us to be resolved by God and only therefore possible to him; or again Descartes might have denied mathematical truths (and eternal truths in general) the status of logical truths. The first interpretation would be detrimental to Descartes's intentions, the second meaningless, the third perhaps too good to be true.

Alexander Koyré, who chose the first interpretation, painted a very attractive picture of the development that Descartes's position in these matters underwent—from utter voluntarism to the subsumption of God's operations under at least the conditions of logical possibility.[3] If this were true, then the early Descartes would have the whole Scholastic tradition against him, to the point of becoming vulnerable to the accusation of heresy. A God not subject to the law of contradiction could not only annihilate everything created, but even himself to boot; for which reason already Anselm of Canterbury had insisted that such an omnipotence would, in effect, be a weakness.[4] If God's self-annihilation were a real possibility, however unthinkable, God could not be an *ens necessarium* either.

Unless we admit what Thomas named *per se impossibilia*[5] into the

---

[2] The distinction between them goes back to Aristotle; but Aristotle, and most logicians till our century, saw them as coextensive, and equally valid, and therefore would sometimes use the former as a shorthand for both. Exceptions are the discussion of future contingents that were sometimes exempted from the principle of excluded middle—if not by Aristotle (against Lukasiewicz, cf. Anscombe, "Aristotle and the Sea Battle," pp. 1–15; Rescher, "An Interpretation of Aristotle's Doctrine of Future Contingency and Excluded Middle," in *Studies in the History of Arabic Logic,* pp. 43–54) then by some medieval authors; likewise, the distinction between *logica divina* and Aristotelian logic could be so construed (above III.B.5).

[3] Koyré, *Descartes und die Scholastik,* pp. 21–26 (pp. 25f., Development), 85–86 (Scotus). Descartes did not lean on the Scotistic version of the distinction between God's absolute and ordained power. In fact, we shall argue in the following, many of his basic attitudes are closer to those of the Terminists. Nor can we find any real development in Descartes's formulations in this respect. It is often assumed that Descartes did not distinguish between logical and mathematical (i.e., eternal) truths; see Bréhier, "La création des vérités éternelles dans le système de Descartes," pp. 15–29; Kenney, *Descartes: A Study of His Philosophy,* pp. 37–39. Although this position has some support in the text (cf. below D.4), it ought, as we shall argue, to be reexamined.

[4] Cf. above III.B.n.17 (Anselm).

[5] Above III.B.n.22

horizon of God's omnipotence, God's necessity could at best remain as a necessity for us only. The very foundations of the revived ontological argument would be shaken.[6] And it would remain shaken if we chose to modify this interpretation of Descartes's stand on the question of omnipotence to say not that God must be capable of defying the principle of contradiction, but rather that we do not know whether or not he is capable of doing so. For the *nervus probandi* of the new ontological argument, the feature that made it in Descartes's eye far superior to the Anselmian version, was the circumstance that it commences with a concept of God as *ens necessarium* rather than as *ens perfectissimum*.[7] The slightest possibility of his annihilation, even by himself, would destroy the argument. But none of this is necessary. Nowhere do we find Descartes abrogating the principle of contradiction *as such* with respect to God. Whether we look into his earliest remarks on the matter (in the letters to Mersenne) or into the latest (in a letter to More),[8] Descartes's examples are always the same: mathematical truths, mountains without valleys, actual atoms, or even "creatures independent of God";[9] and Descartes characterizes them even in his more radical moods as "evident contradictions to us," not to God. Put differently, Descartes attributes to God not the creation of the principle of contradiction, but the determination as to what should constitute a contradiction (or, conversely, a necessity). He seems to distinguish between absolute necessity and a necessity for us.

But is this distinction (and with it our second interpretative suggestion) not meaningless; meaningless at least as a logical distinc-

---

[6] Cf. above II.A.2. In both its medieval and modern version, the ontological argument establishes God as an *ens necessarium* (or, epistemologically, as a *notum per se ipsum*). Yet, while Anselm started his proof from the notion of God as *ens perfectissimum*, Descartes founded his proof on the very concept of God as *ens necessarium* itself, which relieved him from the necessity to interpret existence as an attribute or a perfection. Cf. Henrich, *Der ontologische Gottesbeweis*, pp. 10 ff. It may perhaps be said that there is only one way in which Descartes—or the ontotheological tradition following him—could be refuted: namely by denying that *e nihilo nihil fit*; in other words, denying that sui-sufficiency and necessity imply each other. That something may appear without cause literally out of nothing, it may be contended, is not a logical fallacy. Fred Hoyle's steady-state cosmology was defeated on empirical rather than logical grounds.

[7] This shift, Henrich shows (ibid., above note), became clear to Descartes himself only through the *objectiones*.

[8] Descartes to Mersenne, 15 April 1630, AT, 1:135ff.; 6 May 1630, AT, 1:147f.; Descartes to More, 5 February 1649, AT, 5:267ff.

[9] Descartes to Mesland, 2 May 1644, AT, 4:110 ff. It reminds us of the ancient paradox whether God can create a stone he could not upheave.

tion? For in one way or another, if asked to explain what precisely this distinction distinguishes, we are forced either to abandon the concept of an absolute (or logical) necessity altogether or to construe, alongside it, a necessity that is less than logical. It is not impossible, *pace* Quine, to attack the uses and abuses of analyticity.[10] A theory may be construed that erases the notion of analyticity altogether from our active vocabulary. Another theory is likewise conceivable that concedes the impossibility of eliciting analytical sentences from any given language, but insists nonetheless on the important function of analyticity as a regulative ideal. It is a matter of but minor significance whether, in such a theory, we let analytical sentences stand as regulative ideals, that is, as never totally realizable limiting cases of absolute clarity of speech, or whether we rather let them inhabit God's mind. Both ways, we may distinguish between analytical and semi-analytical sentences. Yet whether or not one fancies such a distinction, it is certain that it was not entertained by Descartes, who never made the examination of language his business. There is, for him, no shadow of a doubt that analytical sentences (or logical contradictions) are clearly recognizable.

We are left with only one direction in which to seek the difference between the necessary and the really necessary. It may resemble the difference between the analytical and synthetic a priori: Kant, we remember, wanted also the latter to be characterized by its "*Allgemeinheit und Notwendigkeit.*" Perhaps, then, Descartes's eternal truths (and with them, mathematics) are not reducible to purely logical principles. If so, their epistemological and ontological status calls for a reexamination in Descartes's own terms.

2. Creation and Validity

What could the option of not creating eternal truths (ideas) mean? Are they not "clear and distinct" and their negation inconceivable?[11]

---

[10] Above I.D.n.2 (Quine).

[11] Descartes, *Meditationes*, AT, 7:436: "Nec opus etiam est quaerere qua ratione Deus poterisset ab aeterno facere, ut non fuisset verum bis 4 esse 8, etc.; fateor enim id a nobis intelligi non posse." Cf., however, *Meditationes 6*, AT 7:71: "Non enim dubio est quin Deus est capax ea omnia efficiendi quae ego sic percipiendi sum capax; nihilque unquam ab illo fieri non posse indicavi, nisi propter hoc quod illud a me distincte percipi repugnaret." Is this a mere tautology to the effect that whatever I judge to be impossible I cannot conceive as possible? Or is Descartes arguing rather from the point of view of God's ordained power? Or is it rather the case that no matter what God can do—including the

If the necessity of geometry falls short of a logical necessity, then "being inconceivable" is not equivalent to a strictly logical contradiction. Eternal truths are indeed evident truths; simple, immediate, and independent of each other.[12] In Hegel's *Vorlesungen über die Geschichte der Philosophie* we find a rather adequate characterization of Descartes's "eternal truths" as discrete *"Facta des Bewusstseins."*[13] They are referred to as *intuitive* cognitions in a sense differing from the Scholastic use of the term; they are, for Descartes, detached from their object both epistemologically (since sense perceptions may generate "confused" ideas) and ontologically (for he denies throughout a *commercium mentis et corporis*). Their validity is unquestionable. Yet Descartes is willing, in the second move of his experimentation in radical doubt, to conceive of them as a gigantic deception of a *spiritus malignus*—which he is not ready to assume of the *cogito* (a *cogito, sed non sum re cogitans* would be a logical contradiction).[14] This is one indication that eternal truths, albeit self-evident, do not simply owe their evidence to the laws of thought.[15]

But the skeptical analysis of sense perceptions has produced, even prior to the introduction of the deceiving spirit, a positive result. It

---

creation of contradictions to us—he cannot deceive? (cf. below, n. 16). At any rate, this passage does not indicate any developmental phase in Descartes's thought, for it is preceded and succeeded by extreme assertions of God's omnipotence in other writings. At the worst the passage is inconsistent with them.

[12] Descartes, *Regula ad directionem ingenii* 6, AT, 10:383–84; 11, AT, 10:407–10. The mutual independence of "clear and distinct" ideas—and their *spontaneity*—justifies the separation of mind and body.

[13] Hegel, *Vorlesungen über die Geschichte der Philosophie*, in *Werke*, ed. Holdenhauer and Michel, 19–20:147; "Faktum der praktischen Vernunft" is a Kantian term for the ultimate underivability of human freedom.

[14] This does not exclude the interpretation of the *cogito* as being in some ways "performance." Descartes often puts the *cogito* as *veritas aeterna* in the neighborhood of the the law of contradiction, e.g., *Principia Philosophiae*, AT, 8:1, 23–24. Cf. Hintikka, "Cogito Ergo Sum: Inference or Performance," pp. 3–32; *id.*, "Cogito Ergo Sum as an Inference and a Performance," pp. 487–96. Against this position see Frankfurt, "Descartes on His Existence," pp. 329–56, esp. pp. 344ff.: although not a syllogism, the *cogito* is a truth *e terminis*. We note in passing that perhaps its own existence is the very pattern of a truth the negation of which is unthinkable to the own self and yet, even as a matter of fact, is not an impossibility. On the medieval precursors of the evidence of self consciousness see A. Maier (below n. 25), and on the history of the *cogito* since Augustine, Blanchet, *Les Antécédents historiques de "Je pense donc je suis."*

[15] For a similar interpretation of Descartes's eternal truths see Miller, "Descartes, Mathematics, and God," pp. 451–65 (not every necessarily true proposition is analytic); but the author does not link this important distinction to Descartes's theory of substances and their cognition.

has taught us that "matter" (the object of sense perception) is first and foremost extension, for extension is the only determination of matter perceived "clearly and distinctly." Mathematical relations (and geometry, for Descartes, is throughout quantifiable) constitute all that is known and all that can be known about matter. It is strange that this basic Cartesian tenet was kept by his interpreters outside the discussions on his delineation of the limits of the divine omnipotence. In the light of the interchangeability of geometry and matter, Descartes's belief that God could have abstained *de potentia eius absoluta*[16] from creating mathematics may be given a minimal, and most conservative, interpretation: God could have abstained from creating *matter*. This interpretation rests on the assumption that the eternal truths (e.g., mathematics) do not exist Platonically in and of themselves, but are always truths in reference to existent things; their truth lies in their (present or future) reification.

3. Existence and Extension

The key to understanding the ontic status of "eternal truths" is therefore the doctrine of substances. Only substances *exist*—one matter, souls, and God; and only one of them exists necessarily as *causa sui*.[17] In the light of Descartes's doctrine of substances, it is a mistake of interpretation to juxtapose (as Koyré had to) souls, matter, and eternal truths; souls and matter can exist independently; eternal truths exist only inasmuch as they are a reference structure within or between substances. According to how this or that (non-divine) substance was to be created, these or other eternal truths became "eternally" valid. In this sense only some of them are "created" and may be postulated—though not conceived—as possibly not existing; while others, pertaining to God himself (the law of contradiction), do not depend on God's will but on his very existence.

It is not difficult to see how truths, or for that matter anything

[16] Cf. above III.A.n.2. It belongs to the facets of Descartes's doctrine of omnipotence that he does not accept, under any circumstances, a possibility of a "deceiving" God, only of a God creating other truths. Above n. 11. This is why God's existence *suffices* to guarantee our clear and distinct ideas, which were subject to doubt before God's existence was proven. This might also explain the passage quoted above n. 11. This is why Descartes is willing to ascribe mathematical truths a certitude that is more than moral (*plus quam moralis*), *Principia Philosophiae* 4.206, AT, 8.1, p. 328.

[17] Above n. 6.

created, could be both created and eternal. God, Descartes maintains, created the world in time rather than from eternity not because he had to do so, but because he wanted it to be so. There is nothing really new in this figure of thought. Scholastic philosophy had learned from Maimonides that "being eternal" and yet "being created" are not mutually exclusive predicates. Had God wanted it, the world could have been eternal. It would remain created in the sense that at any moment of its existence God would have the option of destroying it—from eternity to eternity.[18]

But we are left with another difficulty. If mathematics and its reification (matter) are interchangeable, then the former is not only created, but also created in time. The eternity of eternal truths would seem to hinge on the eternity of matter. Descartes never really addresses the problem. He could, of course, plant these truths in God's mind prior to their reification, not as Leibniz's "possibles," the ideation of which is forced on God, but as confirmed blueprints of things to be created.

Ours is an almost Nominalistic reading of Descartes; as it was not tried earlier, let us see how far it carries us. Descartes, much as the Conceptualists of the fourteenth century, believes in the primacy (and epistemological immediacy) of substances and some of their attributes, of which each is totally independent of the other because each of them is totally dependent on God's will, that is, created *ex nihilo*. Descartes also inherited from the Terminists the criterion of "singularity," namely, *the method of annihilation*: a substance must be conceivable "in itself" even if we imagine the context of other things in which it is actually placed as destroyed. Only those concepts that stand for one substance without necessarily connoting another can be construed as absolute attributes of a substance. No substance necessitates or implies an *ordo ad invicem* to other substances. Substances, both for Ockham and Descartes, are perceived immediately; Descartes's "intuition" has the same positional value in his system as the *cognitio intuitiva* in the epistemology of the Concep-

---

[18] *Meditationes*, AT, 7:432: "Nempe, exempli causa, non ideo voluit mundum creare in tempore, quia vidit melius si fore, quam si creasset ab aeterno; nec voluit tres angulos trianguli aequales esse duobus rectis, quia cognovit aliter non posse etc. Sed contra, . . . quia voluit tres angulos trianguli necessario aequales esse duobus rectis, idcirco jam hoc verum est, et fieri aliter non potest; atque ita de reliquis." See also Buxtdorf's Latin translation of Maimonides, *Doctor Perplexorum* (Basel, 1629), p. 244.

tualists. Both believe thus to add to the understanding of the divine omnipotence.

The thoroughgoing "rationalism" of Descartes—and the point of difference between his and Ockham's understanding of realities—is anchored not in his ontology (which is easily translatable into Terministic terms), but in his epistemology. Throughout its Nominalistic career, the "principle of annihilation" remained only a negative principle, defining what a thing is *not* rather than what it *is*. For Descartes (as, in another way, for Hobbes)[19] it became a constructive principle, since intuitive cognition meant something other for him than it meant for Olivi or Ockham: the immediate evidence of concepts or images as such rather than the immediate evidence caused by the presence of "things." The Scotistic disjunction of abstractive and intuitive cognition again lost its meaning; nor did Descartes have any difficulty concerning a *notitia intuitiva de rebus non existentibus*.[20] The "intuitive" cognition is not connected by natural causation with the existence of an object *extra animam*, as it was for the Terminists. Intuitive knowledge, for Descartes, is either immediate awareness of images or immediate knowledge of essential attributes: whether existing or not (and for a while Descartes is willing to assume that it does not exist), matter *is* extension. Severing intuitive cognition from existential judgment is the deeper sense of the Cartesian ἐποχή.[21]

---

[19] Hobbes begins his phenomenal analysis of "things" by imagining the whole world destroyed. Left are then (he argues against Descartes) not only the thinking self, but likewise its memories, from which the concept of space as a *phantasma* underlying the memory of things outside us (Kant's "anticipation"!) may be reconstructed: "verum et factum convertuntur." Hobbes, *De Corpore* 2 7.1.2, in *Opera*, ed. Molesworth. Hobbes's analysis of the "state of nature" of society without a sovereign is likewise an exercise in the method of annihilation. Cf. below v.c.3.

[20] Above III.B.3.

[21] Husserl, *Cartesianische Meditationen*, ed. Strasser, pp. 27, 60, and *passim*. The following analogy may elucidate Descartes's epistemological position as against that of the later medieval Nominalists still more. Descartes interpreted the "intuitive cognition" in the same way in which the Nominalists interpreted the "intuitive cognition of non-existents"—namely as caused immediately by God or, at any rate, independent of the actual presence of the intuited object. But unlike the Nominalists, this independence (or spontaneity) of the intuitive cognition was to him not a very exceptional, hypothetical case of the exercise of the divine omnipotence. Nor, of course, was the purely intuitive cognition reserved, as it was to Scotus, for the *visio beatifica* of angels or future life. Immediate intuitive cognition irrespective of sense perception and even irrespective of the actual presence of the object became, for Descartes, the essence of intuition, the rule rather than an exception. There is another instance (Specht, *Commercium mentis et corporis: Über Kau-*

Two immediate consequences result from this doctrine: (i) matter, if it exists, is only *one* substance; the Nominalists, on the other hand, had to postulate an indefinite number of singulars, nor did they limit the number of qualities (attributes) of a singular to one; (ii) intuitive notions are spontaneous, mere *entia rationis*. One cannot "deduce" the existence of matter (or other souls) logically, but neither is it empirical knowledge. That whatever we conceive "clearly and distinctly" exists is based on Descartes's version of the principle of sufficient reason: God's "goodness" and "consistency." Throughout the seventeenth century, the principle of sufficient reason will be invoked in various forms to account for or guarantee "physical necessities"—judgments that are neither logical nor contingent but factual, Kant's "synthetic a priori" judgments. The philosophy of the seventeenth century continued to secure the principle of sufficient reason with different versions of a necessary being, from which physical necessities were not to be emancipated (without being altogether destroyed) until Kant.

In a short formula, Descartes agrees with the medieval Nominalists as to the total *independence* of every single intuitive cognition; he disagrees with them as to the origin, or causation, of such cognitions. They are not necessarily caused by objects *extra animam*, and therefore do not imply existential judgments in themselves. They are spontaneous, and the principle of spontaneity of our conceptual network will acquire ever more prolific formulations in one branch of modern philosophy of science, culminating in Kant's transcendental unity of the apperception. But in the measure in which the principle of spontaneity will become clearer, the Cartesian-Terministic emphasis on the total independence of primitive truths (intuitions, clear and distinct ideas) of each other will become more difficult to maintain, and will be exchanged for a new concept of *context*.

---

salvorstellungen im Cartesianismus pp. 7–28, esp. pp. 12ff.) in Descartes's thought where he took properties that medieval theology ascribed to angels only and bestowed them on man, or, more generally, on all thinking substances. Angels, in the medieval understanding, cannot have a body: they are pure intelligences, each a species in itself, as against man whose soul *informs* the body and whose matter is therefore a principle of individuation. An angel or other spirit who chooses to appear with—i.e., to "assume"—a human or another body for either honorable or unclear purposes can only be conceived as using the assumed body in the manner of humans handling *automata* of their creation. Descartes, thus, made all of us closer to being angels (Maritain); at any rate, the exceptional in the eyes of medieval theologians becomes the rule in his eyes.

For Descartes, however, the absolute independence of each intuitive cognition remained axiomatic; thus he separates "laws of nature" far beyond the actual heuristic necessities. The first and general "law of nature" is the principle of sufficient reason itself: the
uniformity of nature is grounded in God's goodness and consistency. In physical terms, the *quantitas motus* in the universe is constant. The first and second *leges naturae secundariae*—particular laws
of nature—formulate the principle of inertia: each body tends to remain in its state (of rest or motion), and it tends to keep its given
direction. The separation between "motion" and "direction," so
consequential for Descartes's mechanics, breaks the principle of inertia into two distinct laws, each of them specifically grounded on
God's consistency. The inner logic of this separation is clear. Each
of the "secondary laws of nature" can operate, that is, determine
matter in motion, without the other, and therefore each needs a special proof (i.e., sufficient reason).[22]

4. Some Final Doubts

God could have abstained from creating matter (extension); having
created matter, he could have created it without motion; or having
created motion, he could change the *quantitas motus* (bulk × speed)
at every minute, or again the direction of motions. The making of
the universe is thus a process of separate, discernible divine deci-
sions. Each of them could have fallen differently. How much differently? Could God only have refrained from reifying mathematics
(i.e., matter), or could he also create another mathematical world?
In other words, can God only make it untrue that $2 + 2 = 4$, or can
he make it also true that $2 + 2 = 5$?

Descartes's position is not very clear on this matter. He argues
only the negative case (of God invalidating mathematical theorems), but nowhere the positive possibility of a different mathematics. This may be intentional and significant, though the difference is never argued. Since, however, our interpretation already

---

[22] Descartes, *Principia Philosophiae*, AT, 8.1, pp. 62–63. Descartes's laws of motion have
been also interpreted as a priori synthetic by Buchdahl, *Metaphysics and the Philosophy of
Science*, pp. 147–55. But Buchdahl does not recognize the importance that the principle
of sufficient reason already has for Descartes (as the very ground of such "synthetic a
priori judgments"), nor does he try to interpret Descartes as a Nominalist. The separation between motion and direction had a profound impact on his physics; above II.E.1.

maneuvers Descartes into a difficult position (though not a position as difficult as the one he would be in if he exempted God from the principle of contradiction), we might as well consider the worst alternative. To argue that God could create other mathematical truths might mean that, whenever an inhabitant of a surrealistic world of another mathematics adds to the segment AB a segment BC in a straight line, AC is greater than $BC + AB$, since God would constantly add (*ex nihilo*) to the whole something which has not been in the parts. Assume, however, that (as in our world) God is a lazy gentleman, then $AC = BC + AB$. The "sufficient reason" for the truth of mathematics turns out to be similar to the sufficient reason for the preservation of motion and direction, namely that there is no sufficient reason to assume a divine intervention. The guarantee for laws of nature is, as a matter of fact, a *negative* rather than positive "sufficient reason." In both cases, we assume nevertheless a primacy of our mathematics over any other: God does not "decide" to create ours, but must decide to invalidate it. If this interpretation is viable, then Descartes assumes, after all, a primacy of geometry as such over its reification (matter)—though in vague terms indeed. For which reason Descartes is willing to ascribe to mathematical truths a certitude that is more than moral (*plus quam moralis*).[23] We discussed already the primacy of mathematics over its reification and considered the possibility that although it is conceivable clearly and distinctly only as mathematical relations, matter is created in time while mathematical relations are eternal if contingent. A tension lies here in Descartes's thought that he never resolved.

Since Descartes never formulated his principle of sufficient reason, he could not distinguish its uses, either. I intend to show that even Leibniz, in whose methodology the principle of sufficient reason acquired such prominence, was unaware of the source of its ambiguity. Neither were his interpreters.

To sum up, we must concede that our last suggestion—the possible primacy of mathematics over its reification—is ambiguous. Is the rest of our interpretation better founded? It is one thing to show, as I believe I did, that it makes sense in Descartes's own terms to interpret his eternal truths as intuitive rather than analytical. But it is another matter to claim that Descartes actually intended this solu-

[23] Above n. 15.

tion, however vaguely. Some of the relevant passages suggest it, others are hard to reconcile with this view.

Particularly the following: in his letter to More of 5 February 1649, which Koyre mistook as a sign of his mitigated initial position, Descartes in fact reiterated his radical concessions to the divine omnipotence. Under attack is his denial that a vacuum could exist without matter, for Descartes a *contradictio in adiecto*. "But you are quite ready to admit that in the natural course of events there is no vacuum: you are concerned about God's power, which you think can take away the contents of a container while preventing its sides from meeting. . . . And so I boldly assert that God can do everything which I conceive to be possible, but I am not so bold as to deny that he can do whatever conflicts with my understanding—I merely say that it involves a contradiction," as if we wanted to construe an unextended extension. Since the basic equation of matter and extension was won by "intuition," we may still accommodate the passages with our interpretation. But Descartes continues: "I confess that no reasons satisfy me even in physics unless they involve that necessity which you call logical or analytical (*contradictoria*), provided that you except things which can be known by experience alone, such as that there is only one sun and only one moon around the earth and so on."[24] Descartes, it seems, foreshadows Leibniz's distinction between *vérités de raison* and *vérités de fait*—not in itself an astonishing achievement, since it is a heritage of the later medieval theories of evidence.[25] But Descartes seems to count among the former physical laws—all the more mathematics. He seems to imply that eternal truths are analytic; if he does, we are left only with the doubtful comfort of having perhaps understood Descartes better than he did himself.

But then again consider his examples: extension as the only essential attribute of matter; physcial laws. They are not, as Descartes himself shows, won by analysis of terms; they are intuitions or based on intuitions. They are defended by the principle of sufficient reason. Even in our passages, Descartes might have wanted to say that since we intuit matter as extension, an empty space is *for us* a

[24] Descartes to More, 5 February 1649, AT, 5:267, trans. Kenney, *Descartes: Philosophical Letters* (Oxford, 1970), pp. 237–45, esp. pp. 240–43.

[25] Cf. A. Maier, "Das Problem der Evidenz in der Philosophie des 14. Jahrhunderts," in *Ausgehendes Mittelalter*, 2:367–418; Weinberg, *Nicholas of Autrecourt*.

contradiction, and that God could create (how, we do not know) unextended matter—remember that Ockham so explains transubstantiation—but not, once he created extended matter, create it in an empty space. I believe, in other words, that Descartes did not see the principle of contradiction as a basic intuition, but as a *condition for intuitions* and their connection. Even in the quoted passage, we are not forced to ascribe to Descartes the opinion that God could actually create absolute contradictories,[26] although admit we must that his language is ambiguous.

Precisely this ambiguity puts Descartes in the mainstream of the history of the distinction between physical and logical necessities, and the discussion of whether mathematics belongs to the one or to the other. Kant's separation of the analytic from the synthetic judgments a priori was prepared by centuries of discussions, since the Middle Ages, on the true limits of the divine omnipotence. It seems that Descartes forces God's hand after all, in that at every stage of creation he allows God only two alternatives—a reasonable and an unreasonable one. God could have created or not created matter (extension). Once created, he could have left it without motion (and with it differentiation) or with both. Once motion was given, it could be regular or arbitrary and unpredictable. If regular and predictable, then—*nulla alia ratione interveniente*—vortices had to be formed, our solar system, and finally and inevitably all other mechanical constellations which are the inorganic and organic compounds of matter in motion. The meaning of laws of nature changed. They became blueprints for the construction and reconstruction of nature, *more geometrico*, out of a homogeneous substrate. God constructed it; Descartes tries to *reconstruct* it; and only by so doing will he have understood creation.

And mathematics? Against what he believed was the Cartesian position, Leibniz included mathematical theorems among the logical necessities. Curiously, a closer look reveals that nevertheless he defended the calculus rather with the principle of sufficient reason. Kant returned to the position that may also have been Descartes's,

---

[26] Such, in essence, was the opinion of Suarez, *Disput. metaphysicae* 30.17 a.13, in *Opera*, 26:210: "Deinde non potest naturale lumen intellectus nostri esse regula objecti possibilis, vel repugnantis omnipotentiae Dei." Only context and unnecessary caution made Descartes's answer appear more irrational than it is—and perhaps intentionally so, in order to eliminate all other vestiges of contingency.

better equipped to defend it. Better, but not well enough. Not until Gödel proved the theorem of incompleteness did we find good reasons to abandon the vision of mathematics as a grand enfoldment of tautologies.

E. NEWTON AND LEIBNIZ

1. Newton and Boyle on Possible Worlds

Throughout the seventeenth century, a new concept of "laws of nature" gave a new urgency and vigor to the old, almost exhausted, distinction between God's absolute and ordained power. Methodological discussions about the theoretical foundations of science became, in the seventeenth century, a protracted exercise in the reification of modal categories. To the theological sense in which this was true in the Middle Ages, the seventeenth century added new concerns: its laws often referred to ideal-abstract conditions and entities which, while not describing reality exhaustively, were claimed to be absolutely constitutive to any description of reality. To say, with Cassirer, that functions took over the former place of substances is both true and false.[1] It is true that the notion of laws, or relations, by now governed the manner in which the *relata*—elements, particles, or subsystems—were conceived. But Boyle, Huygens, and Newton maintained the existence of nonrelational physical entities, while Leibniz did not; the "laws," "rules," and "principles" that govern these entities were won by a new mode of abstraction which called for thorough justifications of their "necessity," justifications both reminiscent and different from their medieval counterparts. To say, in the Middle Ages, that God could create other orders of nature meant mainly that he could create other things or other species and genera of things. But early modern physics, and later chemistry, and even later biology, ceased to view the classification of things, however precise, as the ultimate goal of the knowledge of nature. The new sciences sought general—perhaps even a priori—conditions of all possible entities of nature. And these conditions, whether seen from the vantage point of a physicist

---

[1] Cassirer, *Substance and Function*, pp. 21ff., 162, 168ff. and *passim*. It stands in greater proximity to Cohen's Neokantianism than later books—see pp. 99, 355, especially in the employment of the *Ursprungsprinzip*; cf. A. Funkenstein, "The Persecution of Absolutes: On the Kantian and Neokantian Theories of Science," pp. 51–58.

or that of the theologian, had to be conceived as necessary and contingent at once. To employ Kant's terminology, laws of nature had to be grounded on synthetic a priori judgments.

Newton and Leibniz solved this dilemma each in his own way—but still theologically. Common to both is the conversion of God into a methodological guarantee of the rationality and intelligibility of the world. In a famous passage of the *Queries*, Newton defended the infinity of space with reference to God's omnipotence: "And since Space is divisible *in infinitum*, and Matter is not necessarily in all places, it may be also allow'd that God is able to create Particles of Matter of several Sizes and Figures, and in several Proportions to Space, and perhaps of different Densities and Forces, and thereby to vary the Laws of Nature, and make Worlds of several sorts in several Parts of the Universe. At least, I see nothing of Contradiction in all this." How similar yet different are Boyle's words:

> But if we grant, with some modern philosophers, that God has made other worlds besides this of ours, it will be highly probable, that he has there displayed his manifold wisdom in productions very different. . . . In these . . . we may suppose that the original fabric, or frame, into which the omniscient architect at first contrived the parts of their matter, was very different from the structure of our system; besides this . . . we may conceive, that there may be a vast difference between the subsequent phenomena and productions observable in one of those systems . . . though we should suppose no more, that two or three laws of local motion, may be differing in those unknown worlds . . . God may have created some parts of matter to be of themselves quiescent [i.e., an Aristotelian world with an *inclinatio ad quietem* side by side with an atomistic universe of bodies "restlessly moving themselves"]. . . . And the laws of this propagation of motion among bodies may not be the same with those . . . in our world.[2]

---

[2] Newton, *Opticks* 3.1 (query 31), pp. 403–404; *The Works of the Honorable Robert Boyle*, 3:139. Oakley, *Omnipotence*, pp. 72–77, shows clearly Boyle's indebtedness to the *potentia absoluta–ordinata* terminology. He is correct in calling him a voluntarist; but it does not follow ipso facto that Newton was one, as he, and Burtt, *The Metaphysical Foundations of Modern Science*, p. 294 (and *passim*), assume—*pace* Leibniz. Newton's insistence on the arbitrariness-cum-perfection of creation resembles rather Aquinas (above III.B.3). The distance between him and Leibniz on this count is smaller than assumed. Burtt, at any rate, does not belong to those historians who "have failed to discriminate this more voluntaristic understanding of natural law"; he speaks (ibid.) of the "voluntaristic British(!) tradition in medieval and modern philosophy" which "tended to subordinate in God the intellect to the will." Cf. also Dick, *The Plurality of Worlds*, p. 146.

Newton speaks of a mere possibility; Boyle of a probable state of affairs. Boyle stresses the difference between possible worlds, Newton that which is common to them. Newton only permits God to "vary" laws of nature, Boyle to change them. In the atomic structure of matter Newton recognizes a physical necessity across all possible worlds; Boyle, in comparison, was much more of a "voluntarist."

Newton, so it seems from the passage just quoted, was willing to consider only such worlds as possible that can coexist with ours, though "in other parts of space." The atomic structure of matter is shared by all possible systems. Is this also true of gravitation? It stretches to infinity. Another system in which $F = a \cdot m_1 \cdot m_2 / r^n$, but $a \neq g$ or $n \neq 2$, would have to be infinitely distant from ours; otherwise, the gravitational forces between this system and ours (or between any two bodies, one in another system and the other in ours) would have to assume two contradictory values. But perhaps the "density" of matter can be changed by creating smaller atoms, thus changing the measure of mass and the strength of gravitational forces without changing the law itself; in this case, other possible systems could coexist with ours in finite, variable distances from it. Newton, we have seen, believed that the mass of a body (in our world) depends on the number of atoms times its volume. He also believed that God is indeed capable of splitting atoms,[3] wherefore space must be "infinitely divisible." Atomic matter and forces are, at any rate, the bearers of both contingency and necessity: all possible systems necessarily consist of them, although they permit an infinite range of variation. They are necessary matrices for any possible world with any quantifiable magnitudes.

Atoms and forces guarantee the necessity-cum-contingency not only of the world as a whole, but also within every world. Newton knew as well as Leibniz that "Bodies which are either absolutely hard, or so soft as to be void of Elasticity, will not rebound from each other. Impenetrability makes them only stop. If two equal Bodies meet directly *in vacuo*, they will by the laws of Motion stop where they meet, and lose all their Motion, and remain in rest, unless they be elastick. . . . if they have so much Elasticity as suffices to make them re-bound with a quarter, or three quarters of the

---

[3] *Opticks* 3.1, p. 400.

Force with which they come together, they will lose three quarters, or a half, or a quarter of their Motion."[4] Newton does not shun the consequence that "the variety of Motion which we find in this world is always decreasing."[5] In fact, God must add periodically to it so as to keep the world-watch ticking. Newton had no use for a conservation law of force spent over a distance à la Leibniz,[6] which rested on the assumption that all bodies are truly elastic. Even were the same forces to obtain without change, the system, rational as it may be, is doomed of itself to collapse. Newton's matter and forces are, at one and the same time, homogeneous and unequivocal, necessary and contingent. His employment, or reification, of modal categories is strictly spatio-temporal: "possible" is that which can be realized somewhere at some time. Only absolute space and absolute time seem to be absolutely necessary—so to say *de potentia Dei absoluta*.

2. Leibniz: Necessity and the Unity of Sufficient Reasons

The same dialectics of the necessity and contingency of laws of nature, to which Newton gave a spatio-temporal, almost pictorial interpretation, were given a purely conceptual interpretation by Leibniz. If our universe is "the best of all possible worlds"—which the optimists maintain, and the pessimists fear is true—then no other possible universe can or will ever be realized. There are, however, other possible worlds that God could have chosen but never did or will choose by virtue of the "principle of the better."[7] On the surface, this simple formula permits Leibniz to assume, as a scientist, the thorough rationality of the universe and to believe, at the same time, in the free choice of an omnipotent deity. But the advantage appears to have been lost as soon as it was gained. What, we

[4] *Opticks* 3.1, p. 398. This is the most crucial difference between the Cartesian laws of motion and those of Newton and Leibniz (above II.H.6–7).

[5] *Opticks* 3.1, p. 399; cf. *Leibniz-Clarke Corresp.*, GP, 7:370.

[6] Above II.G.2. It is therefore (*Opticks* 3.1, p. 402), "unphilosophical to seek for any other origin of the world [than the divine arrangement of hard particles], or to pretend that it might arise out of a Chaos by the mere Laws of Nature; though being once formed, it may continue by those laws for many ages [but not indefinitely!]." The adversary is Descartes's cosmogony (below V.B.6).

[7] Principe de Meilleur: *Leibniz-Clarke Corresp.*, GP, 7:390; *Théodicée*, GP, 6:44. Principe de la perfection: *Tentamen anagogicum*, GP 7:272 (compared to method of maxima and minima).

may ask, as Aureoli and Ockham once asked Duns Scotus, is the exact status of mere logical "possibles"?[8] Is the difference between a mere logical possibility and a real one such that the former cannot be realized even by God once he gave existence to our universe? Again Leibniz is torn by contradictory impulses, and again he tries to bridge the contradicting poles with a continuum. A "real possibility," or better, the reality of a possibility, is not an unequivocal concept. There are, we saw, infinitely many degrees of reality, each assigned to a different possible world. Purely logical, perhaps, is the possibility of the least possible world. The degree of the "reality of a possibility" is determined, as we saw, by the compossibility or interconnection of the largest number of possible singulars. Time and space are only derivative, not constitutive expressions of this compossibility. Perhaps a universe in which only one enitity exists—say, God—is possible even now; but it is least possible by virtue of the principle of sufficient reason (PoSR).

We have seen earlier[9] that, side by side with this *extensive* interpretation of the degrees of reality (which cover different possible worlds), Leibniz holds to another, *intensive* interpretation: within *our* universe, there is a continuum of degrees of reality from pure relations to real substances. Force is more real than space and time; material substances (*vinculum substantiale*) are more real than forces. But are these two orders of reality truly different? The more abstract a relation is, the more worlds it can fit: Leibniz does not say so, but it almost follows directly from his conception of laws of nature as valid for many possible worlds, while the laws of logic (which, for Leibniz, are always bivalent) are valid for all possible worlds. But if he ever were to draw the equivalence between both orders of reality, he would have inevitably become a Spinozist. In order to avoid such a horrible fate, Leibniz, we saw, separated reality and existence; but this separation would come to naught if the reality of possible worlds and the reality of relations were inter-

---

[8] Above III.B.4. Since Leibniz's notion of the possible is not equivalent to that which is not demonstrably impossible by finite steps of proof (above II.H.1), it is very questionable whether modern possible world semantics such as Kripke's can be applied to him without reservations (cf. R. M. Adams, "Leibniz's Theories," pp. 32–36). It seems more profitable to approach him with the medieval interpretations of modal categories in mind. The distinction between consistency and ω-consistency, which Adams brings to bear on Leibniz's notion of contingency, is most valuable.

[9] Above II.H.4; 7.

changeable expressions. Leibniz probably developed the intensive interpretation of relations later than the extensive interpretation and never addressed the problem directly. Yet, he could resort to a theory, which again he never fully articulated, even though its elements are clearly present: I mean the difference in the derivation of principles and laws from the PoSR. The principle, we saw, lurks from every corner of Leibniz's thought; it is the embodiment of his belief in contingent-cum-necessary orders, since it "inclines God without necessitating him." Our attempt to understand the various impulses and layers of Leibniz's system must turn, at last, to that principle.

The principle of sufficient reason, which accounts for all extralogical necessities, has many faces; they led some of his interpreters to assume that it stands for two or more independent heuristic devices, or to place at its side other principles, for example, perfection.[10] Leibniz himself spoke only of two "great principles"—noncontradiction and sufficient reason. At least from Leibniz's point of view, Couturat was correct in admittitng no more.[11] Only if the PoSR is construed very narrowly—as standing only for the predicate-in-subject postulate—can one argue for the independence of the "principle of the better." But there is no textual evidence for so narrow a construction, and it can be easily shown to undermine Leibniz's intent. The "principle of the better" is deduced in the same way as that of the identity of indiscernibles: "If two incompatible things are equally good, and no one of them—in itself or by its combination with others—has an advantage over the other, God will produce neither."[12] Possible worlds do not differ from monads in

---

[10] Rescher, *Leibniz*, pp. 22–34, esp. p. 33; pp. 47–57, esp. p. 57, in which a *stemma* of principles is presented, whereby only the principle of identity of indiscernibles and identity derive from the PoSR; the principles of plenitude, harmony, and continuity derive from the (independent) principle of perfection. Only the latter is, says Rescher, the *radix contingentiae*. But cf. GP, 6:612: "Mais la raison suffisante se doit aussi trouver dans les vérités contingentes ou de fait"; GP, 2:56 (PoSR inclines without necessity). It is this "inclining without necessity" that is "the root of contingency." Others who tend to split the PoSR: B. Russell, *A Critical Exposition*, pp. 53f., split the PoSR into an existential and possible domain. Cf. also *The Leibniz-Arnauld Correspondence*, ed. Parkinson, ed. and trans. Mason, introduction pp. xxiii–iv (but both identify perfection with the PoSR).

[11] Couturat, *La Logique de Leibniz d'après des documents inédits*, p. 224; a similar position G. Martin, *Leibniz*, pp. 8–16. Cf. also Belaval, *Leibniz*, p. 387.

[12] *Leibniz-Clarke Corresp.*, GP, 7:374 (§19). Cf. n. 14 below. It is the principle *nulla in rebus est indifferentia* and that of perfection in one. Rescher was perhaps also misled by

them in their degree of reality or the reason why they exist. Indeed, degrees of perfection, we saw, are degrees of reality. The PoSR, it is true, advances different *kinds* of reasons, depending on the level of discourse. On the ontological level, it is both a criterion of reality (possibility *and* compossibility) *and* a criterion of God's choice (existence). The latter, inasmuch as it is reasonable (but absolutely speaking it need not be), has the same reasons as the former. On the epistemological level, it is a criterion of objectivity of the phenomenal world, by which we assess the relative degree of reality (in the sense discussed) of our mental representations and constructs. On the methodological, scientific level, it is a criterion of choice among hypotheses. Accordingly, the PoSR appears either as the predicate-in-subject principle, or as the principle of the better, or of harmony, of plenitude, of continuity, of parsimony. Different levels of discourse employ different reasons.

3. Leibniz versus Spinoza: Sufficient Reason and Possible Worlds

The following observation may help to cut through this maze of reasons. While the positive formulation of the PoSR is uniform but trite—"that a reason can be given to every truth or, as it is commonly said, that nothing happens without a cause"[13]—it splits into two forms of negation that almost seem to contradict one another. If there are no reasons why *P* should not be (or be true), *P* will be: being itself is a reason, it is "better than" nothingness. But if there were as many reasons why *P* should be as there are for *P* not to be, neither will be; or, if God has to choose between equally good but incompatible possibilities, he will choose neither.[14] Leibniz's crite-

---

Leibniz's contention that the PoSR *underlies* the PoC (ibid., p. 420 §130). What Leibniz meant is not altogether clear—he does not prove, or argue, the position—and it leads to a long discussion, among eighteenth-century Leibnizians, whether the PoSR could indeed be deduced from the PoC. But if so, then also the principle of the better, even if it were the only root of contingency. It is the PoSR that is invoked in the refutation of atomism (GP, 7:420 §128). And elsewhere (*Demonstratio contra atomos*, GP, 7:288) Leibniz adds to the same argument a most interesting remark concerning God's omnipotence: "Quodsi quis Atomos saltem decreto DEI fieri posse arbitretur, ei fatemur posse DEUM efficere Atomos, sed perpetuo miraculo opus fore, ut divulsioni obsistatur, cum in ipso corpore principium perfectae firmitatis intelligi non possit. Potest DEUS praestare quicquid possibile est, sed non semper possibile est, ut potentiam suam creaturis transcribat, efficiatque ut ipsae per se possint quae sola ipsius potestate perficiuntur."

[13] *Specimen inventorum*, GP, 7:309.

[14] Above n. 12; cf. *De rer. orig.*, GP, 7:303: "primum agnoscere debemus eo ipso [!],

rion, then, in the assessment of sufficient reasons cannot have been by way of simple arithmetical calculus; arithmetically there is no difference between having (n) or zero reasons for either P or $\sim P$. The cases differ, however, in the nature of the negation involved in each: whether it is determinate or indeterminate. If the affirmation or assertion of P (a subject, a state, an event, a property, a proposition) does not entail the negation of any P_i, then the very consistency of P_1 is reason for its being, since every possibility harbors a drive to exist (*exigentia existentiae*). If, however, $P_1 \equiv P_2$—as in the law of the lever—then neither one can be affirmed or asserted, since (another version of the PoSR) *nulla in rebus est indifferentia.*[15] In other words, the PoSR gives Leibniz a confidence bordering on absolute (but not logical) certainty that reality knows no two things or situations that are identical in all respects.

Spinoza, it has been shown by Curley,[16] may have been Leibniz's source for this minimal construction of sufficient reason:

> Everything must be assigned a reason or cause why it exists or why it does not. E.g., if a triangle exists, one ought to give a cause or reason why it exists; if it does not exist, a cause or reason must be given which impedes or takes away its existence. This cause or reason must either be in the nature of the thing or outside it. E.g., the reason why a square circle does not exist is indicated by its very nature, namely in that it involves a contradiction. . . . The reason why a triangle or a circle exists or does not exist does not follow from their nature, but from the order of the natural corporeal universe [in which, contrary to Leibniz, matter is extension only]. . . . It follows that this exists necessarily for which no reason or cause can be given that somehow impedes its existence [id necessario existens, cuius nulla ratio nec causa datur, quae impedit quo minus existat].[17]

---

quod aliquid potius existit quam nihl, aliquam in rebus . . . esse exigentiam existentiae. . . ."; and further: "semel ens praevalere non enti, seu rationem esse cur aliquid potius existit quam nihil." Cf. GP, 7:289: "ratio est in natura, cur aliquid potius existat quam nihil. Id consequens est magni illius principi, quod nihil fiat sine ratione."

[15] And yet, all possibilities, taken by themselves, have the same right and tendency to exist: "omnia possibilia pari iuri ad existendum tendunt pro ratione realitatis" (GP, 7:303); the "quantitas essentiae" of GP, 7:303 is, then, the same as "ratio realitatis." It is the context (compossibility) which gives an edge to one possible over another. Cf. above II.H.3.

[16] Curley, *Spinoza's Metaphysics*, pp. 83–117; *id.*, "The Roots of Contingency," in *Leibniz: A Collection of Critical Essays*, ed. Frankfurt pp. 69–97, esp. pp. 90f. (Leibniz). Cf. already Lovejoy, *The Great Chain of Being: A Study of the History of an Idea*, pp. 152–53.

[17] Spinoza, *Ethica* I prop. 11, Van Vloten-Land, 1:44 (as proof for God's existence).

Spinoza's proof for God's existence hinges on this minimal construction of the PoSR, which principle also ensures the contingency-cum-necessity of finite modes.[18] Leibniz's PoSR has a similar construction and similar functions.

That two identical things exist in reality is a physical impossibility (by virtue of the PoSR). But it is not an impossibility in physics. The scientific abstraction sometimes needs the aid of assumptions contrary to reality, such as the law of the balance or, for that matter, all laws governing the phenomenal world. *If* two equal bodies were equidistant from the fulcrum, neither will descend: in the phenomenal world, for the sake of mathematical abstraction, we do assume a non-identity of indiscernibles (space, we remember, is not sufficient to distinguish between entities).[19] The homogeneity of matter is another such abstraction.[20] The isolation of a thing or a group of things from its context so as to study its behavior *ut per se est* is permitted and necessary—but only as an auxiliary to be disposed of when the work is done.[21] On further consideration, this must also be true on the metaphysical level. Indeterminate negations are abstract aids for reasoning. In reality, to assert P_1 is to exclude some concrete P_2—a monad, a state, another possible world. If so, then every determination is *eo ipso* a negation. Leibniz holds to a paradoxical methodology: in physics we construct incomplete notions and treat them as if they were complete; in metaphysics we assume complete notions but, since we know we do not have them, treat

---

But we must keep in mind that Spinoza does not envisage impossibility by infinite analysis, which Leibniz does, wherefore the latter's notion of "possibility" is not only that which cannot be (finitely) proven to be contradictory. Cf. also Schepers, "Zum Problem der Kontingenz bei Leibniz," in *Collegium Philosophicum*, pp. 326–50.

[18] As possibilities. Cf. also Van Vloten-Land, 1:45: "Perfectio igitur rei existentiam non tollit, sed contra ponit; imperfectio autem contra eandem tollit; adeoque de nullius rei existentia certiores esse possumus, quam de existentia Entis absolute infiniti seu perfecti, hoc est Dei."

[19] Above II.H.5.

[20] Leibniz to de Volder (1703), GP, 2:252–53: "Tantum nempe interest inter substantiam et massam, quantum inter res completas, ut sunt in se, et res incompletas, ut a nobis abstractione accipiuntur, quo definire liceat in phaenomenis quid cuique parti massae sit adscribendum." Cf. ibid., pp. 277, 282 (ideality) and *NE* pref., GP, 5:57 (*matière logique* versus *réel*; the latter either *métaphysique* or *physique* = *une masse homogène solide*).

[21] Leibniz to de Volder (n.d.), GP, 2:190: "Etsi autem revera nulla actio in natura sit sine obstaculo, abstractione tamen animi separatur quod in re per se est, ab eo quod accidentibus miscetur, praesertim cum hoc ab illo accipiat aestimationem tanquam a priori." Cf. also above II.E.n.3 (*in se*), III.C.n.10 (Clauberg).

them as incomplete. The difference between these modes of abstraction accounts also for the problem we raised in the beginning of this section: how can Leibniz avoid identifying the two orders of reality? Reality, of course, is one. But we assign "degrees" of it to one *or* the other order of conceiving reality, and their terms cannot be matched in a one-to-one correspondence.

Only God guarantees the validity of Leibniz's PoSR. Leibniz's God is a methodological guarantee for the utter rationality of the world. Hardly can we find a better expression for the distance between medieval theology and seventeenth-century philosophies of nature than in the employment of the very same figure of thought—the distinction between physical and logical necessities. To the medieval theologians it was ultimately the source of the utter contingency of the world. Even Thomas, for whom everything created by God must bear some order, could not conceive of a best order: the decision which possible world would come to exist must be totally arbitrary.[22] But this, says Leibniz, is impossible by virtue of the PoSR. There must be a reason why our universe was created; it therefore must be the best of all possible worlds.[23] From the source of all contingency, God became the source of all rationality—a methodological guarantee that nature is thoroughly intelligible. *Mutatis mutandis*, even Newton was not far from this position.

The position had its price. Spinoza could employ the PoSR to prove God's existence without begging the question. Leibniz could not afford to avoid circularity. He needed the PoSR to prove God's existence, but the validity of the principle, and with it a host of other principles that ensure the intelligibility of nature, rests on God's choice. There seems no other way out than to dismiss God and have "reason" posit itself as sufficient reason for itself. No one before Hume did dare say so, and no one before Kant tried to show how this can be done.

[22] Above II.H.3.

[23] And since it requires an infinite number of steps, it is contingent though a priori. In a speculative vein one could, perhaps, apply the criterion of compossibility not only to singulars within possible worlds, but also to possible worlds themselves (Leibniz, to my knowledge, does not do so explicitly). That world is best which is at any given phase (state) still compossible with the greatest number of other possible worlds, i.e., has the largest degree of freedom. Transition from phase to phase within this best world would mean navigating toward that next state that excludes the least number of other possibilities. It is only the sum total of all steps that excludes all other worlds as less perfect.

IV

DIVINE PROVIDENCE AND THE
COURSE OF HISTORY

A. THE INVISIBLE HAND AND THE
CONCEPT OF HISTORY

1. Vico's Providence

A respectable family of explanations in social and economic thought since the seventeenth century is sometimes known by the name "invisible-hand" explanations, a term borrowed from Adam Smith. In many variations, we are taught how "private vices" turn, of themselves, into "publick virtues"; how the individual pursuit of self-interest contributes ipso facto to the common wealth and welfare. Spinoza based his political theory on this mechanism; Mandeville popularized it with his Fable of the Bees.[1] Likewise since the seventeenth century, versions of the invisible-hand explanation were employed to illuminate the course of history, the evolution of society. Giambattista Vico described at length the slow process by which man created his social nature out of his initial brutish existence; a spontaneous process, even if unintentional and "occasioned" by outer necessities. Vico named this process "providence" and stressed time and again the oblique nature of its operation—unintended by individuals and unknown to them.

> For, though men have themselves made this world of nations, it has without doubt been born of a mind often unlike, at times quite contrary to, and even superior to, the particular ends these men had set themselves. . . . Thus men would indulge their bestial lust and forsake their children, but they create the purity of marriage, whence arise the families; the fathers would exercise their paternal powers over the

[1] On the origins and career of Mandeville's *topos* see Euchner, *Egoismus und Gemeinwohl: Studien zur Geschichte der bürgerlichen Philosophie*, pp. 82–125 and *passim*. On Mandeville and Vico see Goretti, "Vico et le hétérogenèse de fins," pp. 351–59. We met already a similar idea, relevant to the interpretation of society and nature alike, in Bernardino Telesio (above, II.D.3). The rationale that some Christian thinkers adduced for private property is not very far from this sentiment: Thomas Aquinas saw the *divisio possessionum* already anchored in natural law and in man's state prior to the Fall. It encourages his productivity. Cf. Thomas, *Summa theol.* I q.2 a.105.

clients without moderation, but they subject them to civil power, whence arise the cities; the reigning orders of nobles would abuse their seigneurial freedom over the plebeians, but they fall under the servitude of laws which create popular liberty; the free people would break loose from the restraint of their laws, but they fall subject to monarchs . . . by their always acting thus, the same things come to be.[2]

And again: "Divine providence initiated the process by which the fierce and violent were brought from their outlaw state to humanity. . . . It did so by awakening in them a confused idea of divinity, which they in their ignorance attributed to that which it did not belong."[3] If Mandeville's constructive private vices generate social stability, Vico's constructive collective errors of the first men—"stupid, insensate, and horrible beasts"[4]—generated early social systems. These evolved, propelled again by constructive errors and self-interests, into ever more humane orders. To Hobbes's contention that "we make the commonwealth ourselves,"[5] Vico adds that it happens unintentionally. It sounds like a historical version of the "fortuitous original sin" of which Ambrose once said that it bears "more fruit than innocence."[6]

Later versions of the same figure of thought are better known.

---

[2] Giambattista Vico, *Principi di scienza nuova*, 3rd ed. (1744) §1108 (hereafter *SN*), in *Opere*, 4, ed. Nicolini; *Vico, Selected Writings*, ed. L. Pompa (Cambridge 1982), p. 265; Löwith, "Vicos Grundsatz: *Verum et factum convertuntur*," in *Aufsätze und Vorträge 1830-1970*, pp. 169–70.

[3] *SN* §§178–79; cf. Vico, *De universi iuris uno principio*, in *Opere* 2:55: "Non igitur utilitas fuit mater iuris et societatis, sive metus, sive indigentia, ut Epicuro, Machiavellio, Hobbesio, Spinosae. Baylaeo adlubet; sed occasio fuit, per quam homines, natura sociales et origines vitio divisi, infirmi et indigni ad colendam societatem sive adeo ad colendam suam socialem naturam raperentur." On the historical appropriation of Occasionalism see below IV.F.2.

[4] *SN* §374; Vico, like Hobbes, characterized the dominant passion of his primordial man as fear—but fear of gods rather than of his fellow man. Others emphasized timidity: so Montesquieu, *De l'esprit des lois* 1.2, in *Oeuvres*, ed. Callois, p. 235, who argued against Hobbes's alleged view of the brutality of man in his natural state. Cf. also E. Leach, "Vico and Lévi-Strauss on the Origins of Humanity," in *Giambattista Vico: An International Symposium*, ed. Tagliacozzo, pp. 309–12. Cf. below IV.F.n.5.

[5] Thomas Hobbes, *Six Lessons of the Principles of Geometry*, in *EW*, ed. Molesworth, 7:184. Cf. Watkins, *Hobbes' System*, p. 69. This dictum also ties to the *verum-factum* principle; cf. Löwith, "Vico's Grundsatz" (above n. 2). Verene, *Vico's Science of the Imagination*, pp. 36–64, shows, with G. Fasso, how scarce Vico's explicit use of the principle was before reintroducing it to the second edition of the *SN* (esp. pp. 57ff.).

[6] Ambrosius of Milan, *De Iacobo*, 1.6.21, p. 18. Cf. Ladner, *The Idea of Reform*, p. 146 and n. 67. Cf. also Funkenstein, *Heilsplan*, p. 34 (Ambrose and Theophilus of Antiochia).

Kant, like Vico, combined the synchronic with the diachronic aspects of the invisible hand in his remarks about "the hidden plan of nature" (*verborgene Plan der Natur*),[7] which Hegel transformed into the doctrine of the "cunning of reason" (*List der Vernunft*). He brought all the connotations of invisible-hand explanations of history to bear on his theory. "This is to be called the cunning of reason, that it lets the passions do its work."[8] The individual is propelled by his subjective perception of his self-interests; he has an "infinite right" to pursue his egotistic freedom; bending it directly to any higher goals would violate that right, would violate the (Kantian) categorical imperative never to use man as means to an end, and always to consider him as an end-in-itself. Reason, therefore, should not, and could not, implement the objective goals of history against the subjective desires of its agents, for otherwise history would not be "the progress in the consciousness of freedom" (*Fortschritt im Bewusstsein der Freiheit*). The objectives of reason are realized obliquely. Without *being* an instrument, the historical agent *acts* as one by following his will. Only to the subjective consciousness do subjective freedom and objective necessity appear to be in conflict. In the *Zeitgeist* of each phase they coincide; the growing insight into their coincidence constitutes the progress in the objective consciousness of freedom. This very mediation of freedom and necessity is the "cunning of reason."

A strong sense of the absolute autonomy and spontaneity of human history is common to all historical constructions of the invisible hand. From Vico to Marx, they envision the subject of history—human society—as capable of generating all of its institutions, beliefs, and achievements of itself. Whether they speak of providence, nature, or reason as acting indirectly and invisibly,

---

[7] Kant, *Idee zu einer allgemeinen Geschichte in weltbürglicher Absicht* prop. 8, in *Werke*, ed. Weichschedel, 11:45 (cf. pp. 34, 47); written (1784) a year before the *Grundlegung der Metaphysik der Sitten*, Kant does not (yet) avoid speaking of nature using social antagonism as "means" to further its noble ends (prop. 4, p. 37).

[8] Hegel, *Philosophie der Geschichte*, ed. Brunstädt, pp. 61, 65, 69, 78. Wundt was later to speak of the "heterogeneity of ends." Cf. Stark, "Max Weber and the Heterogeneity of Purposes," pp. 249–64. I believe that if Hegel's sense of "meaning in history" were to be traced backwards, then it can be found in the idea of divine accommodation rather than in other ancient or medieval traditions. In his important book *Meaning in History*, Löwith neglected to distinguish traditions and their setting, which led him to some erroneous comparisons.

in all of these constructions the "finger of God" disappeared from the course of human events. When Vico named that historical necessity by which "the nature of people is rude at first, solemn thereafter, benevolent later, gentle after that, and finally desperate"[9] both "nature" and "providence," or "ideal eternal history" (*storia ideal eterna*),[10] he evidently twisted the medieval notion of special providence, perhaps even turned it, like Spinoza, on its head. Providence, for Vico, stands for the immanent mechanism by which "the age of Gods, of heroes, and of men"[11] follow from each other. Perhaps only caution caused Vico to exclude from his scheme of *corsi e ricorsi* the chosen people, that is, Jews and Christians. No such caution prevailed in later versions of "reason in history." They seem to be an antithesis of the medieval modes of eliciting the meaning of history from a transcendent premise or promise; none of them could say, as Bonaventure once had, that "faith moves us to believe that the three periods of law, namely that of natural law, of the Scripture, and of Grace, follow each other in the most harmonious order."[12]

2. History as Contextual Reasoning

There is yet another, more fundamental aspect to the denial of God's direct intervention in history to which we ought to attend.

---

[9] *SN* §242.

[10] *SN* §245; see below IV.F.2

[11] *SN* §31; the direct classical reminiscence may be Censorinus, who recalls that Varro distinguished three ages (*tria discrimina temporum*), namely, "primum ab hominum principio ad cataclysmum priorem, quod propter ignorantiam vocatur adelon, secundum a cataclysmo priore ad olympiadem primam, quod multa in eo fabulosa referentur mythica nominatur, tertium a prima olympiade ad nos, quod dicitur historicon quia res in eo gestae veris historiis continentur": Censorinus, *De die natali* c.21, ed. Hulfsch, pp. 44–45. Cf. Spranger, "Die Kulturzyklentheorie und das Problem des Kulturverfalls," esp. p. 22 n. 5 (reference to Vico); Scholz, *Glaube und Unglaube in der Weltgeschichte: Ein Kommentar zu Augustins De civitate Dei*, p. 164 (reference to Eusebius and Augustine). Christian authors mixed this *topos* with the *tria tempora* (*ante legem, sub lege, sub gratia*; below IV.D.n.17). Vico may have thought of both traditions. Censorinus's distinction between "forgetful" and "historical" time testifies to the concept of history I shall discuss below.

[12] "Nam fide credimus, aptata esse secula verbo dei; fide credimus, trium legum tempora, scilicet naturae, scripturae et gratiae sibi succedere ordinatissime decurisse; fide credimus, mundum per finale iudicium terminandum esse; in primo potentiam, in secundo providentiam, in tertio summi principii advertentes": Bonaventura, *Itinerarium mentis ad Deum* 12, *Opuscula varia theologica* in *Opera omnia*, 5:298b; cf. Augustine, *Enchiridion*, ed. Scheel, p. 73. Note, in Bonaventura, the correspondence between periods and divine attributes—a modified Joachimitic interpretation.

The many versions of reason in history from Vico to Marx are only speculative byproducts of a profound revolution in historical thought in the sixteenth and seventeenth centuries, namely the discovery of history as *contextual reasoning*. A new concept of historical facts, and of the meaning of historical facts, emerged in the seventeenth century; a conception of every historical fact, be it a text, an institution, a monument, or an event, as meaningless in itself unless seen in its original context. This new sense gradually replaced the medieval perception of historical facts as simple, so to say atomic, entities, understandable and meaningful in and of themselves.

History writing, in the classical and medieval perception, was "simplex narratio gestarum": the simple story of things that happened as they really happened (*ut gestae*).[13] The terms "historical sense," "simple sense," and "literal sense" were synonymous to the medieval exegete, who recognized a deeper sense (*spiritualis intelligentia*) only on the theological level.[14] And because historical facts were viewed as immediately given and their meaning as immediately recognizable, the eyewitness was regarded as the ideal historian—if only he kept to the truth, which is the *officium* of the historian.[15] "History is the narration of events by which we learn what

---

[13] For a detailed reasoning of the following remarks see my *Heilsplan*, pp. 70–77 and nn. 187–92. Since its publication, my interpretation of the medieval view of "historical facts" and the writing of history has tacitly or explicitly been accepted: cf. Koselleck, *Vergangene Zukunft: Zur Semantik geschichtlicher Zeiten*, pp. 311–13; Melville, *System und Diachronie: Untersuchungen zur theoretischen Grundlegung geschichtsschreiberischer Praxis im Mittelalter*, pp. 33–67, 308–41; Gurewitsch, *Kategorii srednevekovoie kulture* (trans. Lossak, *Das Weltbild des mittelalterlichen Menschen*, pp. 156–57).

[14] "Si enim huius vocabuli significatione largius utimur . . . non tantum rerum gestarum narrationem, sed illam primam significationem (didicimus)": Hugh of St. Victor, *Didascalicon* I.vi.3, Migne, *PL* 176: 801 (cf. *id., De scripturis* 3, Migne, *PL* 175: 12A; *De sacramentis* prol., c.4, Migne, *PL* 176: 185); John of Salisbury, *Polycraticus* 8.12, ed. Webb, 2:144; Lubac, *Exégèse médiévale*, 2:425, 428 n. 6, 474. "Historialiter facta sunt, et intellectualiter Ecclesiae mysteria per haec designantur" (Isidore of Seville, *De ordo creaturarum*, Migne *PL* 83: 939–40) is a recurring formula. Even Campanella, in the sixteenth century, could still say of history writing: "dicitur simplex, hoc est 'pura,' quoniam non habet sensum alium, nisi quem verba primo exprimunt, et in hoc differt a parabola. . . . Solius tamen sacrae historiae est alios sensus admittere mysticos" (*Rationalis philosophiae* pt.5 c.1 (historiographia), in *Tutte le opere di Tommaso Campanella*, ed. Firpo, 1:1226.

[15] Truth as "proprium" or "officium": *Cnutonis regis gesta . . . auctore monacho sancti Bertini*, ed. Pertz, p. 1; Otto of Freising, *Chronica sive historia de duabus civitatibus*, ed. Hofmeister p. 5; cf. M. Schulz, *Die Lehre von der historischen Methode bei den Geschichtsschreibern des Mittelalters*, pp. 5ff.; Simon, "Untersuchungen zur Topik der Widmungsbriefe mittelalterlicher Geschichtsschreibung bis zum Ende des 12ten Jahrhunderts," pp. 52ff.

happened in the past. The word is derived form the Greek ἀπὸ τοῦ ἱστορεῖν, that is, to see or recognize [!]: among the ancients, namely, no one wrote history unless he was present and saw the events which were to be written down." So begins the definition of history in Isidore of Seville's *Etymologies or Origins*, the most widely used encyclopedic reference book between the seventh and twelfth centuries.[16] Since the historical fact is self-evident and the eyewitness the best historian, every generation can be trusted to have committed to writing those events that are "worthy of memory."[17] The *annales* were regarded as the ideal form of history writing.[18] The historian was merely to continue, in a straight line, the work of his predecessors;[19] and the whole of history could be viewed as a continuous, unbroken chain of one historical narrative: *erat enim con-*

Love for truth means nonpartisanship, e.g., *Gesta abbatum Trudonensium*, ed. Köpke, p. 250. The historian stands between Scylla and Charybdis: William of Tyre, *Historia rerum in partibus transmarinis gestarum*, p. 1. Regino of Prüm withdraws from the actual duty of a historian to record what he saw and leaves contemporary history to his successors: Regino, *Chronica*, ed. Knopf, p. 1.

[16] Isidore of Seville, *Etymologiarum sive originum libri XX* 1.41.1, ed. Lindsay. For his source see M. Schulz, *Historischen Methode*, p. 20 n. 2; Keuck, "Historia: Geschichte des Wortes und seine Bedeutung," pp. 12ff. Isidore goes as far as to deny history where there is no record: *Etym.* 5.39 ("Helius pertinax ann. I nihil habet historia"). Isidore disregards the original Greek meaning of history—search and judgment (in contrast to mere logography, e.g., Herodotus 2.45): cf. Cochrane, *Christianity and Classical Culture: A Study of Thought and Action from Augustus to Augustine*, pp. 458ff.

[17] Isidore, *Etym.* 1.41.2: "haec disciplina ad Grammaticam pertinet, quia quidquid dignum memoria est litteris mandatur." Similar passages in M. Schulz, *Historischen Methode*, p. 66. "Historical times" are those that show continuity of historical record. Varro's "historical times" (above p. 205 n. 11) carried this connotation.

[18] Isidore, *Etym.* 1.44.1 (Annales). Lampert of Hersfeld, in the eleventh century, still pretended to write "annales" only. Flavius Josephus once derived the superiority of Oriental historiography (Jewish, Chaldean, and Egyptian) over the Greek from the circumstance that, while the former records are continually written and kept by priests—he was a priest himself!—in temple archives, the Greeks write their history anew every generation: Flavius Josephus, *De Iudaeorum vetustate sive contra Apionem* 1.6–7, in *Opera*, ed. Niese, 5:7; 1.4, p. 6; it may be an allusion to Plato, *Timaeus* 21e–25d. cf. Rohr, *Platons Stellung zur Geschichte*, p. 108.

[19] John of Salisbury, *Historia pontificalis* praef., ed. Chibnall, pp. 1–2: "Lucas . . . nascentis ecclesie texit infantiam; cui succedens . . . Eusebius Caesariensis. . . . Cassiodorus quoque . . . sicut previos in cronocis descriptionibus habuit, sic illustres viros huius studii reliquit successores. Versantur in hoc Orosius, Ysidorus, et Beda, et alii . . . etatis quoque nostre quam plurimi sapientes"—he names the chronicle of Hugh of St. Victor. "Secutus est enim Sigebertus Gamblensis monachus." His own history is a *continuatio* of Sigebert's. Cf. Lampert of Hersfeld, *Annales*, ed. Holder-Egger, p. 304. See previous note, and Funkenstein, *Heilsplan*, pp. 74–76, for popular images of chains of testimonies from the creation of the world.

tinua historia mundi, we still hear from Melancthon.[20] Some of the prejudice that equates history with the writing of history still lives in our distinction between history and prehistory, or in Hegel's remark that it is not a coincidence that the word history stands both for the *res gestae* and for the *historia rerum gestarum*: there are no significant social events without record.[21]

Against this admittedly simplified medieval historical methodology stands our sense of history, rooted in the "historical revolution" of the seventeenth century: that historical facts are meaningful only in their context; that this context must be reconstructed painstakingly, often by alienating words or institutions from their present connotation or function, lest we fall into anachronism; that the eyewitness is not at all the best historian, because he is, even if subjectively sincere, captive of his vantage point; that, indeed, every period inevitably reinterprets history from its own vantage point, and approaches the past with a unique canon of questions born out if its own experiences.

The first historians to speak of the unique historical "point of view" of each age were the German historians of the eighteenth century, such as Gatterer and Chladenius.[22] They borrowed the term

[20] Philipp Melancthon, *Chronicon Carionis,* in *Opera omnia,* ed. Bretschneider, 12:714. Cf. Klempt, *Die Säkularisierung der universalhistorischen Auffassung,* p. 131 n. 29. In this sense one should understand the insistence of Ambrosius of Milan that Moses was, of sorts, an eyewitness to the creation of the world: *Hexameron* 1.7, ed. Schenkel, p. 6. An old apologetic theme of Christian literature is the comparison between the reliable, continuous biblical tradition from the beginning of the world as against the pagan historical accounts that are fragmentary and, the further away in time, the more mythical: Theophilus of Antioch, *Ad Autolycum* 3.6, ed. Otto, p. 8; Tertullian, *De pallio* 2, pp. 734–36; Hieronymus, *Chron. Euseb.* praef., ed. R. Helms, GCS 24 (1913) pp. 7–8; Orosius, *Historiarum adversus paganos libri VII,* ed. Zangemeister, pp. 3–9; Frechulf of Lisieux, *Chronicon,* Migne, *PL* 106: 919; Frutolf of Michelsberg, *Chronicon,* ed. Waitz, p. 34. Cf. Von den Brincken, *Studien zur lateinischen Weltchronistik bis in das Zeitalter Ottos von Freising,* p. 137.

[21] Hegel, *Vorlesungen über die Philosophie der Geschichte,* ed. Glockner, pp. 97–98. Against this identification (here without mentioning Hegel) Heidegger, *Sein und Zeit,* p. 396: "So ist denn auch die Herrschaft eines differenzierten historischen Interesses . . . an sich noch kein Beweis für die eigentliche Geschichtlichkeit einer 'Zeit' . . . unhistorische Zeitalter sind als solche nicht auch schon ungeschichtlich." On pp. 405ff. the difference between his understanding of temporality and Hegel's is diagnosed as the difference in the notion of *time.*

[22] Koselleck, *Vergangene Zukunft,* pp. 176–207, esp. pp. 183ff.; Reill, "History and Hermeneutics in the *Aufklärung*: The Thought of Johann Christoph Gatterer," pp. 24–51; Reill, *The German Enlightenment and the Rise of Historicism,* pp. 125f. (Gesichtskreis).

from Leibniz's *Monadology*, in which each "metaphysical point" reflects in its unique way the entire universe in which it is embedded. But the revolution in historical reasoning was, by that time, long under way, nourished by many sources. Ever since the sixteenth century, philologists, jurists, and biblical critics had developed methods of understanding through alienation and reconstruction; they severed past monuments, institutions, events from their actual connotation and association, and interpreted them in the light of their remote original setting as if they were details of a strange new continent. History became *eo ipso* interpretation.

Common to both "invisible hand" interpretations *of* history and the new, contextual reasoning *in* history was a sense of immanent structures that have to be unearthed. Historical sources reveal their information indirectly; saints' lives may be replete with superstitions, but they tell us about the times of their authors "sans le vouloir." Monuments should not be seen "détachez et isolez," but in context and development.[23] The "spirit of the people," the "genius of the times" does not announce itself in the sources; it has to be reconstructed from them.

Again it was Vico who gave the first systematic expression to most facets of this methodological revolution. A new concept of historical periods as dynamical contexts emerges from his writings: it consists of the demand and of the serious attempt to determine historical periods from within, through some internal, integrating principle rather than, as hitherto, in mere contraposition to earlier and later segments of history. Vico's key terms in this respect are "harmony," "convenience," "correspondence," or "accommodation."[24] All human affairs (*cose uomani*) of a society at a given phase correspond to and reflect each other; they form a harmonious whole and are shaped by the very same "mode of the time."[25] Each of the

---

[23] Gossman, *Medievalism and the Ideologies of the Enlightenment*, pp. 111–12 (quotation from Lévesque de la Revalière, Mabillon).

[24] Vico, *SN* §32: "Convenevolmente a tali tre sorte di natura e governi, si parlarono tre specie de lingue . . ."; §311, p. 112 ("tra loro conformi"); §348, p. 125 ("necessaria convenevolezza delle medisme umane cose").

[25] *SN* §979: mode of the time; on the medieval term *qualitas temporum* see, e.g., Hildebertus Cenomanensis, *Sermo in Septuagesima*, Migne, *PL* 177: 1073; or *ratio temporum* (in a noncomputistic sense): Beda, *Super acta apostolorum*, Migne, *PL* 92: 953. On the term in the legal tradition see the excellent article of Kelley, "Klio and the Lawyers," pp. 24–49.

ideal successive ages—the divine, the heroic, the human—brings
forth its own characteristics, its own significations and achievements, its own language, poetry, jurisprudence, social institutions,
and religious imagery. All manifestations of an age are facets of one
and the same collective imagination, the "common sense."[26]
Knowledge of the common sense is acquired not by deductive reasoning or mechanical, anachronistic analogies. "It is another property of the human mind that whenever men can form no idea of distant and unknown things they judge them by what is familiar and at
hand."[27] The historian's guide, as we shall see later, is his own imagination and empathy. Vico's *New Science* draws its strength from the
insight that no historical fact is understandable unless we reconstruct the mentality, the context that endows it with significance.

3. Vico's Forerunners

Not all of this was Vico's discovery. By the time of Vico it had become almost a truism to warn against "those who mold their notion
of antiquities after their resemblance to the present."[28] The injunction against anachronisms—not only textual, but also historical—
became almost commonplace. Another author could note, as a matter of course: "Je le prie de quitter les idées particulieres de nôtre pais
et de nôtre temps, pour regarder les Israëlites dans les circonstances
des temps et des lieux où ils vivoient; pour les comparer avec les
peuples qui ont été les plus proches d'eux, et pour entrer ainsi dans
leur esprit et dans leurs maximes."[29] Vico was an heir to generations
of humanistic scholarship since the sixteenth century. Sixteenth-
century philologists returned to the level once achieved by the ancients and surpassed it. Porphyry attributed to Aristarch of Sa-
mothrake the demand "to interpret Homer by Homer only," not to

[26] Vico, *SN* §142: "Il senso commune e un giudizio senz'alcuna riflessione, commu-
namante sentitio da tutto un ordine, da tutto un populo, da tutta una nazione o da tutto
il genere umono." Cf. below IV.F.4. (comparison with Ibn Khaldun's *asabiya* and Mannheim's *totaler Ideologiebegriff*, the shifts in the meaning of "common sense"). On the history of the *topos* of the three ages see Funkenstein, *Heilsplan*, pp. 129–32 nn. 27–29 and
passim; cf. also above n.11 and below IV.D.n.17.

[27] Vico *SN* §122; see below IV.F.3.

[28] "De rebus antiquissimis secundum sui temporis conditionem notiones forment";
Budde, *Historia ecclesiastica*, praef.; quoted by Diestel, *Geschichte des Alten Testaments in
der christlichen Kirche*, p. 463. A history of the notion of anachronism has not yet been
written.

[29] Fleury, *Les Moeurs des Israelites*, p. 8.

impose later connotations on earlier words.[30] Porphyry himself turned classical philology to potentially good use when he proved that the Book of Daniel could not have been written earlier than the Hasmonean revolt to which its symbols allude.[31] The methods by which Lorenzo Valla proved the inauthenticity of the Donation of Constantine were no different.[32]

Soon, however, humanistic philology surpassed its ancient paradigms; and it did so by moving from textual criticism and textual exegesis to the reconstruction of history. Vico was also an heir to the *mos docendi Gallicus*, the humanistic interpretation of Roman law.[33] It was a reaction against the elevation of the *Corpus iuris civilis* to the status of a universal, inexhaustible paradigm of legal wisdom, as if it were an ideal law valid for all times. The historical school of legal interpretation explains the Roman institutions within their now obsolete circumstances; Hotman, the contemporary of Jean Bodin, even denied that the Justinian Code ever reflected an existing society: in its own time it was an abstract ideal, never actualized.[34] The tedious dispute between the adherents of the *Loi écrit* and the adherents of the *coutumes* generated the insight that there could never be an ideal law valid for all times. From the interpretation of Roman law, humanistic jurisprudence advanced to the historical reconstruction of non-Roman legal institutions, the origins and career of "feudal" law.[35] There was, it seems, no real precedent for their work in the tradition they inherited. True, they may have remembered Thucydides' "Archeology," the short description of the Greeks' slow development from primitive warlike conditions to a

[30] R. Pfeiffer, *Geschichte der klassischen Philologie von den Anfängen bis zum Ende des Hellenismus*, trans. Arnold, pp. 276–78. Of the intensive philological work of Antiquity the Middle Ages inherited but the general theoretical principles of *accessus ad auctores*, among which was the admonition to pay attention to time and circumstances: Wolters, *Artes liberales: Studien und Texte zur Geschichte des Mittelalters*, ed. Koch, pp. 66ff.

[31] Above III.B.n.7. Pagan biblical criticism was often the starting point of later critical endeavors—Moslem, Jewish, and early modern.

[32] Kelley, *Foundations of Modern Historical Scholarship: Language, Law, and History in the French Renaissance*, pp. 19–50, esp. p. 38.

[33] *Id.*, "Klio and the Lawyers," above n. 25. Kelley alerts us to the roots of modern historical thought in medieval legal texts: as I hope to show, biblical exegesis and the various other occurrences of the idea of accommodation are no less significant.

[34] On the origins of the new legal hermeneutics see Dilthey, *Weltanschauung und Analyse*, pp. 11ff., 113; Kelley, *Foundations*, pp. 106–12; Franklin, *Jean Bodin and the Sixteenth-Century Revolution in the Methodology of Law and History*, esp. pp. 48–58 (Hotman).

[35] Pocock, *The Ancient Constitution and Feudal Law*, esp. ch. 1.

state of law.[36] And surely the tradition of rhetorical and legal scholarship in Antiquity and in the Middle Ages also contained references to changing times as one of the *circumstantiae* to be explored by the exegete. But all of these hints were now fleshed out with living details. Vico must have likewise been aware of the beginning of biblical criticism, which also advanced from the wish to recover the "pure" text to a reconstruction of the *mens auctoris*.[37]

Legal scholarship, biblical criticism, and classical philology were the main bearers of the new historical method: history writing lagged considerably behind. But that the trust in the "simple narrative" of events eroded in the sixteenth century can also be shown in other spheres, far removed from the religious or legal contests. In a famous passage of his *Essais* Montaigne says: "Others shape man: I narrate about him" (*Les autres forment l'homme; je le récite*). And again: "I do not teach; I just narrate."[38] "Man" refers primarily to the man Montaigne himself, his inconsistencies, changeable moods, and unique moves. There can be hardly any doubt as to the ironic undertone of his words. Montaigne is very aware of the fact that his description of himself and of others is saturated with prejudices, that his judgments depend on time, place, and mood. Because of this awareness he does not claim to offer systematic reflections but rather "rhapsodic" impressions, "essays." And he is likewise aware that his very preoccupation with himself, his very act of writing in voluntary solitude, is also an act of self-education, "shaping" rather than merely "narrating." A simple story does not exist, nor do the simple facts of history.

Vico's impact was negligible. But his main themes—the contextual harmony within each period, the necessary regularity in the succession of periods (nature), and the growing spontaneity of the social endeavor (freedom)—maintain a regulative role in the formation of modern historical reasoning. The meaning of history (as

---

[36] Fritz, *Die griechische Geschichtsschreibung*, 1: 575–617 and nn. 263–80. On the Sophistic affinities of Thucydides, Jaeger, *Paideia*, 1:479–513, esp. pp. 483–85. Cochrane, *Christianity*, pp. 469–74 stresses the proximity to Hippocrates. On "nature" against "convention" or brute force below v.C.2.

[37] Bentley, *Humanists and the Holy Writ*, pp. 294–314; Spinoza: below IV.B.3.

[38] Montaigne, *Essais* 2.2 (2:222); Auerbach, *Mimesis: Dargestellte Wirklichkeit in der abendländischen Literatur*, pp. 271–75. On Vico's notion of *vera narratio* see Mali, "Harehabilitatsia shel hamythos: Vico vehamada haḥadash shel hatarbut" [The Rehabilitation of Myth: Giambattista Vico's New Science of Culture], pp. 68–101.

a whole) and the meaning in history (of its facts) underwent a revolution of no less significance than the revolution in the natural sciences. Again we wonder: how radical was this break, what precisely
was new in this "New Science"? Evidently, the ways of seeking
signs for the divine providence working in history have changed;
but "harmony," "correspondence," "concordance" within historical periods—the notion of change in the *qualitas temporum*—were
likewise not altogether alien to the medieval historical reflection; we
encounter them, in particular, in the medieval notions of *divine accommodation*.

Medieval Jewish and Christian exegesis shared the hermeneutical
principle of accommodation: the assumption that the Scriptures are
adjusted to the capacity of mankind to receive and perceive them.
Out of this exegetical *topos*—which I will discuss first—grew various explanations of the less palatable and less understandable biblical precepts and institutions as the adjustment of God's providence
to the primitive religious mentality of the nascent Israel. Out of
these explanations, or side by side with them, grew grand historical
speculations, which saw in the whole of history an articulation of
the adjustment of divine manifestations to the process of intellectual, moral, and even political advancement of mankind. It is astonishing that so little has been written about a principle that was so
fundamental to the medieval reflections on God and mankind, nature and history.

B. "SCRIPTURE SPEAKS THE LANGUAGE OF MAN": THE EXEGETICAL PRINCIPLE OF ACCOMMODATION

1. A Legal Principle Turns Exegetical

The *exegetical* career of the medieval principle of accommodation (I
mean its function within the interpretation of the Bible) is often tied
to a phrase: "The Scriptures speak the language of man." The Latin
phrase—*Scriptura humane loquitur*[1]—is a literal translation from the
Hebrew—*dibra tora kileshon bne 'adam*. In Jewish sources it appears at
first in a legal context, and has little to do with its later employ-

---

[1] E.g., Thomas Aquinas, *Summa theol.* 1–2 q.98 a.3: "secundum opinionem populi loquitur Scriptura"; Oresme, *Le livre du ciel* 2.25, p. 530: "L'en diroit que elle se conforme
en c'est partie à la manier de commun parler."

ment.[2] R. El'azar ben Azaria, the first *tana* to invoke the rule, refused to read into the laws concerning the discharge of Hebrew
slaves the provision to endow the slave, whether or not he profited
the household, with a gift, just because the biblical verse redupli-
cates the verb: "you should donate a donation" (*ha'anek ta'anik*).
The reduplication has no specific legal meaning, but it is a *rhetorical*
phrase only. If the narrative passages of the Bible contain colloqui-
alisms, as for example "two by two" (*shnayim shnayim*), so do the
legal parts. Similar differences arose between R. Akiba, who
searched for (*darash*) the legal meaning of every seemingly redundant particle of speech, and R. Yishmael, who was much more willing to admit that rabbinical provisions cannot be deduced from the
Scriptures. At best they can be related to a hint.

What to the ancients was primarily a *legal* hermeneutical principle
became in the hands of medieval exegetes a general rule to justify or
to limit the philosophical allegoresis. In this new sense it is employed in the Geonitic literature as well by Sa'adia or other early
medieval philosophers. The numerous anthropomorphic expressions of the Bible could more or less easily be translated into a less
offensive idiom; the right [hand] of God, *yemin adonai*, could be
made to mean God's power. Even those who deny that God can be
spoken of with positive attributes could still claim that all scriptural
predicates of God are reducible to attributes of action or negations
of a privation. Still, the very original presence of prima facie anthropomorphism in the Bible was embarrassing and called for a justification. The reason they are employed is to accommodate the lesser
capacity for abstraction of the masses. The law was given to all in a
language to be understood by all (Maimonides).

Gradually, as the heuristic horizon of the principle broadened, it
came to explain more than anthropomorphisms. Evidently the cosmology of the Bible differed from the last word of scientists—in the
Middle Ages no less than today. But Scripture cannot be mistaken;

---

[2] *Entsiklopedya talmudit* [Talmudic Encyclopedia] 8, s.v. "dibra tora." Lauterbach,
"The Saducees and Pharisees," in *Rabbinic Essays*, pp. 31ff. n. 11, believed in the Sadu-
cean origin of the formula; but he himself drew attention to the fact that, against their
distorted tanaitic image, the Saducees possessed oral traditions of their own in their domain. The origin of the adverse position of Yishmael and Akiba may rather go back to
the traditions of Hillel and Shamai—whether or not one should try to derive oral from
written law. This suggestion may also be supported by an inverted analogue: Hillel used
hermeneutical principles, it says, even on secular texts: *haya doresh leshon hedyot.*

rather, it speaks the language of everyday man, or of primitive man. At this very point in the career of the principle, "the Scripture speaks in human terms" splits into two possible approaches: a *maximalistic* and a *minimalistic* employment of the formula.

The maximalist will see the whole body of science and theology—needless to say, *his* science and *his* metaphysics—epitomized in the Bible. The Bible may not *read* as a general encyclopedia, but it is one to him. The scientific information is clothed in metaphors so as to remain understandable to the masses. The task of the interpreter is to decode the biblical phrases and show that nothing worth knowing evaded the notice of the revealed text. This was done by the mainstream medieval Jewish exegetes *mutatis mutandis*: Sa'adia, Ramban, Sforno. Ramban (Nachmanides) went as far as to claim that the philosophical translation actually constitutes the simple, literal sense of the Scriptures, while allegory is the mystical, Kabbalistic dimension of understanding, in which the whole Scripture is nothing but a continuous name God.[3] The literal sense embraces, on the other hand, the whole range of rational science rather than merely colloquial speech.

2. A Minimalistic versus Maximalistic Construction

The merits of Abraham Ibn Ezra's exegesis can be partly measured on the basis of this, the maximalistic extension of the Scriptural principle of accommodation. Ibn Ezra himself polemicizes against this approach—the first among his list of five exegetical methods (of which the first four are wrong or useless).[4] It is neither true nor false, but often irrelevant. "If you want to learn the sciences, go to the Greeks." The *Geonim* in their philosophical allegorization invoke the *results of*, for example, astronomy, not its proofs; and likewise unscientific, by implication, would be the Bible itself if it were to be read as an encyclopedia. This would be a far cry from real science.

Ibn Ezra suggests, instead, a minimalistic approach. It may be that he was preceded in it by some of the extreme rationalists in Spain, such as R. Isaak, but we know him only through Ibn Ezra.

---

[3] Cf. A. Funkenstein, "Nachmanides' Typological Reading of History," pp. 35–59, esp. pp. 43–47; also in Dan and Talmage, *Studies in Jewish Mysticism*, pp. 129–50.

[4] Abraham ibn Ezra, *Perush hatora*, ed. Weiser, 1:1ff. (text).

"The Scriptures speak a human language"[5] means simply that Scriptures adapt themselves to the point of view of the multitude. They do not contradict science, but neither do they contain all of it. Indeed, nowhere is this minimalistic interpretation of the principle of accommodation more evident than in Ibn Ezra's exegesis of Genesis 1. To quote a few examples, on Genesis 1:1, he explicates:

> "The Heavens": with a definite article, to indicate that he speaks of those [heavens] seen. ["Heaven" and "earth" he will later interpret as referring to sublunar elements only.] "And void" [*vabohu*]: For Moses did not speak about the world of Celestial Bodies [*olam haba*: the otherwise eschatalogical term is used here in an astronomical sense, spatial rather than temporal] which is the world of angels [*hamal'ahim*, here in the sense of intelligences], but only about the world of generation and corruption [*olam hahavaya vehahashkhata*—the medieval, Aristotelian equivalent for the sublunar realm].[6]

Time and again, Ibn Ezra emphasizes that Genesis is not a scientific, comprehensive account of the creation of the world *ex nihilo*, but rather the account of the formation of the sublunar realm through natural processes, that is, laws. Genesis only tells the facts immediately pertinent to the formation and status of *man*. Even the celestial bodies appear in the narrative of creation *only from the vantage point of the average man*, not with any reference to their essence or true nature:

> "And should one ask" [he explains in Genesis 1:16], Did not astronomers [*hachme hamidot*] teach that all planets excepting Mercury and Venus are bigger than the moon, and how could it be written [in the Scriptures] "the big ones." The answer is that the meaning of "big" is not in respect to the bodily size but in respect to their light, and the moon's light is many times [stronger] because of its proximity to the earth.[7]

That the moon is called a great luminary, while the bigger planets are only called stars—this mode of speech corresponds merely to

[5] E.g., to Gen. 1:26, ibid., 1:18: 've'aḥar sheyada'nu shehatora dibra kilshon bene 'adam, ki hamedaber adam gam ken hashome'a."

[6] Ibid., 1:13 (text). In some of its boldest points, Abaelard's interpretation of Genesis 1 is astonishingly similar: Petrus Abaelard, *Expositio in Hexaemeron*, Migne, *PL* 178:733c (Heaven = air and fire), 735a–b, 737b. The possible link deserves further study, though I failed to notice verbal correspondences.

[7] Abraham ibn Ezra, *Perush hatora*, 1:15.

our point of view. "The speaker is a man [Moses], and so are the listeners," he says elsewhere.[8] Further hermeneutical devices were not even needed to reconcile the Copernican theory with the Scriptures. Already Nicole Oresme discarded the exegetical arguments against the motion of the earth as the least disturbing parts of the geodynamic hypothesis (which he eventually rejected). It may be that "Scriptura humane loquitur," even where it appears to hold to a geostatic cosmology. Galileo would later use a similar argument to defend the Copernican, heliocentric system in a most ingenious way. If time were to stand still, he argues in his letter to the Duchess Christina, it is not enough that "The sun in Gideon stand still" (Jos. 10:12); even according to Ptolemy, the sphere of the fixed stars must cease to move.[9] Whether we accept Ptolemy or Copernicus, there is no way to interpret the Scriptures literally; under both systems the Scriptures speak in a human way. Having to allegorize anyway, why not adopt the astronomically sounder hypothesis? Yet, the Church was not out so much to save the literal meaning of the Scriptures as it feared the undermining of its authority. Centuries of *patres* and theologians held to the geocentric system and advanced reasons for it; their involved, meticulous arguments could not be taken as an accommodation to a lesser understanding. If the Copernican hypothesis describes reality, they were simply wrong. That was the true danger of Galileo, increased by his apt employment of theological arguments. And even Cardinal Bellarmine had to admit that if anthropomorphisms in the Scriptures could be allegorized away, so could the seemingly geocentric references. He just called for the armistice until empirical evidence was adduced.

The narrative of creation is, according to Ibn Ezra, the narrative of the creation of objects immediately perceived in proportion to the way in which they are perceived. If not to give an adequate cosmology, what, then, is its purpose? For one thing, the sublunar world, of which only it speaks, was created for the sake of man, unlike the supralunar, about which the story of Genesis is mute:

[8] Cf. above n. 5 (Ibn Ezra), n. 1 (Oresme).

[9] Galileo, *Opere*, 5:281–88, 309–48; Drake, *Galileo*, pp. 224–29; Sambursky, "Three Aspects of the Historical Significance of Galileo," pp. 1–11; Westman, "The Copernicans and the Church," in *God and Nature: The Encounter of Christianity and Science*, ed. Lindberg and Westman (forthcoming); Wallace, *Galileo and His Sources*, pp. 291–95, and above, II.D.n.41.

And now let me pronounce a principle. Know that Moses our Master did not give the laws to the philosophers [*khachme halev*] only, but to everybody. And not only to the people of his generation, but for all generations. And he did not refer in the story of creation to anything but the sublunar world, which was created for the sake of man.[10]

Moreover, the story of Genesis 1 shows how man is at one and the same time subject to necessities of matter and above them: man represents the material universe and participates in the realm of the intelligences (Ibn Ezra seems to have endorsed a *unitas intellectus*). Man is a microcosm, just as God is the macrocosm—this is Ibn Ezra's Neoplatonic, almost pantheistic interpretation of "in our image and likeness" (Gen. 1:26).[11]

All this is not to say that Scripture does not contain metaphysical allusions, but that the exegete should be careful when, where, and how to look for them *or* to refrain from it. Ibn Ezra, just like Spinoza at the beginning of early modern biblical criticism, established a most fruitful methodological principle. Whether a biblical image has to be interpreted literally or metaphorically cannot be decided arbitrarily from a point of view outside the text, but rather immanently. In other words, Ibn Ezra delimits the borderline between permissible and impermissible allegorization. As we shall see later, it is with this principle more than in any detail of his interpretation that Ibn Ezra influenced Spinoza's exegetical approach. Against Sa'adia he sees in references to the word of God—"and God said"—not a substitute for "God's will," but the image of a king commanding his servants: the work of creation was effortless, since God operated through "servants"—laws of nature or natural elements.[12] In other words, Ibn Ezra, when allegorizing, does not always look for the most abstract ("scientific") substitute, but for a middle level of abstraction as demanded by the context. Ibn Ezra, in his grammatical as well as in his allegorical interpretations, looks for the *context* of the *explanandum*.[13]

With the exception of the proper name of God, Ibn Ezra looks for a deeper meaning (*sod*)—astronomical or metaphysical—not in the

---

[10] Abraham ibn Ezra, *Perush hatora*, 1:14.

[11] Ibid., 1:18.

[12] Ibid., 1:14.

[13] Thus, e.g., he accepts Rashi's interpretation to Gen. 1:1 (*bereshit*) as a construct case, but refuses to see it as a general rule: the context ought to decide the issue in each case.

biblical formulations, but in the things—objects and events—to which they refer, a principle reminiscent of the exegetical revolution that Christian exegesis underwent in the thirteenth century.[14] And, just as Maimonides and Thomas Aquinas after him, Ibn Ezra founds his doctrine of permissible allegorization on the properties of language. Language is, by its nature, ambiguous and analogical: we project the familiar onto the unfamiliar "above us and below us." Indeed, Ibn Ezra develops an exegetical doctrine of *analogia entis* to explain the creation of man in "God's image and likeness." In short, as an exegetical principle, "the Scripture speaks the language of man" eventually referred to a body of theories concerning the properties of the sacred language. The language of revelation uses elements of the familiar and natural in order to transcend them; and this procedure is in itself a property of the language of man, which operates through analogies and metaphors.

3. The Principle Secularized: Spinoza

Ibn Ezra's influence on Jewish exegetes was considerable; not until Nicholas of Lyra did many of his interpretations become known to Christian students of the *veritas hebraica*. But neither Jewish nor the Christian medieval exegesis ever fully understood, let alone adopted, Ibn Ezra's method of minimal-contextual allegorization. The exegetical principle of accommodation served for a long while mostly to reconcile reason with revelation. Ibn Ezra's moment of true impact came with the beginnings of biblical criticism in the seventeenth century. Let me hasten to add that nothing was further from Ibn Ezra's intentions than to question the authenticity or revelatory origin of the Scriptures. Most of his remarks used by biblical critics for their purposes are instances in which he refuted—and thus preserved—arguments against the consistency and authenticity of the biblical narrative. Biblical criticism did not start *de novo*; most of the questions it raised were already asked by the traditional exegetes, and some of the answers were already given by the classical pagan anti-Jewish and anti-Christian polemicists.[15] More important

---

[14] The recognition, namely that the *sensus litteralis* may include allegory and metaphor if they were the *intentio auctoris*; and that spiritual intelligence is not an interpretation of the text but rather of that which the text refers to. See above II.C.n.51.

[15] The different names of God used in the first and second story of creation were interpreted as different attributes (Justice versus Mercy). The question how "the kings that

than critical remarks gleaned from Ibn Ezra's writing was his use of the principle of accommodation. It was easily susceptible to secularization. It was put on its head—or, if you wish, on its feet—by Spinoza.

Spinoza's *Theological-Political Treatise* contains one of the earliest documents of biblical criticism. The borderline between exegesis and criticism is not always sharp. For a preliminary definition, it suffices to say that biblical criticism is indifferent if not hostile to the authentication of the Bible as a superhuman document. Spinoza, here as elsewhere, does not oppose outright the theological terms and principles of the Middle Ages. His strategy is more subtle: to use them in an adversary meaning. "General" and "special" providence, he says, are legitimate terms—but only if understood as two kinds of universal laws of nature![16] Likewise, "the Scripture speaks the language of man" is a legitimate principle—but only if understood so that, since the author is human, the contents of the Scriptures are his language.[17]

---

ruled Edom before there was a king to the children of Israel" (Gen. 36:31) could be mentioned—if it is a prophecy, it should be cast in the past tense—was answered by R. Yitshale Hasefaradi to the effect that it is a later interpolation: Abraham ibn Ezra, *Perush hatora*, though mentioning this radical solution, rather wishes to interpret "king" as referring to Moses. On pagan biblical criticism see Anastos, "Porphyry's Attack on the Bible," in *The Classical Tradition: Literary and Historical Studies in Honor of Harry Kaplan*, ed. Wallach, pp. 421–50 and Rembaum, "The New Testament in Medieval Jewish Anti-Christian Polemics," pp. 17–61, and above III.B.n.7.

[16] Spinoza, *Tractatus* c. 3, Van Vloten-Land, 2:123–25 (Gebhardt ed., 7:45–47): "auxilium Dei externum" is the chain of causation that determines the actual course of a body; "auxilium Dei internum" is the law that determines individuality, internal balance of motion ($m \cdot v$), or, in simple bodies, the law of inertia. So also in states: the *auxilium externum* determines their actual fate, the *auxilium internum* their constitution. Cf. also below v.c.5.

[17] Ibid., pp. 152, 242, esp. p. 248: "Nam, ut iam etiam monuimus, sicuti olim fides secundum captum et opiniones Prophetarum et vulgi illius temporis revelata scriptaque fuit, sic etiam iam unusquisque tenetur eandem suis opinionibus accommodare, ut sic ipsam absque ulla mentis repugnantia sineque ulla haesitatione amplectatur." On Spinoza's use of the principle of accommodation see also Hassinger, *Empirisch-rationaler Historismus: Seine Ausbildung in der Literatur Westeuropas von Guiccardini bis Saint-Evremond*, pp. 141–43; Scholder, *Ursprünge und Probleme der Bibelkritik im 17. Jahrhundert: Ein Beitrag zur Entstehung der historisch-kritischen Theologie*, p. 168. In the *Ethics*, the term is used in the almost opposite sense—our need to adjust to the course of nature that is indifferent toward our happiness or unhappiness: *Ethics* 4 prop.4, corol.; cf. Walther, *Metaphysik als Anti-Theologie: Die Philosophie Spinozas im Zusammenhang der religionsphilosophischen Problematik*, p. 111 ("natürliche Ethik als Theorie der Anpassung"). Both senses of accommodation are, of course, one.

Whether Moses or Ezra, the author of the Scriptures was a man reflecting the world view of his age. The exegete should not assume a priori what this world view is, or force it to conform to true metaphysics. Take, for example, Moses' image of God.[18] In Deuteronomy 4:24 we read: "For Yahweh your God is consuming fire, a jealous God." Should this verse be interpreted literally or allegorically? Is it anthropomorphic or not? No external philosophical viewpoint should guide us. We rather ought to establish, from the context, an internal principle of permissible allegorization. We know that the Pentateuch rejects bodily images of God. "A consuming fire" (*esh ochla*) could be allegorized, all the more so since fire stands elsewhere as a metaphor for jealousy and vengeance; "a jealous God" (*'el kana*) refers to a psychical attribute, and nowhere do we find scriptural objections to psychical attributes. It has to be interpreted literally. Moses' image of God is the image of a God without a body, but with a soul—an unphilosophical image indeed, since *ordo et connexio idearum idem est, ac ordo et connexio rerum.*[19] There can be no soul without a corresponding body, for both denote one and the same constellation of acts. The Bible is a book written by primitive man in his own language, which he could not escape. It is a historical rather than a perennial document: this is Spinoza's use of the exegetical principle of accommodation. The theological language, before being abandoned, was vacated of its content; or better, was turned on its head.[20]

[18] Spinoza, *Tractatus*, Van Vloten-Land, 2:174–75. The portrait of Spinoza as a "critic of religion," is somewhat misleading. In the preface to the *Tractatus* he argues for the relative validity of orientational concepts which enable persons to function in normal times; such concepts are vague, incomplete, and unphilosophical, but Spinoza, unlike Descartes, did not cast unclear and inadequate ideas out as if they had no foundation in reality (cf. above II.F.I and the notion of self, below V.C.5). For the peasant the horse is a work animal; the knight has another idea of it. Religious ideas in any given society are of that category. Only in times of social tumult and breakdown of norms do inherited modes of explanation collapse, and people seek refuge in pseudo-truths and "superstitions," while their rulers use the chance to impose tyranny. Spinoza's attitude toward average, non-philosophical orientational concepts is similar in many ways to a recent anthropological explication of "common sense as a cultural system": Geertz, *Local Knowledge*, pp. 73–93.

[19] Spinoza, *Ethica* p. II prop.7, Van Vloten-Land, 1:77 (above II.A.n.6). The demand for context was broadened into a system by Leibniz, *Commentariuncula de Iudice Controversarium seu Trutina Rationis et Norma Textes, SB*, 1:548–59.

[20] It became, however, hotly debated in Protestant theological circles, especially in Holland in the seventeenth and eighteenth centuries. Cf. G. Horning in *Wörterbuch der Philosophie*, ed. Ritter and Gründer, s.v. Akkommodation. It is remarkable that this article starts the history of the concept only with the seventeenth century.

C. ACCOMMODATION AND THE DIVINE LAW

1. Sacrifices as Divine Accommodation:
Early Christian Literature

The divine accommodation to "the language of man" became, we have seen, an almost indispensable exegetical figure of thought. It also inspired—without in the least implying them—a host of speculations about the divine accommodation to the history of mankind. The former defended the truth of revelation, the latter, its wisdom and justice. That God adjusted his acts in history to the capacity of men to receive and perceive them presupposed a notion, however vague, of a relatively autonomous social and cultural evolution of some or all men; attempts to link the divine plan of salvation with the intrinsic evolution of humanity through the mediation of a divine pedagogy comprised the backbone of Christian philosophies of history since the second century. They ascribed, at least on the level of theological interpretation, a different *qualitas temporum* to different periods of history.[1] In so doing they aided in the formation of categories of historical reasoning that have been used since the seventeenth century. Yet before turning to the larger, and perhaps more interesting, speculation about the working of providence throughout the whole of history, it would be proper to trace the career of the earliest, most persistent, and most elaborate accommodational interpretation of an institution of the past—the sacrificial and ceremonial laws of the Bible. It served, in a way, as a bridge between the merely exegetical and the historical employment of the principle; it certainly served as a paradigm.

The first hints of an accommodational interpretation of the sacrificial law within the Jewish tradition may have been no more than a sour grapes reaction to the cessation of sacrifices after the destruction (A.D. 70) of the Temple. *Vajikra Rabba*, commenting on Lev. 17:7, attributes to R. Pinhas ben Levi the opinion that the sacrifices were but a divine concession to polytheistic customs; God used them to eradicate idolatry all the more forcefully. "[A simile to] a prince whose heart has forsaken him and who used to eat carcasses

---

[1] E.g., Hildebertus Cenomanensis, *Tractatus theologicus* 2, Migne, *PL* 171: 1073A; sometimes *ratio temporum* is also used in a noncomputistic, similar meaning, e.g., Beda, *Super acta apostolorum*, Migne, *PL* 92: 953; for parallels in the legal literature, Kelley, "Klio and the Lawyers."

and gory meat [trefot]. Said the king: 'Let these be always on my table, and of himself he will get weaned.' So also: since Israel was eagerly attracted to idolatry and its sacrifices in Egypt . . . God said: 'Let them always bring their sacrifices before me in the tabernacle and thus they will separate themselves from idolatry and be saved.' "[2] Christian Church fathers employed similar interpretations, whenever it served apologetic or polemical purposes;[3] the Antiochian exegetes, in particular Theodoret of Cyrrhus, made it into a systematic hermeneutical principle, well suited to combat the excesses of allegorical-pneumatic interpretations of Alexandrian provenance. In his commentary to Leviticus Theodoret sums up that which he "explained in many places" in words very close to the Jewish *Midrash*.[4] One element of his theory appeared much earlier than all of these references, Jewish or Christian, in the arguments of anti-Jewish polemicists. The Jewish cult and law, this was the essence of Manetho's counter-biblical reconstruction of Jewish history, had nothing original or authentic about them: they were but an inverted mirror of the Egyptian cult and laws.[5]

In one of Augustine's better-known letters we find an explicit and precise expression of the link between such interpretations and the idea of accommodation:

It befitted [aptum fuit] God to request sacrifices in earlier times; now, however, things are different, and he commands that which befits this time. He, who knows better than man what pertains by accommodation to each period of time [quid cuique tempori accommodate adhibeatur], commands, adds, augments, or diminishes institutions . . . until the beauty of the whole history [saeculi], whose parts these periods are, unfolds like a beautiful melody [velut magnum carmen].[6]

[2] *Leviticus Rabba* 22.6 (ed. Margulies). On this, and the following, see Funkenstein, "Maimonides: Political Theory and Realistic Messianism," pp. 81–103; Benin, "The 'Cunning of God' and Divine Accommodation," pp. 179–91, esp. p. 183; *id.*, "Thou Shalt Have No Other God Before Me: Sacrifice in Jewish and Christian Thought."

[3] Beginning with Justin the Martyr; Benin, "Sacrifice," pp. 10ff.

[4] Theodoret of Cyrrhus, *Questiones in Octateuchum, in Leviticum*, Migne, *PG* 80: 300; cf. *In Isaia* 1.2, *PG* 81: 226; *Graecorum affectionum curatio* 7 (De sacrificiis), *PG* 83: 991ff., esp. 995ff. For other correspondences with the Jewish *Midrash* see Funkenstein, "Gesetz und Geschichte: Zur historisierenden Hermeneutik bei Moses Maimonides und Thomas von Aquin," pp. 147–78, esp. p. 165 and n. 71.

[5] Below IV.C.5 (Spencer), E.2. (Manetho).

[6] Augustine, *Epistulae* 138.1.5, ed. Goldbacher, p. 130: "aptum fuit primis temporibus sacrificium, quod praeceperat deus, nunc vero non ita est, aliud enim praecepit, quod

In question was the wisdom of the sacrificial rituals in ancient Israel. The pagans ask: If they were not good, why were they instituted? And if they were good, why were they abolished by a new dispensation? Again, as in the matter of divine omnipotence, pagan polemics are directed not at the idea of one God, but at the notion of God as a busybody, acting arbitrarily in history, changing, as it were, his mind (*concilium*). We remember how and why Celsus spoke of the notion of special divine providence as the "frog and worm perspective."[7] Augustine tells such pagan contenders that the process of history, far from being arbitrary, is as beautiful, if seen as a whole, as the cosmos is, and for the very same reasons: the parts fit into the whole. In accord with his Platonic aesthetic theory, Augustine distinguishes here, as in other instances, between the "fitting" (*aptum*) and the "beautiful" (*pulchrum*). The parts of whole need not be, in themselves, beautiful; they must, however, be fitting to each other if the whole is to be beautiful. In other places, Augustine conceded even to each single period in the life of a person or in the process of history "its own beauty" (*pulchritudo sua*),[8] a relative beauty, since the signs and institutions of each period in history also fit with each other and are adjusted to the capacity of humanity to perceive them and to live by them.

huic tempori aptum esset, qui multo magis quam homo novit, quid cuique tempori accommodate adhibeatur, quid quanto imperiat, addat, detrahat, augeat minuative immutabilis mutabilium sicut creator ita moderator, donec universi saeculi pulchritudo, cuius particulae sunt, quae suis quibusque temporibus apta sunt, velut magnum carmen . . . excurrat." Cf. also Augustine, *Adversus Judaeos* 3.4, Migne, *PL* 42: 53: "ut rerum signa suis quaeque temporibus conveniant"; *Contra Faustum* 6.7, Migne, *PL* 42: 417; Lubac, *Exégèse* 2.1, p. 347 n. 7. On the motive of history as a harmonious melody cf. Augustine, *DcD* 11, CCSL 30.1, pp. 537–38; *Contra Seceundinum Manichaeum* 15, Migne, *PL* 42: 577. Augustine names in this context *Jes. Sir.* 33:14 (*DcD*, ibid.); other possible sources: Marrou, "Das Janusantlitz der historischen Zeit bei Augustin," in *Zum Augustin—Gespräch der Gegenwart*, ed. Andresen, p. 379. Later employment: e.g., Bonaventura, *Breviloquium*, in *Opera omnia*, ed. Quaracchi; 5:204; Lassaux, *Philosophie der Geschichte*, ed. Thurner, pp. 65–66 and n. 8 below.

[7] Above III.B.1. and n. 6.

[8] Augustine, *De diversis quaestionibus* 44, Migne, *PL* 40: 28: "Quia omne pulchrum a summa pulchritudine est . . . temporalis autem pulchritudo rebus decedentibus succedentibusque peragitur. Habet autem decorum suum in singulis quibusque hominibus singula quaeque aetas. . . . Sicut ergo absurdus est, qui iuvenilem tantum aetatem vellet esse in homine temporibus subdito . . . sic absurdus est qui in ipso universo genere humano unam aetatem desiderat." On the equation ages of the world = ages of man = days of creation see Funkenstein, *Heilsplan*, pp. 38–40 and below IV.D.3. The enfoldment of beauty also: *De vera religione* 21.41 (113), p. 213; Spitzer, *Classical and Christian Ideas of World Harmony: Prolegomena to an Interpretation of the Word "Stimmung,"* pp. 28ff.

If, indeed, Christian authors often tended to downgrade or deny the value of sacrifices, they never completely forgot this tradition in which the sacrifices were seen as a positive preparation for Christ; eventually it found its way also to the *Glossa ordinaria*.[9] In his treatise "On the Origin and Development of Some Ecclesiastic Customs," a Carolingian author, Walahfrid Strabo, wishes to defend the diversity of customs and liturgy in the *one* Church; the Church adjusts itself to diverse exigencies of times and places. God himself has set the example for such flexibility. Before the law (*ante legem*) the cult of demons filled the earth; the sacrificial law was both a concession to that cult and the best means to combat it: "The omnipotent and patient creator, always willing to aid his creation, knew that, due to the weakness of mortals, he could not remove at once all of their habits."[10] Anselm of Havelberg, whose reflections on the variety of religious orders I shall discuss later, based them on similar arguments.[11]

The law given to Israel was "good for its times"[12] (*bonum in suo tempore*);[13] the tragedy of the Jews, says Joachim of Fiore,[14] is that they

---

[9] *Glossa ordinaria*, Migne, *PL* 113: 344–45: "Lex ergo, quasi paedagogos eorum, praecipit Deo sacrificare (Exod. 32), ut in hoc occupati abstinerent se a sacrificio idolatriae. Talia tamen sanctivit sacrificia, quibus mysteria significantur futura." For a different, negative tradition of evaluating sacrifices see *Glossa ord.*, Leviticus, Praef., Migne, *PL* 113: 295–97; cf., e.g., Hysechius, in Lev. 1, Migne, *PG* 93: 792, 1002f. (to Lev. 17:7).

[10] "Omnipotens et patiens creator facturae suae volens undecunque consulere, quia vero propter fragilitatem carnalium omnes consuetudines pariter tolli non posse sciebat": Walahfrid Strabo, *De exordiis et incrementis quorundam in observationibus ecclesiasticis rerum*, ed. Boretius and Krause, p. 476.

[11] Below IV.C.4. and n. 52; D.4 and nn. 71–73.

[12] "Oportuit enim ut et illa quae finienda erant nequaquam subito vel precipitanter, sed paulatim et quasi cum quadam reverentia dimitterentur, ut ostenderentur bona fuisse tempore suo. Et similiter quae incipienda erant non subito in auctoritatem assumerentur, sed cum mora et gravitate incoharentur, ne velut aliena et praeter rem aliunde inducta subito putarentur": Hugh of St. Victor, *De sacramentis Christianae fidis* 2.6.4, Migne, *PL* 176: 450a. Hugh argues for the wisdom of an "intermediate time" of overlapping systems of law. On this idea in early Scholastic thought see Gössman, *Metaphysik und Weilsgeschichte: Eine theologische Untersuchung der Summa Halensis (Alexander von Hales)*, pp. 280–81.

[13] The expression "Bonum in suo tempore" was seen by Grundmann (*Studien zu Joachim von Floris*, pp. 99–100) as characteristic for the difference between Joachim and the Catholic tradition (though conceding that it may be older). Hugh employs it throughout (cf. also *De vanitate mundi*, Migne, *PL* 176: 740c). As possible sources in Jewish exegesis see Funkenstein, *Heilsplan*, p. 165 n. 5 (Rashi to Gen. 6:9).

[14] "Noluerunt . . . ipsi Judaei mutari cum temporae": Joachim of Fiore, *Super quattuor Evangeliarum*, in Lubac, *Exégèse* 2.1, p. 144 n. 2. At about the same time, I tried to show elsewhere, a radically new form of polemic against Jews appeared in the west: the contention that, indeed, they did change; that they have acquired a new, man-made law and

"refused to change with the times." The *processus religionis*[15] did not cease even in Christian times: more miracles were needed to persuade people in the beginnings of Christianity than are needed now. The appearance of heresies necessitated the generation of dogmas—wherefore even heretics had a function in the divine plan of history. Religion, then, has progressed in accord with the refinement of human capacities. Knowledge, even in matters religious, increased with time.[16] Does it mean that faith itself, even among the true worshippers of God that always existed, has changed? The twelfth century debated the question repeatedly.[17] Of special interest is Hugh of St. Victor's solution. His systematic exposition of the sacraments—sometimes seen as the first theological *Summa*—is historical throughout; the sacraments are discussed in the order in which they were introduced during "the work of restoration" of mankind. They were adjusted to the changing conditions of mankind's "disease," and were introduced slowly (*paulatim*) rather than suddenly (*subito et precipitanter*), so that there always were periods of transition.[18] Faith increased, but not in substance (*materia fidei*), rather in effectiveness. In each of the three periods of salvation one

merely pretended to hold to the Old Testament literally—that they are, in a way, heretical even from the point of view of genuine Judaism (as the Church understood it). The campaign against Jewish post-biblical writing was based on this new propaganda. See my "Changes in the Patterns of Anti-Jewish Polemics in the 12th Century," pp. 125–45 and J. Cohen, *The Friars and the Jews: The Evolution of Medieval Anti-Judaism*, pp. 51ff., 129ff. (who concurs in the evaluation of the nature of the change). Cf. below n. 63.

[15] Abaelard, *Dialogus inter Philosophum Judaeum et Christianum*, Migne, *PL* 178: 1614: "Quid enim? mirabile est, cum per seriem et temporum successionem humana in cunctis rebus creatis intelligentia crescat, in fide, cuius errori summum periculum imminet, nullus est profectus?" Cf. Alanus ab Insulis, *Contra haereticos* 3.2, Migne, *PL* 210: 402c.

[16] Augustine, *De vera religione* 25.128, pp. 216ff.; Tajo, *Sententiae* 2.12, Migne, *PL* 80: 794b; Odo of Cluny, *Collationum libri tres* 1.25, Migne, *PL* 133: 536: "De signis vero illud sciendum, quia iuxta scripturam: 'unicuique rei tempus suum est sub coelo (Eccl. 3:17).' Unde sancta Ecclesia signis ad corroborandum suorum fidem in primordiis suis indiguit. Nunc vero constante iamdudum fidei statu signa ad modum non requirit." Heresies: cf. Grundmann, "Oportet et haereses esse: Das Probleme der Ketzerei im Spiegel der mittelalterlichen Bibelexegese," pp. 129–64.

[17] On the "questio scholastica de fide antiquorum et modernorum"—namely "an secundum incrementa temporum mutata sit fides" (e.g., Hugh of St. Victor, *De sacramentis* 1.10.6, Migne, *PL* 176: 355ff.) see Grabmann, *Die Geschichte der scholastischen Methode*, 2:276ff.; Grundmann, *Studien*, pp. 123–24; Beumer, "Der theoretische Beitrag der Frühscholastik zu dem Problem des Dogmenfortschritts," pp. 209ff., 220ff.; Lubac, *Exégèse* 2.1, p. 356.

[18] Above n. 12.

finds all three groups—the "openly bad," the "deceivingly good," the "truly good"; the last may be and may always have been a minority, but now they can act *publicly* and therefore more effectively.[19]

Inspired by Jewish or Christian sources—or by both—a well-known convert of the twelfth century, Petrus Alfunsi, transformed the idea of accommodation into an anti-Moslem argument. In his "Dialogues," his old self, Moses, debates Petrus, his new self: if he wanted to join the most rational and progressive of religions, why not embrace the newest of them—Islam? Because, answers Petrus:

> People in the time of Mohammed, without law, without scripture, ignorant of any values except arms and agriculture, desiring luxury, given to gluttony, were easy to preach to [only] what they desired. Had he done otherwise, he would not have drawn them to his law.[20]

All of these traditions of accommodation, especially those centered around the historical relativization of the sacrificial law, found their way into later Scholasticism. The schoolmen were also acquainted with a much more radical and elaborate interpretation of this sort—Maimonides' *Guide to the Perplexed*.

2. Maimonides on the Indeterminacy of Nature

In the *More Nebuchim* III.26–56, Maimonides unfolds his philosophy of law, the doctrine of "reasons for the commandments."[21] Against the Sa'adianic disjunction between commandments of obedience (*mitsvot shim'iyot*) and of reason (*sikhliyot*), a disjunction that combined the Kalam terminology with Midrashic reminiscences,[22] Maimonides holds that every single precept has a dual structure and

---

[19] Funkenstein, *Heilsplan*, pp. 52–53, p. 167 n. 12; A. Funkenstein and J. Miethke, "Hugo von St. Viktor"; Schneider, *Geschichte und Geschichtsphilosophie bei Hugo von St. Viktor*, pp. 54–55.

[20] Petrus Alfunsi, *Dialogi* 5, Migne, *PL* 157: 667b; on his polemics see my "Changes in Patterns," pp. 133–37. Cf. Vives, *De veritate fidei Christianae* 4.12, *Opera Omnia* 8:402.

[21] Henceforth *MN*; ed. Kafih, *Dalalat el Hairin*, trans. S. Pines, *The Guide of the Perplexed*. The following remarks on Maimonides are a modified version of my articles in *Miscellanea medievalia* and *Viator*. See also Twersky, *Introduction to the Code of Maimonides (Mishne Tora)*, pp. 380–406, 430ff., 450–59, 473ff.

[22] The *Midrash* furnished the name for the discipline (*ta'ame hamitsvot*, e.g., Numeri Rabba 16.1:149a) and some of the paradigms (the red heifer). Cf. Heinemann, *Ta'ame hamitsvot be-safrut Yisrael*, 1.22–35; Urbach, *Ḥazal*, pp. 320–47.

may be seen as both a commandment of reason and a commandment of obedience. Every commandment serves a rational design:
"The law of God is perfect" (*torat hashem temima*). But the right obedience to every commandment should not be dictated by insight
into its purpose: it must be based on the *potestas coactiva* of the law,
the fact that it is the will of the sovereign.[23] Maimonides is thus
forced to look for a specific rationalization of those commandments—the ceremonial and dietary laws—to which Sa'adia assigned only a generic rationale. A perfect constitution, Sa'adia held,
must include some irrational commandment as an opportunity for
the subjects to profess blind loyalty; and Sa'adia, in the endeavor to
demonstrate that the written and oral law form a perfect constitution, valid for all societies and all times, had to limit the number of
such pure "commandments of obedience" to a minimum. Maimonides, who questioned this very axiom of Sa'adia's legal philosophy, needed a new starting point. He started, as so often, by trying
to define anew the meaning of old questions.

What do we really look for when we ask for the *reason* of a commandment? Must a rationale for a specific law cover every part and
detail of that law? In a preliminary answer, Maimonides draws a
strict analogy between laws of nature and social laws.[24] In the second part of the *Guide*, Maimonides developed one of the most original philosophies of science in the Middle Ages. There he proved
that not only are laws of nature (the ordering structures of nature)
in themselves contingent upon God's will; but that each of them
must include, by definition, a residue of contingency, an element of
indeterminacy. No law of nature is completely determining, and no
natural phenomenon completely determined, not even in God's

---

[23] Even in the domain of obligations pertaining to non-Jews (*sheba mitzvot bne Noah*)
Maimonides insists that insight into their rationality (*hekhra hada'at*) does not suffice to
characterize an obedient gentile, a "pious from among the nations," but only the fulfillment of these commandments because they are the will of God (*Hilchot Melachim* 8.11).
See also Levinger, *Darche hamachshava hahilchatit shel harambam* [Maimonides' Techniques
of Codification], esp. pp. 37ff.; J. Faur, "The Basis for the Authority of the Law According to Maimonides," *Tarbiz* 38.1 (1969):43ff. (Hebr.).

[24] *MN* 3.26 (Pines trans., p. 509): "this resembles the nature of the possible for it is
certain that one of the possibilities will come to pass . . ." i.e., which necessitates the actualization of one of the possibles within a material substrate. Cf. *MN* 2.25, as well as my
following notes.

mind.[25] To illustrate the matter, allow me to invent an example. Assume that tables should all be made out of wood; assume that the kind of wood most suitable for tables is mahogany, and that the best mahogany can be found only in a remote forest in Indonesia. A carpenter who wishes to make a perfect table has good reasons to choose mahogany and to travel all the way to the said forest. But there and then he will ultimately be confronted with two or more equally reasonable possibilities. Should he choose the tree to his right or to his left? He must choose one, and both are equally suitable. The purpose can never determine the material actualization in all respects, down to the last particular; a "thoroughgoing determination" is ruled out by the very material structure of our world. In the very same way, there may (indeed must) be a purpose to the universe, but it does not govern all particulars. The purpose of the universe may require the circular orbit of the celestial bodies. But it does not account necessarily for the different velocities or colors of the planets.[26]

Technically, Maimonides seems to have recognized that the Aristotelian concept of matter carried two different explanatory burdens.[27] It was both a principle of potentiality and a *principium individuationis*. Maimonides deemphasized the second connotation of matter; matter becomes for him mainly the source of contingency throughout the universe, and not only in the sublunar realm. Between essential forms (laws, necessities) and matter qua mere potentiality (contingency, possibility) lies a hierarchy of contingent structures—*causae finales*—that account for the individuation (i.e., particularization) of all singulars. The natural world is thus a *contin-*

---

[25] Maimonides does not say so explicitly, but it follows clearly from his discussion of the particularization of percepts and of natural phenomena. The Maimonidean theory of nature, and in particular his doctrine of contingency, have not yet received due emphasis, but see Julius Guttmann, "Das Problem der Kontingenz in der Philosophie des Maimonides," pp. 406ff. Cf. Twersky, *Introduction*, pp. 397–98.

[26] *MN* 2.19 (Pines trans., pp. 302–14). On similar examples in the Kalam, Davidson, "Arguments from the Concept of Particularization," pp. 299ff., esp. pp. 311f., 313 n. 50 (Maimonides). On the Aristotelian concept of contingency (e.g., *De generatione animalium* Δ 3.778b, pp. 16–18); cf. Hintikka, *Time and Necessity*, pp. 27–40, 93–113, 147–75.

[27] *MN* 2.19 discusses Aristotle's *failure* to account for the particularization of terrestrial as well as celestial bodies; the failure is then converted into a maxim—namely that matter can never be "omnimodo determinatum," since it is, by definition, a principle of potentiality (cf. 3.26, n. 24 above). Of prime importance for the understanding of this chapter is the distinction between necessity and purpose.

uum of instances of the accommodation of divine planning to indifferent if not resilient substrates. The influence of parts of this doctrine on Scholastic philosophy was considerable. Some of it we discussed earlier.[28] In a sense, Maimonides' principle of indeterminacy is closer to modern than to classical physics: modern physics likewise assumes a principle of indeterminacy not as a limit to our knowledge, but as an objective indeterminacy within nature itself.[29]

His principle of indeterminacy and the corresponding principle of accommodation allowed Maimonides to rephrase that which Kant later was to call the "physico-theological argument," the proof for God's existence from the order of the universe. If the universe throughout were to be thoroughly well ordered, it would be of itself necessary and would not imply an ordering hand. The physico-theological argument assumes neither that the universe is completely ordered nor that it is completely disconnected (in the manner of the extreme Nominalism of the *Isharia*), but that its order is imposed on the heterogeneous elements, which of themselves do not demand or imply this particular order.[30] The argument from particularization had been used already by the Kalam; Maimonides gave it the balanced form in which it was to remain effective until Kant.

The principle of indeterminacy allowed him to introduce most miracles—or, more generally, instances of special providence—without violating laws of nature.[31] Miracles are mostly, but not always, taken from the reservoir of the remainder of contingency on

[28] Above III.B.2.

[29] Niels Bohr, "Discussion with Einstein on Epistemological Problems in Atomic Physics," in *Albert Einstein: Philosopher-Scientist*, ed. Schlipp, 1:199–241. Here and there, "indeterminacy" is not a limit to our understanding, but a limit within nature itself.

[30] Kant, *Kritik der reinen Vernunft*, in *Werke*, ed. Weischedel, 4:552 (B654 = A626): "Den Dingen der Welt ist diese zweckmässige Anordnung ganz fremd und hängt ihnen nur Zufällig an, d.i. die Natur verschiedener Dinge konnte von selbst, durch so vielerlei sich vereinigende Mittel, zu bestimmten Endabsichten nicht zusammenzustimmen, wären sie nicht durch ein anordendes vernünftiges Prinzip . . . dazu ganz eigentlich gewählt und angelegt worden."

[31] MN 2.48 and *Ma'amar techiyat ha metim* 10, in *Iggrot harambam*, ed. Kafiḥ, pp. 98–101. The words "shekol ze taluy behiyuv hokhma seen anu yod'im ba me'uma, velo od ela she'anu hizkarnu kevar ofen ha'hokhma bekhakh," whose meaning eluded the translator and editor, may be taken as reference to the divine "cunning," i.e., to "purpose" (rather than necessity). Cf. Twersky, *Introduction*, pp. 473ff.

all levels of nature. Maimonides calls such miracles "miracles of the category of the possible (*moftim misug . . . ha'efshari*).[32]

3. Indeterminacy, Accommodation, and "the Reasons for the Commandments"

Precisely the same figure of thought is used by Maimonides to clarify what we search for in "reasons of the commandments" (*ta'ame hamitsvot*). Take, for example, the sacrifices. We may be able to explain, in view of their purpose, why sacrifices should have been instituted in the first place; "but the fact that one sacrifice is a lamb and another a ram; and the fact that their number is determined—to this one can give no reason at all, and whoever tries to assign a rationale enters a protracted madness."[33] Rather than looking for an always determining principle for each law, we should look for a contingent rationale. Maimonides found such a contingent rationale in the concrete historical circumstances under which these laws were given to the nascent Israel. Sacrifices and the bulk of the dietary laws are not in themselves beneficial for every society at every time. The former are in particular suspicious, because they invoke anthropomorphic associations of a smelling or an eating deity. Considering the vigor with which Maimonides eradicated even the most abstract positive attributes of essence from the concept of God,[34] the institution of sacrifice must have been to him unworthy of a truly monotheistic community. And indeed he interprets it as a remnant of the universal polytheistic culture of the Sa'aba which prevailed in the times of Abraham and Moses. So deep rooted and pervasive were its abominable creeds, that they could not be eradicated altogether in one sweeping act of revelation and legislation.[35] Human nature does not change from one extreme to another suddenly ("Lo yishtane teba ha'adam min hahefekh el hahefekh pit'om: natura non facit saltus"). Had anyone demanded of the nascent Israel to cease the practice of sacrifices, it would be just as impossible a demand as if "someone

---

[32] Maimonides, *Ma'amar*, p. 98.

[33] Maimonides, *MN* 3.26 (Pines trans., p. 509; mine is a translation from the Hebrew).

[34] Above II.C.n.44.

[35] Maimonides calls these practices and beliefs "an abomination (*to'eba*) to human nature ('*altaba 'alanasani*)" (*MN* 3.29), "against nature" (*MN* 3.37). On the other hand, he describes how mankind lapsed gradually, i.e., almost naturally, into such a universal error (*Sefer hamada, hilchot Avodat kohavim*).

demanded today (of a religious community) to abandon prayer for the sake of pure meditation." Only a miracle could have transformed the polytheistic mentality immediately into an altogether monotheistic one, but God does not wish to act "contra naturam." He rather prefers to act with the aid of nature, to accommodate his plans to existing, contingent circumstances, to use contingent elements within nature in order to change it. Rather than eradicating all polytheistic inclinations among the emerging monotheistic community from the outset in a miraculous act, he preferred to use elements of the polytheistic mentality and culture in order to transform this very mentality by degrees. Sacrifices were conceded with maximum restrictions and changed intents. They are turned into a fruitful error.

Just as Hegel's *objektiver Geist* uses the subjective, egotistic freedom of man to further the objective goals of history,[36] so also Maimonides' God fights polytheism with its own weapons and uses elements of its worship as a fruitful deceit. Maimonides spoke of the "cunning of God" (*'ormat hashem utebunato; talattuf fi 'allahu*)[37] where Hegel will speak of the "cunning of reason" (*List der Vernunft*). Their point of agreement is at one and the same time the point of their difference. Hegel's *List der Vernunft*, much as its forerunners—Mandeville's "private vices, publick benefits" or Vico's "providence"—articulate a sense of the absolute autonomy of human history and its self-regulating mechanisms. Maimonides, as all other medieval versions of the divine economy, allows at best a relative autonomy to the collective evolution of man.[38]

Maimonides demonstrates with considerable detail how every single allegedly "irrational" precept is a countermeasure to this or that Sabean practice. Now it matters little that the Sabeans, of whom Maimonides speaks with the genuine enthusiasm of a discoverer, were actually a small remnant of a gnostic sect of the second or third century A.D. rather than a polytheistic universal community[39]—note that Maimonides uses for it the Moslem self-

---

[36] Above IV.A.I.

[37] *MN* 3.32: "talattuf alalla wahakhmatah." Cf. *MN* 3.54 where "talattuf" stands for "practical reason." The Quran attributes God with "cunning": Goldzieher, *Vorlesungen über den Islam*, p. 23 (*kejd* and *makr*, "Kriegslist").

[38] Below IV.F.5.

[39] *MN*, Pines trans., Intro., pp. cxxiii–iv.

denomination " 'umma." The mistake in the identification of the background of the Mosaic law led Graetz to discard the Maimonidean explanations as "flat."[40] But it is still possible that the argument of Maimonides is new and reliable in its method rather than in the actual validity of his historical reconstruction.

Yet, we have already seen that the interpretation of sacrifices as a divine concession to polytheistic usages in order to eradicate idolatry all the more forcefully was not altogether new, not even in the Jewish tradition. A few remarks that remind us of the passage quoted above from Leviticus Rabba—which Maimonides, oddly enough, never mentions—can be gathered from Jewish authors before Maimonides. One of his direct sources may have been the observations of the Karaite Qirquasani.[41] It seems as if Maimonides' doctrine is just another variation of the medieval principle of accommodation. But consider the following. None of these traditions is actually concerned with the reconstruction of the original meaning of biblical legal and ritual institutions out of their forgotten historical background. Maimonides raised such a reconstruction to a methodical level. His theory not only explains, in detail, how the "forgotten" culture of the Sa'aba accounts for opaque parts of the law; it explains at one and the same time why these original "reasons for the commandments" were forgotten and must now be recon-

---

[40] H. Graetz, *Die Konstruktion der jüdischen Geschichte*, pp. 85–86 and the note. In her famous study, Douglas, *Purity and Danger: An Analysis of Concepts of Danger and Tabu*, pp. 41–57, refers to Maimonides' theory of *ta'ame hamitsvot* as a paradigm of wrong methodology (looking for external causes for taboos). She already takes for granted Maimonides' methodological breakthrough—if perhaps on the wrong object—in trying to reconstruct the original setup in which taboos were meaningful; she takes it for granted that Maimonides *could* make errors on the same level as Robertson-Smith. That Maimonides' extreme rationalism obscured his view of the polarity of taboo laws is even more evident in his inability to explain why sacred books "defile the hands" (*metam'in et hayyadayim*). But this should not reduce our appreciation for his achievement. As to Douglas's powerful thesis itself, I am not competent to judge its merits. I have but one specific question, leading to a more general observation. Jewish law lacks any prohibitions concerning the consumption of vegetables; surely, in any conceivable primitive taxonomy, some plants will resist neat classification. One may answer that the dietary prohibitions stem from an older, nomadic, cattle-raising society; but mixed "weaving and sowing" (*sha'atnez reki-l'ayim*) were likewise prohibited, as was the sowing with two kinds of animals. Perhaps the thesis suffers from overprecision. Granted that taboos originate in the contraposition of order and disorder, culture and wilderness (chaos); the forbidden belongs to the latter, the undomesticated, but it need not *defy* attempts at classification. Its classification may not be even always *attempted*. It suffices that it is at odds with the familiar.

[41] Qirquasani, *Kitab al Anwar* I, 44, ed. L. Nemoy.

structed so painfully. The very intention of the lawgiver was to eradicate all the reminiscences of the abominable rites and opinions of the Sabean *'umma*. The fact that the reasons for certain commandments were forgotten is in itself testimony to the success of the divine "cunning" or pedagogy. Not only among the Jews: the whole inhabited world, Maimonides believes, is by now monotheistic.[42]

Indeed, Maimonides did not shy away from the employment of similar structures of interpretation to explain the polytheistic residues within Islam—in a far more sophisticated manner than Petrus Alfunsi. In his famous letter to Obadia the Proselyte he denies that the cult of the Quaaba is of the category of "throwing stones for Mercury": Moslems, it is true, inherited the cult from their idolatrous fathers, but endowed it with a new, monotheistic meaning.[43] His attitude toward Christianity was more ambiguous; but he viewed both Christianity and Islam as a necessary, though negative, preparatory stage for the messianic age.

4. *"The Cunning of God" in the Course of History*

Scattered passages in Maimonides' writings add up to a distinct view of the course and phases of human history seen as a growth of monotheism. It is a gradual process, which shall be succeeded by an indefinite period of unchallenged, universal monotheism, and was preceded by a likewise gradual process of polytheization. From Enos to Abraham, the original monotheism of Adam degenerated through polyatrism into polytheism, which then enabled a priestly class to exploit and terrorize the superstitious masses.[44] If this sounds like an outright inversion of the evolutionary models of an-

---

[42] *MN* 3.51; only a few nomads on the fringe of civilization (some Turks, some Africans) are still "people without a religion" (*mibene adam she'en lo emunat dat*). This is possibly the source of Hameiri's *'umot hagedurot bedarche hadatot*. On him see Katz, *Ben yehudim legoyyim*, pp. 116–28.

[43] It seems as if Maimonides implies a somewhat similar structure of understanding to explain the polytheistic residues within Islam. Here as at the outset of Israel, pagan cults are *reinterpreted*. Maimonides, *Teshuvot ha Rambam (Responsa)*, ed. Blau, 2:726–27. Cf. Lazarus-Yafe, "The Religious Problematics of Pilgrimage in Islam," pp. 222–23, 242–43. A different, but explicit usage of the principle of accommodation to explain the origins of Islam can be found in Petrus Alfunsi, *Dialogi* (above n. 20); cf. 603a, 667b: "Quoniam in mundi exordio quasi silvestres erant adhuc homines et bestiales . . ." (slow introduction of the divine law).

[44] *MT, Sefer hamada, Hilkhot avodat kokhabim*, ch. 1, pp. 1–3.

thropologists of the nineteenth century, it is due to one basic agreement and another basic disagreement. The medieval and modern rationalistic views of the development of (true or false) religions share a dislike of radical mutations; they only disagree as to the starting point of the evolutionary process. To the Middle Ages, the knowledge of God's unity was part of the *lumen naturale*. Not its presence, but any deviation from it called for a historical explanation; all the more so since Adam, as it were, encountered the Almighty frequently and directly, if not always on friendly terms. Schmidt's anthropological arguments for the primacy of the *Urmonotheismus*[45] are but old theologoumena in a modern guise, for example, Eusebius of Caesarea's description of the gradual corruption of man's "kingly nature" through polytheism and polyarchy and its restitution through universal monarchy and monotheism.[46] Similar questions already bothered the author of the *Wisdom of Salomon*;[47] and of similar scope is the Maimonidean attempt to reconstruct the prehistory of monotheism.

The second period in the essential history of mankind begins with the establishment of a monotheistic community. The "feeble preaching" of Abraham[48] did not suffice to guard against a relapse of his followers: the masses were, and still are, prone to superstition, and can only be held within the boundaries of religion by laws. These laws, we have seen, were construed by the "cunning of God" so as to utilize polytheistic images and rites with the intent to abolish them. The emergence of a monotheistic mentality was slow and difficult: "tanta molis erat Romanam condere gentem." Gradualness and slowness, we noted already, are the formal marks of natural change—here as in the Christian versions of the principle of accommodation since Irenaeus of Lyons.

If the transformation of a small nation into a monotheistic community was a slow and difficult process, all the more so the monotheization of the entire *oikoumenē*. This is a dialectical and highly dramatic process, guided again by the operation of the divine ruse.

[45] Cf. Pettazzoni, *L'essere supremo nelle religioni primitive: L'omniscienza di Dio*, ch. 1.
[46] Below IV.D.3.
[47] *Sapientia Salomonis* 14:12–17, a Euhemeristic interpretation.
[48] *MN* 3.32. In his placing of the role of Moses above that of Abraham, Maimonides may also have intended to invert the Moslem historical scale of values, which placed Abraham above Moses.

Time and again "the nations of the world" wish to destroy the people of Israel, whose election they envy (even if, one may add, they deny it).[49] They generate successively destructive ideologies—Maimonides calls them "sects"—each of greater sophistication than the former, though all of them exist at present, wherefore they correspond loosely to the "four monarchies" of the Book of Daniel.[50] Having failed in their attempt to extinguish the true religion by force or argumentative persuasion (Hellenization), the nations of the world resort to a ruse. A third sect emerged that imitated the basic idiom of the monotheistic, revelatory religion in order to assert a contradictory law, so as to confuse the mind and thus cause the extinction of both the original and its imitation. "And this is of the category of ruses which a most vindictive man would devise, who intends to kill his enemy and survive, but if this is beyond his reach, will seek a circumstance in which both he and his enemy will be killed." Yet inasmuch as this latter sect and those similar to it— Christianity and Islam—*do* imitate a monotheistic mentality, they help to propagate and prepare the acceptance of the true religion against their will; their stratagem turns, by a divine ruse, against them; or better, their ruse turns out to have been a divine ruse from the outset. The effect of their resistance to the truth is a negative *preparatio messianica* (or, in the fortunate phrase of H. H. Ben Sasson, a *preparatio legis*). It is in this sense, I believe, one has to interpret the phrase that Christianity and Islam are "road pavers for the king messiah."[51]

Our attention was drawn repeatedly to some analogies between Maimonides' historical employment of the principle of accommodation and its Christian counterparts. The broad role that Maimonides assigned to the divine "ruse" also reminds us of one of the most original pieces of historical speculation in the twelfth century, Anselm of Havelberg's *Dialogi*.[52] The *spiritus sanctus* accommodates its

---

[49] *IT*, ch. 1, p. 21.

[50] *IT*, ibid. Maimonides, unlike some Jewish and most Christian philosophers of history, did not pay specific attention to detailed periodization. Nor was he interested in history as such. Cf. Baron, "The Historical Outlook of Maimonides," in *History and Jewish Historians*, pp. 109–63, esp. pp. 110–13.

[51] Ben-Sasson, "Yihud 'am yisrael le'daat bne hame'a hastem esre," pp. 212–14.

[52] Anselm of Havelberg, *Dialogi* 1.10, Migne, *PL* 188: 1152ff. Cf. Kamlah, *Apokalypse und Geschichtstheologie*, p. 64; Berges, "Anselm von Havelberg in der Geistesgeschichte des 12. Jahrhunderts," pp. 38ff., esp. p. 52 (reference to Hegel's "List der Vernunft");

historical operations not only to the degree of perception of man, but also to the ever more refined stratagems of Satan: each of the seven successive *status ecclesiae* is characterized by a less obvious and therefore more dangerous opposition of the adversary. In his own, fourth, *status ecclesiae* Anselm sees Satan penetrating the Church with pretension and imitation, "sub praetextu religionis," through "falsi fratres," a move that the Holy Spirit counters by a variety of new, fresh turns of religiosity. Needless to say, such analogies do not suggest direct mutual influence; their interest lies precisely in the circumstance that these figures of thought belong to so disparate cultural horizons. The search for the theological meaning of history was much more a part of Judaism and Christianity than Islam. A similarity of the problem led, at times, to somewhat similar patterns of answers.

Returning to Maimonides, we note that even though the scheme of each of the "sects" is doomed to failure, they still inflict on Israel severe physical and mental blows. It is the lot of Israel to endure in spite of dispersion and deflection. Among the current types of historical theodicies—that is, attempts to invest meaning in the discrepancy between being God's chosen people and the present humiliation in dispersion—Maimonides occupies a unique position. His explanation is neither of the cathartic, nor of the missionary, nor again of the soteriological type.[53] Neither the purification and punishment for old sins, nor the propagation of the seeds of the *logos*, nor again suffering for the sins of the nations so as to redeem the world are for Maimonides the essential rationale of the *galut*. His language is rather sacrificial-martyrological. Israel is constantly called to bear witness. Time and again it brings itself as a sacrifice, *korban kalil*,[54] throughout this long phase of world history.

The last period, namely the messianic age, will finally transform the hostile and implicit recognition of the spiritual primacy of Israel, which most nations share already against their will and word, into a more or less voluntary, explicit recognition of the community of Israel as a most perfect and paradigmatic society. It will be a time

---

Funkenstein, *Heilsplan*, pp. 60–67, esp. p. 66; Chenu, *Nature, Man, and Society*, pp. 174–75.

[53] I have explained this classification in "Patterns of Christian-Jewish Polemics in the Middle Ages," p. 376.

[54] *IT*, ch. 1, p. 30.

of material affluence and security,[55] but not of total equality either among men or nations. The messianic age of Maimonides is in all aspects a part of history, the concluding chapter in the long history of the monotheization of the world. In the Christian medieval horizon there is only one eschatological vision that seems to resemble Maimonides' in this respect: Joachim of Fiore's version of the *tempus spiritus sancti*. But the similarities are only superficial. Joachim's Millennium, even though it is within the boundaries of history, altogether transcends historical processes.[56]

Of course, this strongly contingent interpretation of the vast portions of the revealed law was bound to be challenged. Nachmanides, who by and large speaks of Maimonides with deference, rejects his theory of sacrifices! "I have seen his words . . . they are nonsensical."[57] Did Maimonides relativize the validity of those precepts that he interpreted against the background of a concrete and now bygone historical situation? Maimonides himself never addressed this problem directly, and the problem was to become one of the main issues in the anti-Maimonidean controversy.[58] Should laws be changed? Maimonides, we have seen, insists on the validity of every iota of the law even in the messianic age. He includes explicitly the restoration of the Temple and its sacrifices in the schedule of messianic deeds. Then, as once before, the law will save the masses from a relapse into the superstition to which they are and will remain prone. Maimonides was no *Aufklärer*, and he did not believe in an essential *Erziehung des Menschengeschlechts*, that is, in the capability of the masses to rise to the level of the philosopher.[59]

[55] *HM*, ch. 12, 4: "The sages and prophets did not desire the days of the Messiah in order to rule the entire world, nor in order to tyrannize the nations, nor again so that they be elevated by all people, nor in order to eat, drink, and be merry—but in order to be free for the to'ra and its wisdom, and so that there be no tyranny over them to cause distraction." Cf. *Perush hamishnayot* loc. cit. and *IT*, ch. 9, p. 2.

[56] See Grundmann, *Studien*, pp. 56–118.

[57] Nachmanides, *Perush haramban al hatora*, ed. Ch. D. Chawel, to Lev. 1:9: "and the master said in the Guide to the Perplexed that the reason for the commandments is because the Egyptians and the Chaldeans used to worship the cattle . . . these are his words and he went to great length and they are nonsense (*divre havay*)." Ramban sees in the sacrifices theurgical magical acts; cf. Gottlieb, *Mehkarim besifrut hakabala*, pp. 93–95 and below IV.D.n.70 (typologies).

[58] D. J. Silver, *Maimonides' Criticism and the Maimonidean Controversy, 1180–1240* (Leiden 1965), pp. 148ff., 157ff.; for criticism of the book, Davidson, *Jewish Social Studies* 30.1 (1968): 46–47. It was, of course, part of the controversy over the "hagshama."

[59] Against L. Strauss see my remarks in "Gesetz und Geschichte," pp. 147–78, 162 n.

The respect of the masses before the law is founded on their belief in the law's immutability. Which is not to say that the law cannot be modified at all. Again we have to resort to his doctrine of contingency. A good law, this was already the essence of the Aristotelian doctrine of equity (ἐπιείκεια),[60] must be formulated so as to remain flexible enough to meet changed conditions. It must be precise in its "core" and allow for a "penumbra" of indeterminacy. The absolute immutability of the law may be a necessary fiction for the masses, but the legal experts of every generation have the right and the duty to adjust the law *in casu necessitatis.*[61]

5. Scholastic and Early Modern Parallels

Scholastic thought made ample use of the core of Maimonides' theory, all the more since it corroborated older Christian traditions. The interest in the theory may have been sharpened by the emergence of recent heresies which, as Marcion and the Gnosis once had, abrogated the divine origin of the Old Testament.[62] Maimonides could also be called as a witness to prove that Jews themselves recognize that this law was "good for its times" only; Raymundus Martini argued in this vein.[63] And yet, in spite of its continuous ci-

---

60. Maimonides, I argue there, depicts, e.g., Abraham as already on the height of wisdom; if there is a relative progress, it consists in the taming of superstitions among the masses. For a similar view of the question "an secundum mutationes temporum mutata sit fides" within the Christian horizon (Hugh of St. Victor), see my *Heilsplan,* pp. 52–53.

[60] Cf. Kisch, *Erasmus und die Jurisprudenz seiner Zeit,* pp. 18–26, for the Aristotelian origin of the demand to complement law through equity (to cover the necessary residue of indeterminacy in any legislation). My knowledge of the Arabic sources does not suffice to trace the possible vehicles through which Maimonides might have received the doctrine. Yet Jewish law had similar constructs (*Lifnim mishurat hadin*).

[61] This interpretation is given by J. Levinger, "Al tora shebe'al pe behaguto shel harambam," pp. 282ff. For a different intepretation, Twersky, *Introduction,* pp. 430–70.

[62] Thomas Aquinas, *Summa theol.* 1. q.2 a.98.

[63] Raymundus Martini, *Pugio fidei adversus Mauros et Judaeos* 3.12, pp. 809–10. J. Cohen (above n. 13) accepted my claim that a radical change of pattern occurred in Christian polemics, yet dates it only from the thirteenth century onwards. Raymundus Martini, he claims, advanced this view particularly. I still maintain that the accusation that the Jews adhere to a *nova lex* started to crystallize in the twelfth century, and that Raymundus Martini employed a mixture of two strategies: on the one hand, the claim that the Jews abrogated their law; on the other, the claim that their postbiblical sources implicitly confirm the messianity of Jesus. The latter, too, was a strategy first employed in the twelfth century (Alanus ab Insulis). Cf. my "Patterns," pp. 141–43 and Merchavia, *Hatalmud bir'i hanatsrut* [The Church versus Talmudic and Midrashic Literature], pp. 214–17 (my discovery of the *vaticinium Eliae* in Alanus ab Insulis).

tation, the theory was accepted only with reservations—reservations not unlike those it met in Jewish quarters. William of Auvergne buttressed the theory, on the one hand, with many details from classical sources concerning pagan usages. On the other hand, he insisted, combating idolatry could not have been the only cause for sacrifice: "This cause has no place in Cain and Abel."[64]

Thomas, likewise, accepted the Maimonidean explanation as a partial explanation only, even on the purely literal level of exegesis. Figuratively, the Old Testament, its history and institutions foreshadow the New Testament; literally, the ancient law must be shown to have been adequate for its time. As divine law, it consisted of *moralia, caerimonialia*, and *iudicialia*.[65] The first group coincides with the postulates of natural law, and its demands—the Decalogue—are equally valid at *all* times. The third group specifies and determines those forms of social order (*communicatio hominum ad invicem*) which natural law establishes in principle, but leaves room for the ingenuity of humanity to give them a time- and place-bound concretization. Natural law, for example, dictates the necessity for a division of property (*distinctio possessionum*) and its abrogation for the sake of the common good (*commune quoad usum*); it leaves to the definite law—positive or divine—the exact terms of concretization; in this respect, the biblical institutions are one source of advice and experience among others, but by no means a substitute for the "ingenuity of the human mind" (*adinventio humanae rationis*). This is why Aegidius Romanus could say of the political state that it is partly by nature, partly an artifact. As for the middle group of divine precepts—the *caerimonialia*—they do have, beyond their figurative implications, a general and a particular reason: the right attitude toward God entails the recognition that everything man has, he has from God as first and last principle. This is represented by the

---

[64] *Guilelmi Aluerni episcopi . . . Opera* 1.2, p. 29b: "Septem de causis ante legem, et etiam sub lege sacrificia huiusmodi sibi offeri voluit Deus, non solum propter consuetudinem idolatriae, ut quidam opinati sunt. Haec enim causa in Cain et Abel non locum habet." Cf. Jakob Guttmann, "Der Einfluss der Maimonideischen Philosophie auf das christliche Abendland," in *Moses ben Maimon, sein Leben, seine Werke und sein Einfluss*, ed. Guttmann, pp. 144–54. Nachmanides (Ramban) raised the same objection.

[65] Cf. my "Gesetz und Geschichte," pp. 170–73; on Thomas's systematics see M.-D. Chenu, "La Théologie de la loi ancienne selon S. Thomas," pp. 485ff. It should be borne in mind that the biblical text itself distinguishes between *mitsvot, huqqim*, and *mishpatim*.

sacrifice—remember that Thomas does include allegories, metaphors, and symbols even on the level of literal-historical exegesis inasmuch as they are part of the *intentio auctoris*.[66] But this could have
happened in many ways. The individual, concrete character of the
sacrifices within the Old Testament was dictated by the concrete
historical circumstances of their institution, namely to pull Israel
away from idolatry.[67] In this way Thomas used Maimonides' theory
while removing its sting.

That laws and customs differ according to times and places was
ancient legal wisdom. That mankind slowly accumulates social and
cultural experience has likewise been, of old, a commonplace. Such
topoi gained vigor and color in the reality of the later Middle Ages.[68]
The idea of accommodation could have added a new dimension to
such insights; it could have guided the search for correspondence
and concordances of legal, religious, and political institutions that
express the *qualitas temporum*. Whether the legal schools of the sixteenth century that called for a historical interpretation of Roman
law were actually inspired by the examples of historical explanation
of the sources of sacrifices is hard to determine.[69] The finest hour of
Maimonides' theory came not in the Middle Ages, but in the seventeenth century: the humanists recognized the affinity between
their outlook and his. The first comparative studies of religion were
inspired by him; notably John Spencer's voluminous treatise "On
the Ritual Laws of the Hebrews."

Hardly any ancient or medieval instance of the accommodational

---

[66] Cf. above II.C.3.; IV.B.2.

[67] Funkenstein, "Gesetz und Geschichte," pp. 169–70 (there also a comparison with
Alexander of Hales and William of Auvergne).

[68] Sometimes even in direct leaning on classical accounts of the progress of culture: cf.
Giraldus Cambrensis, *Topographia Hibernica* 3.10, in *Opera*, ed. Dimock, 5:149–50; Funkenstein, *Heilsplan*, p. 54: The ferocious, independent nature of the Irish, their propensity for extremes, is due to their primitivism, to their isolation from culture and civility.
They live "in mundo quodam altero."

[69] Above IV.A.3. Another instance of the secularization of the principle of accommodation during the Renaissance merits attention. Panofsky, *Meaning in the Visual Arts*, pp.
212ff., discusses Vasari's distinction between absolute beauty, the property of recent art,
and "the admirable," a relative beauty that depends on the aesthetic norms and limitations of this or that period, say medieval art and architecture. Panofsky looks for the
origins of this historicizing perspective in the Scholastic distinction between the assertion
per se (*simpliciter*) and *secundum quid*. It seems to me that Vasari's doctrine owes much
more to the historical interpretation that Augustine gave to the Platonic distinction between *aptum* and *pulchrum* (above n. 6). Vasari could have known it directly or indirectly.

interpretation of the Israelite law escaped Spencer's notice. To the work of his predecessors he added a host of classical references of his own. If Maimonides admitted that not all precepts can be given a precise, even if only historical, rationale, Spencer believes that all precepts can be so interpreted.[70] By demonstrating the time-boundedness of every biblical institution he hoped not only to increase knowledge, but also to combat Jews, Catholics, and "fanatics."[71] The Jews of today do not wish to understand their laws, because it will prove them to be anachronistic; the bold beginnings of a historical interpretation made by Maimonides and Abarbanel were, therefore, neglected. The Catholics, it can be shown, took over many of the institutions which had meaning only in their original context and are forced to endow them with fantastic, hairsplitting figural or mystical meanings. The fanatics wish to revive ancient Israelite laws, or even the ancient Israelite "theocracy" in its totality. They have to be taught about the temporal character of the Old Testament as against the divine-eternal character of the New Testament.[72]

But there was more to Spencer's book than erudition and polemics. He differs from his medieval predecessors in method and scope. Instead of mechanical, one-to-one (inverted) correspondences between single precepts and their pagan counter-instances, Spencer tries—but not always successfully—to reconstruct the primitive mentality and the religious imagery that it generates. At his disposal are not only classical authors—most of whom, as he knows well, reflected on the origins of their own religion as an already alien and distant territory[73]—but also the newest accounts from the New World. In this attempt, however feeble, to reconstruct the *mens auc-*

---

[70] John Spencer, *De legibus Hebraeorum ritualibus et earum rationibus libri tres* proleg., pp. 1–18. Cf. Julius Guttmann, "John Spencers Erklärung der biblischen Gesetze in ihrer Beziehung zu Maimonides," in *Festkrift David Simonsens*, pp. 258ff.; Ettinger, "Jews and Judaism as Seen by the English Deists of the Eighteenth Century," pp. 182ff. Spencer, *De legibus* 2.1, pp. 645–49.

[71] Spencer, *De legibus* proleg. 2, pp. 8–12 (p. 10: Judaeos, Pontificos, Fanaticos).

[72] Ibid., p. 12: "Legum Mosaicarum rationes in apricum prolatae multum proderunt ad nonnulla *Fanaticorum* (uti vulgo audiunt) dogmata moresque redarguendos. Ex iis enim nonnulli, Judaeorum sabbatismos docent et ἀεργασίαν; alii, abstinentiam a sanguine et suffocato; sunt et alii, qui Dei (quos ipsi vocant) adversarios, veluti jure zelotarum, perdendi licentiam inferunt, et alia dogmata e mediis Judaeorum lacunis hausta." Like Hobbes, Spencer sees in sectarianism the root of anarchy.

[73] Ibid., proleg. 4, pp. 15–16; 1.4, p. 48.

torum he resembles Spinoza's endeavoring to reconstruct the biblical image of God. Both of them pay particular attention to the political implication of the reconstructed, ancient theology of the Hebrews. The God of the Hebrews had to be cast in the role of a real king, his Temple was a king's abode,[74] the constitution theocratic: only thus could the Hebrews be forced to abandon the Egyptian religion that servitude and oppression had made them assimilate. The Hebrew religion returned to a cult which, albeit more primitive, was also simpler; indeed, the more simple a cult—pagan or nonpagan—the more fundamental the image of the *sacred* it reveals. The concept of the sacred is always associated with the natural, wild, uncultivated, untouched. For this reason the Jews were commanded to build their altar of earth or, later, of untouched, whole stones.[75]

The shift from the mere attempt to reconstruct early institutions, pagan or Jewish, to the attempt to reconstruct the mentality that generated—or called for—such institutions gave the principle of accommodation a new heuristic power. Vico was to exploit it to the limit, and to define it as a new method. There was also, however, another side to the medieval concept of accommodation; it also anticipated the universal-historical usages of invisible hand explanations.

D. ACCOMMODATION AND THE COURSE OF UNIVERSAL HISTORY

1. Providential History: The Bible and Apocalyptic Literature

When and why was history first conceived as a continuous manifestation of God's providence? I apologize in advance for a brief speculative digression into the origins of a distinct sense of history. Most societies of the ancient Near East—indeed most of the ancient societies I know of—traced their origins back to mythical times, to the creation of the world and the human race. *In illo tempore* they, with all of their institutions, were founded by gods or demigods. The

---

[74] Ibid., *Dissertatio de Theocratia Judaica* 1.5, p. 223: "Tabernaculum institutum est, ut Summo regi, palatii et habituculi regi loco, inserviret" (cf. Maimonides, *MN* 3 c. 45). As for gentile temples, Spencer traces their origin to hero worship (p. 828).

[75] Ibid., 2.5, p. 280: "Gentes antiquae, natura vel traditione doctae, naturalia omnia, rudia licet et impolita, sanctiora et Diis suis gratiora, crediderunt." Cf. also 2.6, pp. 281ff.

myth of Ennumma Elish concludes with the foundation of the city-state of Babylon.[1] Very few ancient societies confessed their fairly recent origins; but the ancient Israelites, and to a measure the Greeks, admitted it. Both suffered, in their own, different ways, from an acute historical sense of youth. To the Jahevist, the putative Judean author of the first layer of biblical histories,[2] the founding fathers of the nation lived no more than half a millennium earlier; and the Israelites became a nation (*goy*) barely a quarter of a millennium ago, in Egypt. The origin of Israel took place in history—indeed, in recent history. In and of itself, the consciousness of being a young nation was a burden rather than an asset, a blemish that made a community inferior to others of older pedigree. True nobility is always of oldest vintage. Some of the pejorative connotations of "Hebrew" as persons belonging to a lower class (rather than a distinct ethnic group) were preserved even in the Bible—as, for example, the consistent reference to "a Hebrew slave" against references to the "[free] man of Israel."[3]

But the blemish was turned into a virtue. True, Israel is much younger than most nations, and much smaller. Yet, this circumstance is amply compensated for by the fact that God has *chosen* this particular nation; God has made it into a nation. Historical consciousness and the Israelite version of monotheism went, from the outset, hand in hand. The certainty of being under the continuous, special tutelage of God operating in history compensated for the historical reminiscence of recent origins. All ancient Israelite cultic festivals have a historical meaning grafted onto their original reference to the cycle of nature: the delivery from the Egyptian yoke, the crossing of the desert, the giving of the law.

The pride with which the historian living in the golden times of the united monarchy viewed the God-guided political ascendance of Israel up to his own time contrasts sharply with the sense of doom present between the fall of the northern monarchy (721 B.C.) and the destruction of Judea (586). The discrepancy between the claim of Is-

[1] Pritchard, ed., *Ancient Near Eastern Texts Relating to the Old Testament*, p. 68 l. 47–p. 69 l. 73.

[2] Eissfeldt, *Introduction*, pp. 140–43 and 199–204; North, "Pentateuchal Criticism," in *The Old Testament and Modern Study*, ed. Rowley, p. 81.

[3] Vaux, *Ancient Israel*, p. 83; Alt, *Die Ursprünge des israelitischen Rechts, kleine Schriften*, 1:291–94.

rael to be God's chosen nation and its present powerlessness called for a new justification for the belief in divine providence. The prophets introduced such a new and revolutionary, almost dialectical, theodicy. They inverted the common belief that the measure of the power of a diety is the success of the community obliged to it by the bonds of *religio*. The very powerlessness of Israel is a proof of God's immense power. God's power manifests itself by using the biggest empires, Assyria, Babylon, and Egypt, as a "rod of his wrath" (*mate za'am*; Is. 10:5–8) to purify Israel; yet these world powers are unaware of it (*vehu lo chen yedame*) and attribute their success to their own strength. Here, perhaps, was the earliest, the original version of "the cunning of God" or "the cunning of reason": by following their own, blind urge for power, the nations of the world unknowingly serve a higher design.[4] God still watches over Israel "like an eagle in his nest, hovering over his youngsters"; the whole of history testifies to it; therefore, Israel is called to "remember the days of the world, grasp the years of generation to generation" (Deut. 32:7).[5] Though Israel was made by God into a nation late, God set the boundaries of nation in the beginning of time in view of the (future) Israel; the nations of the world are an instrument in his hand now to punish Israel; and one day in the future he will restore Israel to its pristine glory.

The shift from the pride in past and present achievements to the premonition of future catastrophe and hope of redemption became even more pronounced in the apocalyptic literature and imagery emerging about two centuries before the rise of Christianity. A deep sense of utter alienation, of rejection of this wicked world (αἰών) which is "filled with sorrow and pain"[6] marks the apocalyptic men-

---

[4] Cf. above IV.A.1, C.3.

[5] The song (*ha'azinu*, Deut. 32:1–43) has sometimes been dated early—the "wicked nation" that is to punish Israel for forsaking God identified with the Philistines: Eissfeldt, *Introduction*, p. 227. It seems rather in line with the prophetic theodicy described above, and the catastrophic dimension of the predicted catastrophe fits better to the Assyrians. It seems to be a later didactic-narrative poem, employing, as the song of Moses in Exod., *deliberate* archaisms (32:27, 37, 38). Elements of wisdom literature have been noted by Rad, *Deuteronomy: A Commentary*, pp. 196–200. Rad suggests reading 32:8 *bene elohim* instead of *bene Yisrael* (with LXX and possibly IV Q); but this would suggest a too elaborate myth of the allotment of each nation (except Israel) to a supervising angel.

[6] 4 *Ezra* 4:27, ed. Violet, *Die Esraapokalypse*, 1: *Die Überlieferung*, pp. 37–38; II: *Die kritische Ausgabe*, p. 17, ed. Charles, *The Apocrypha and Pseudepigrapha of the Old Testament* 2:542–624. On the "pessimism" of the apocalyptic literature W. Bousset, *Die Religion des*

tality. Its attitude toward the world is, paradoxically, passive and revolutionary at once. Seldom did it call for violent resistance against the forces of this world, and yet it was the most radical answer to the loss of political autarky since the Seleucid reign: passive resistance against the powers that be, external and internal, was expressed in a new ideology, a new vision of providence. Viewing the entire course of history, from its sinful beginnings to its inevitable end, as the steady though invisible unfolding of a secret, preexisting divine plan "written on divine tablets" had its origin in apocalypticism. A rigid timetable, articulated in distinct periods and subperiods from which Christian historical speculations were later to draw their systems of the periodization of history,[7] left no room to human activity or the notion of a slow, gradual amelioration of the human condition. This version of history underlined the apocalyptical contention that the world "hurries toward its end,"[8] that nothing in it is worth mending or saving. The new aeon, presaged and predicted, will nonetheless be a revolutionary event, coming "like a thief in the night"; but it will come "not by the hand of man." Then, with the new order, "all periods and years will be destroyed."[9] Historicity and change are the marks of the old, bad aeon alone.

The only task of the apocalyptician in this world was the spreading of knowledge of the end. Apocalyptic sects—such as the Qumran community—saw themselves as the avant garde, the representatives of the New World amidst the ruins and tumult of the Old. Their knowledge was their power, their sign of salvation. Indeed, the very knowledge of the end is the surest sign of its imminent ap-

---

Judentums in Späthellenistischer Zeit, ed. Gressmann, pp. 243ff.; Funkenstein, *Heilsplan,* pp. 11–15.

[7] *Ethiopian Ennoch* 81:2, 93:2; *Jub.* 1:29 (Charles, *Apocrypha* 2:163ff). D. R. Russell, *The Method and Message of Jewish Apocalyptic, 200 BC–AD 100,* pp. 108-109. The somewhat similar midrashic image of God as an architect who "saw the *Tora* and created the world" (*Gen. rabba* 1:1) refers to the given canon rather than to a *secret* plan. On passivity as a mark of apocalyptic mentality Rössler, *Gesetz und Geschichte: Untersuchungen zur jüdischen Apokalyptik und der pharisäischen Orthodoxie,* pp. 55ff.; Volz, *Die Eschatologie der jüdischen Gemeinde im neutestamentlichen Zeitalter,* pp. 6, 107, 137. Cf. Daniel 2:45 and Benzen, *Daniel,* ed Eissfeldt, p. 33. Yet the prayer of the just may hasten the end according to some texts (*Eth. Ennoch* 47:1, 97:35, 99:3 and Bousset, *Religion,* p. 248) and at the time of the end the elect will carry "the war of the children of light" against "the children of darkness."

[8] "Quoniam festinans festinat saeculum pertransire," *4 Ezra* 4:26, ed. Violet, *Überlieferung,* 1:36.

[9] *Slavic Ennoch* 65:7–8; *Eth. Ennoch* 91:17; cf. Bousset, *Religion,* p. 244.

proach. Because the apocalytpic visionaries shared with the normative Jewish establishment the conviction that prophecy had ceased
with the end of the ancient monarchy—a conviction that had guided
the process of canonization of biblical books[10]—they were inhibited
from prophesying, and developed, therefore, two alternative
modes of proof for their obsessive expectation of the end. They
either "discovered" old, hidden prophecies or "decoded" the well-
known biblical ones. An apocalypse proper is a prophecy that poses
as a very old prophecy, given to a venerated figure in Antiquity—
Enoch, Moses, Daniel, Ezra—with the instruction to hide it so that
it be rediscovered at the end of days. The fact that most of the events
prophesied in it occurred as predicted is a proof of its authenticity;
the very fact that it was rediscovered is a proof that the end is near,
for only at the end will knowledge about it "multiply."[11] The "decoding" of existing prophecies (*peṣer*) served the same purpose differently.[12] The prophets of old could not understand their own
prophecy; only those who share the right consciousness at the end
of days were shown by the "teachers of righteousness" how to decode each prophecy so as to reveal its real content, namely the
schedule for the last generations and for the end of the world.[13]
Again, the very fact that they possess such a key for the decoding of
prophetic words proves that they, the members of the sect, are in-

[10] Urbach, "When Did Prophecy Cease?" pp. 1–11; *id.*, *Ḥazal*, pp. 502–13.

[11] Dan. 8:26, 12:4–10. The (apocalyptic) prophecy has to be sealed or hidden till the
end; it is not even understood by the prophet himself. Cf. *Eth. Ennoch* 1:2, 108:1; *Ascensio
Mosis* 1:16ff; 4 *Ezra* 4:14, 4:46, 14:6, ed. Violet, *Überlieferung*, 1:28, 46, 406; *Ausgabe*,
2:13, 21, 191. Cf. Rössler, *Gesetz und Geschichte*, pp. 65ff. That "many will roam and
knowledge will multiply" at the time of the end (Dan. ibid.) became a standing *topos*
of Jewish and Christian eschatological readings of history—long after they became
deapocalypticized; cf. Gregory the Great, *Homiliae in Ezechielem*, 2.4.12, Migne, *PL* 76:
980–81; Bernard of Clairvaux, *Ad Hugonem*, Migne, *PL* 182: 1040; Abaelard, *Dialogus*
1.5, Migne, *PL* 188: 1147; Abraham bar Hiyya, *Sefer megillat ha'megalle*, ed. Poznanski,
p. 3.

[12] Elliger, *Studien zum Habakuk—Kommentar vom Toten Meer*, pp. 150ff.; Cross, *The
Ancient Library of Qumran*, pp. 111ff.; D. R. Russell, *The Method*, pp. 178–202. The term
peṣer is the Hebrew form of the Aramaic *pishra* and the older Hebrew *ptr* (Gen. 41:12;
Dan. 2:26). The Bible uses it in the sense of "decoding dreams." That Jesus came "to
solve riddles of old" (κεκρυμμένα ἀπὸ καταβολῆς Matt. 13:35, cf. 13:10), is a mode of
apocalyptic *pesher*.

[13] I *Qp Hab*. 8.1–13 (to Hab. 2:1–3), ed. Heberman, *Megilot midbar Jehuda*, p. 45. It
should be stressed that the "secrets" are not only eschatological-temporal, many of them
are cosmological-uranical. Cf. Gruenwald, *Apocalyptic and Merkavah Mysticism*, pp. 47–
72. This element was taken over by the *Hechalot*-literature and related texts.

deed the "remnants of Israel" and that the transition from the old to the new dispensation is close at hand.

The fascination with the precise succession and structure of historical periods, with the history of the world conceived as a meaningful totality, was an indelible contribution to the Jewish and Christian sense of history.[14] But there was no room in the apocalyptic imagery of the providential course of history for human intiative or evolution, for a *manifest* progress of mankind (as against the *latent* progress of the predetermining divine plan of history). This can also be said of the earliest Christian documents, irrespective of the intensity and importance we may wish to attribute to the apocalyptic elements in them. Christians knew themselves to be already amidst the "New Aeon" anticipated by Jewish apocalypticians. The "kingdom of heaven" had commenced already. But neither the coming of Christ nor his second coming are seen as being prepared by an evolutionary process, not even in the Lukeian version of the history of salvation (*Heilsgeschichte*), which insisted that the community still has tasks in this world. The end may not be as soon; it is still sudden, its exact time unpredictable.[15] The Pauline distinction between the "childhood" under the law as against the "adulthood" in the freedom under the new faith is a juridical rather than a biological-evolutionary metaphor: "because of transgression" the burden of the law was imposed on the Jews so as to enhance their sense of inadequacy and sin, so as to underline the need for salvation. The law, albeit a "tutor toward Christ," is merely a negative preparation;

---

[14] The often-expressed view that the Greek image of history was "cyclical," while the Jewish-biblical or apocalyptic-biblical was "linear" (Mircea Eliade, *Cosmos and History: The Myth of Eternal Return*, trans. Trask, pp. 113ff., 125ff.) has no leg to stand on. The apocalyptic tradition certainly does not exclude the image of an eternal return, at times it even alludes to it, perhaps under the impact of Iranian-Babylonian traditions. Nor indeed does the biblical tradition exclude (in principle) such images—they simply are outside its horizon of discourse. The *uniqueness* of historical events and of history became thematic only in Christianity when, against Origenes' theory of world-successions, Augustine insisted that Christ came only *once* for all times.

[15] Matt. 24:7; 2 Peter 3:10; 4 Ezra 4:34ff.; cf. Norden, *Die Geburt des Kindes: Geschichte einer religiösen Idee*, p. 44 (pagan eschatology); Scholem, "Zum Verständnis der messianischen Idee im Judentum," in *Judaica*, 1:7ff., 19ff., 28ff.; Conzelmann, *Die Mitte der Zeit: Studien zur Theologie des Lucas*, pp. 5ff., 8off., esp. pp. 86, 92 (suddenness). Cf. 1 *Clemens* 23:4, Migne, *PG* 1:259; 2 *Clemens* 11:2, Migne, *PG* 1:343ff.; E. Massaux, *Influence de l'Évangile de Saint Matthieu dans la littérature chrétienne avant Saint Irénée*, pp. 30–31; M. Werner, *Die Entstehung des christlichen Dogmas*, p. 111.

only a new act of grace could make the "slaves" into "heirs."[16] In this sense one ought to interpret the famous distinction, initiated by Paul, of the periods "before the law," "under the law," "under grace"—a distinction of apocalyptic origins.[17]

To the apocalyptic modes of interpretation—discovering prophecies or decoding given texts—the earliest Christian texts added a third mode. From the outset, it seems, the Christian community sought and found symbolic-structural correspondences between the old and the new dispensation. Cain and Abel, Jacob and Esau, Rachel and Lea became such symbolical anticipations of the church and the synagogue. Adam, Moses, and David prefigured Christ— the second Adam, the ultimate deliverer, the true king. So did Melchisedek—king and priest at once, and, like Christ, a priest not from the lineage of Aaron. The Gospels point out time and again how Christ not only fulfilled old prophecies, but also recapitulated, in his life and death, epochal events of the past. The sacrifice of Isaac prefigures the self-sacrifice of Christ, whose high priesthood is antithetical to the Israelite; while the high priest there in Jerusalem expiates the sins of the community by sacrificing something else, Christ sacrificed himself.[18] And the twelve apostles were prefigured

---

[16] Gal. 3:23–25 (παιδαγωγὸς εἰς Χριστόν), 3:19, 4:1ff.; Rom. 6:15ff., 7:1ff.

[17] On the origin of the triad *ante legem, sub lege, sub gratia* cf. *Bab. Talmud, Sanhedrin* 99a; *Avoda Zara* 9a; *Seder Eliyahu Rabba*, ed. Friedmann, p. 6; cf. Strack and Billerbeck, *Kommentar zum Neuen Testament aus Talmud und Midrasch*, 3:826; 4:828; Bousset, *Religion*, p. 247 and n. 1; Hipler, *Die christliche Geschichtsauffassung*, pp. 10–11; Grundmann, *Studien*, pp. 88–89; Schmidt, "Aetates mundi, die Weltalter als Gliederungsprinzip der Geschichte," p. 299; Van der Pot, *Die Periodisierung der Geschiedenis: Ein Overzicht der Theorien*, pp. 43ff. Note that the Jewish tradition often starts "the age of the law" with the fifty-second year of Abraham, while the Christian tradition, at first undetermined, then settles with Moses so as to fit the three *tempora* with the six *aetates*. I have discovered (*Heilsplan*, p. 130 n. 28; "Patterns," pp. 141–42) the first instance in which this Talmudic passage from Sanhedrin was translated into Latin in the attempt to prove, out of the Jewish tradition, the veracity of Christianity: Alanus ab Insulis, *De fide catholica contra haereticos* 3.10, Migne, *PL* 210: 410 ("In Sehale etiam loquitur Elias," etc.). On the career of this *vaticinium Eliae* in the sixteenth century (Reuchlin, Bodin). Cf. Reuchlin, *Augenspiegel, Ratschlag* etc. p. 7b; Bodin, *Methodus ad facilem historiarum cognitionem* 5, pp. 108f., 120; Klempt, *Die Säkularisierung der universalhistorischen Auffassung*, pp. 24 (and n. 45), 67–68. Secularized versions merged with pagan periodizations in Antiquity as well as in the sixteenth and seventeenth centuries; cf., e.g., Bacon, *The Advancement of Learning, Works*, 8.

[18] Hebr. 7:15 (εἰς ὁμοιότητα Μελεχισεδέκ). It has been suggested that the letter to the Hebrews polemicizes against a Christian sect close to the creeds of the Qumran sect, who expected a Messiah from the priestly house of Aaron; Yadin, "The Dead Sea Scrolls and

by the twelve sons of Jacob. The "figure" (τύπος, *figura, praefigura-tio*) and its fulfillment are both historical events, distant in time and place, which stand to each other in a complex relation of identity, contrast, and complementarity, a true *participation mystique*.[19] Though traces of such a symbolical reading (or reenacting) of history can be found in the Jewish, biblical, apocalyptic, or even Midrashic tradition,[20] it became central to the Christian self-understanding to a measure unknown before. The typological reading of history was not a reading of texts and the "decoding" of their symbols, but of history itself. It treated the events themselves—to use the fortunate phrase of Iunilius Africanus—as a "prophecy through things" rather than words: "prophetia in rebus, inquantum res esse noscuntur."[21] Its prominence in the Christian horizon may also be due to the fact that not only Christ's words, but his very life, person, body, and death acquired a central, sacral meaning.

---

the Epistle to the Hebrews," in *Aspects of the Dead Sea Scrolls*, ed. Rabin and Yadin, p. 207. For different opinions see Kümmel, *Einleitung in das Neue Testament*, pp. 349–50 (yet I am not aware of any *historical* typology within the Hellenistic horizon). On typologies in general see Auerbach, *"Figura," Scenes from the Drama of European Literature*, pp. 11–76.

[19] As a genuine symbol, a "figure" is not a linguistic expression or literary metaphor, but a concrete piece of reality—an event, a person, an institution—which, while referring to something else, preserves its own identity and does not dissolve into that which it signifies. For some definitions of "symbol" see, e.g., Wellek and Warren, *Theory of Literature*, pp. 188–90; Fletcher, *Allegory: The Theory of Symbolic Mode*, p. 17. I believe that much of the prevailing uncertainty in distinguishing symbols from metaphors is due to the genetic imparity of the terms. The extraordinary range of connotations of "symbol"—which originally meant just sign—is of fairly recent vintage, while a clear definition of "metaphor" is as old as poetics itself.

[20] The Exodus was always perceived as the pattern of salvation: Is. 11:15–16; cf. Loewenstamm, *Masoret Jetsi'at mitsrayim behishtalsheluta*, pp. 16, 103. The author of 1 Kings 12:28 constructed a negative typological identity between the sacrifice to the Golden Calf at Sinai and in the time of Jeroboam; both generations even used the same formula. The exiles returning from Babylon were supposed to have identified with the last generation of the desert and the first to conquer the land at the time of Joshua: Nehemiah 8:17; cf. 2 Kings 23:21–22 (Josiah). The Dead-Sea-Scroll sect likewise saw itself prefigured in the last generation wandering in the desert (they were "the exiles of the desert"); their eschatological organization was to be according to tribes, families, and camps: Wieder, "The Law Interpreter of the Sect of the Dead Sea Scrolls—the Second Moses," pp. 158ff. and Yadin, "Dead Sea Scrolls." On the Midrash cf. *Gen. Rabba* 40, ed. Albeck, p. 40; *Tanhuma, Leh-Leha* 10; *Vajehi* 10; Heinemann, *Darche ha'agada*, pp. 32–34. The very fact that Christian authors employed the figurative exegesis so vigorously explains, in part, its atrophy in Jewish homiletics and exegesis, but cf. below n. 70.

[21] Junilius Africanus, *Instituta regularia divinae legis* 22, Migne, *PL* 68: 34; Grundmann, *Studien*, p. 37; cf. Thomas Aquinas, *Summa theol.* 1. q.1 a.10.

2. Accommodation and Evolution: Early Church Fathers

Only later was a new sense of gradual, slow, autonomous human progress—a sense, in short, of evolution—grafted onto the interpretation of history as an implementation of a divine plan of salvation. The notion of an accommodation of God's revelation and commandments to the various states in an evolutionary process of mankind was alien to the biblical, and even more alien to the apocalyptic, perspective of history. The notion of evolution was indeed one of those notions with which "Christian philosophy" has "embellished and ornamented"[22]—or, better, transformed—its heritage with an alien conceptualization. Its original home was the Greek traditions of reflections about the origins of culture, law, and society.

Mythical thought, we are assured by Mircea Eliade, is marked by its "denial of historicity." It recognizes, he claims, two modes of time—the changeable, punctual, monochromatic, profane time and the eternally recurring, heroic, or divine time.[23] A genuine sense of history can, therefore, emerge if, and only if, both modes of temporality, the sacred and the profane, collapse into each other. This happened once in ancient Israel, which subsumed past and present events under the providence of God; and it happened once again among the Greeks—perhaps also because they were plagued by the awareness of their relative youth. Greek thought overcame the dichotomy built into the mythical attitude toward time by subsuming the changeable and eternal under *one* notion of time, ascribing to ephemeral events a paradigmatic meaning. Reflecting on the relation of human events to nature, Greek philosophy of culture formulated a variety of evolutionary interpretations of the origins and course of society, culture, religion, law, and language. "Not from the beginning did the gods reveal everything to the mortals; only

---

[22] With these words Origenes (*Contra Celsum* 1.3, ed. Kötzschau, p. 58) defined the intentions of Christian philosophy. Cf. Andresen, *Logos und Nomos*, pp. 69, 206. It is an answer to the contention of Celsus that the barbarians invented theories, but the Greeks are better in defending and employing them.

[23] Eliade, *Eternal Return*, pp. 40ff., 117 (refusal of history). Cf. Dixon, *Oceanic Mythology*, pp. 125ff.; G. van der Leeuw, *L'Homme primitif et la religion* (Paris, 1940), pp. 22ff.; Gelber-Talmon, "The Concept of Time in Primitive Mythus," pp. 201 (Hebrew), 260 (English summary). On the Greek conceptions of history cf. Cochrane, *Christianity and Classical Culture*, pp. 456ff.; H. Arendt, *Between Past and Future: Six Exercises in Political Thought*, pp. 41ff. Cf. also Spranger (above IV.A.n.11).

gradually (χρόνῳ) do these, by searching, find that which is better."[24] Whether law, religion, or even language emerged by nature
(φύσει) or against nature, namely by force or by convention (θέσει);
whether the original state of humanity was happy or brutish—such
were the questions raised systematically by the Sophists.[25] The Stoic
account of the gradual emergence of culture assumed a golden age
in which human capacities developed in accordance with, and by
adjusting to, nature: the division of labor in society to the measure
of what was necessary for survival. Then came the lapse and degen-
eration into luxury.[26] The Epicureans rejected the myth of a golden
age altogether. The gradual formation of the physical world as well
as of society was, in the formulation of Lucretius, a law of nature
(foedera naturae).[27] During the gradual growth of society from wilderness into civilization, propelled by necessity—the struggle for
survival—the balance of happiness and unhappiness remained, according to Lucretius, about the same.[28] Gradual adjustment was the
key concept to many such accounts, such as the adjustment of constitutions to various climates, the adjustment of human goals to human means.[29]

Historical accounts of the origins of Greece or Rome were sometimes enriched by occasional evolutionary insights of this sort. Thucydides traced the gradual development of Athens from a contrac-

[24] Fragm. B 18 (Diels-Kranz); Kirk and Raven, Presocratic Philosophers, pp. 179–80; cf.
Lovejoy and Boas, Primitivism and Related Ideas in Antiquity, p. 194.

[25] Guthrie, A History of Greek Philosophy 3:55–147. On the impact of the "conventional" interpretation of law on Thucydides, cf. above IV.A.n.36.

[26] Reinhardt, Poseidonios, pp. 392ff., esp. pp. 399ff.; id., "Poseidonios über Ursprung
und Entfaltung," in Orient und Occident, ed. Bergesträsser et al.; Pohlenz, Die Stoa, 1:212
(1:79ff., 2:47ff.: cyclical theories). On Varro, De re rustica 2.1.3ff.; cf. Lovejoy and Boas,
Primitivism, pp. 368–69; Reinhardt, Poseidonios, pp. 402–404.

[27] Above III.A.n.22.

[28] Lucretius, De rerum natura 5.826ff. (development of the earth), 925ff. (of mankind),
1454 (paulatim); cf. Farrington, Greek Science, pp. 245ff.; Lovejoy and Boas, Primitivism,
pp. 222ff. (pp. 374–75: influence on Vitruvius, De architectura 2.1). On his concept of (developmental) "laws of nature" Reich (above III.A.n.22), esp. p. 121 (measurability of time).
Already Democritus (according to Diodorus Siculus, Bibliotheka 1.8, Diels, Fragmente 2
[Nachträge] p. XII l. 13–18) demolished the image of an aurea aetas. Cf. Vlastos, "On the
Pre-History of Diodorus." On the balance of happiness and unhappiness Lucretius, De
rer. nat. 5.989ff.

[29] Cochrane, Christianity; Mazzarino, Das Ende der antiken Welt, trans. Joffé, pp. 18ff.
(Polybius). The political climatology of the Ancients—with or without the dynamics of
changes of constitution—was taken over by several Renaissance authors, e.g., Bodin
(Klempt, Die Säkularisierung, above n.17)

tual agreement among pirates to a state of laws; his point of view was Sophistic.[30] Livius described the assimilation of the alien plebs into the republic as the end of Rome's childhood. He seems to say that, had Rome acquired its *libertas* any earlier than it did, the *secessio plebis* would have inevitably occurred earlier and would have destroyed the nascent state.[31] Organological metaphors of growth were used by Roman authors time and again to show how, guided by prudence (*ratio*), the republic progressed from birth to adulthood (*et nascentem et adultam et iam firmam*).[32]

New Christian versions of the economy of salvation since the second century insisted on the accommodation of the divine providence to the *lex humani generis* of slow growth and gradual development (Irenaeus of Lyons),[33] to the *mediocritas humana* (Tertullian).[34] Tertullian even dared to postulate a mutual process of adjustment between God and mankind. During the successive periods of history, God's justice (*aequitas*) manifests itself by his very adjustment (*adequatio*), by his passions and humanlike attitudes—for which, as we remember, God needed a body.

The systematic exploitation of this new figure of thought varied according to polemical or apologetical needs. At times it served to enhance apocalyptic expectations, at times to curb them; at times it served to stress the continuity of the old and new dispensation against Marcion and the gnostics, at times to stress, against Jews, their difference; it was later instrumental in the construction of a political theology, but was also instrumental in refuting any intrinsic link between Christianity and the Roman Empire. It answered a variety of questions: Why did Christ not come earlier (*quare non ante venit Christus*)? Why were anthropomorphic forms of worship such

---

[30] Above IV.A.n.36.

[31] Livius, *Ab urbe condita* 2.1 (ed. Conway and Walters).

[32] Cicero, *De re publica* 2.3 (ed. Ziegler); K. Fromm, "Cicero's geschichtlicher Sinn," pp. 7ff. Lactantius seems to have been the first Christian author to adopt (and reinterpret) the organological images of Rome's growth, i.e., its comparison to the ages of the individual: *De divinis institutionibus* 7.15, pp. 631–35; cf. W. Hartke, *Römische Kinderkaiser: Eine Strukturanalyse römischen Denkens und Daseins*, pp. 393ff. In the life of Carus of the *Historia Augusta*, he sees already a pagan answer to Lactantius: Häussler: "Vom Ursprung und Wandel des Lebensaltervergleichs," pp. 313ff., esp. p. 314 (Seneca, Lactantius). Later adaptations: Paulus Orosius, *Historiarum adversus paganos libri VII*, 2.4, ed. Zangemeister, p. 91; Jordanes, *Romana et Getica*, ed. Mommsen, p. 3; Funkenstein, *Heilsplan*, p. 102 (Otto of Freising).

[33] Below n. 36.

[34] Below n. 44.

as sacrifices permitted? Why are there so many different forms of worship in the Church? Should religious innovation be tolerated or even encouraged? With some of these expressions of the principle of accommodation I have dealt already. I shall confine the balance of the chapter to its role in giving meaning to history at large.

Irenaeus reacted to the attribution of the Old Testament to a bad angel. Marcion may have just drawn the logical conclusion from Paul's insistence that the law given to Moses was a παιδαγωγὸς εἰς Χριστόν in a negative sense only—that the "burden of the law" was meant to increase Israel's sense of inadequacy.[35] Irenaeus invested the evangelical preparation under the law with a positive meaning. The transition from one dispensation to another could not have come suddenly: man had first to get used to the one in order to be ready for the other; his progress is always step by step (ἠρέμα προ-κόποντος).[36] Against the gnostics, Irenaeus used their own weapons, inverted their own images. True, there is evolution—but not in the realm of the divine (plēróma), rather in the history of man. "Irrational in every way are those who do not wait for the fullness of time and ascribe their own weakness to God. . . . transgressing the law of the human race, they want to be similar to God the maker before they [even] became fully human."[37] Man can advance only slowly to become "God's image and likeness," adjusting to successive revelations in time.[38]

[35] Harnack, Marcion, Das Evangelium vom fremden Gott, pp. 30ff., 106ff.; M. Werner, Die Entstehung des christlichen Dogmas, pp. 201–207. For my argument it matters little whether Marcion was, or was not, a gnostic. For Harnack he was a hero before his time: the history of Christianity is the history of its gradual emancipation from the Old Testament.

[36] Sancti Irenaei episcopi Lugudunensis libri quinque adversus haereses 4.63.2, ed. Harvey, 2:296 (lex humani generis); cf. also 2.33.2, 1:330 (in conjunction with his doctrine of recapitulation). He was called, for good reasons, "the first Christian philosopher of history": Kamlah, Christentum und Geschichtlichkeit: Untersuchungen zur Entstehung des Christentums und zu Augustins "Bürgerschaft Gottes," p. 113; see also Funkenstein, Heilsplan, pp. 19–22.

[37] Irenaeus, Adv. haer. 4.63.3, ed. Harvey, 2:297. God "is" while man "becomes": Adv. haer. 4.21.2, ed. Harvey, 2:175; cf. Slomkowski, L'État primitif de l'homme dans la tradition de l'église avant St. Augustine, pp. 53–54; Prümm, "Göttliche Planung und menschliche Entwicklung nach Irenaeus Adversus Haereses," pp. 206ff.; A. Benoit, Saint Irénée: Introduction à l'étude de la théologie, p. 228; Bengsch, Heilsgeschichte und Heilswissen: Eine Untersuchung zur Struktur und Entfaltung des Hl. Irenaeus von Lyons, p. 125.

[38] Man, unlike God, cannot be ἀπ' ἀρχῆς τέλειος (Adv. haer. 4.62; 38.2; 38.3) and the question: could God not have perfected man ab initio? is wrong (4.62.1, ed. Harvey, 2:292–93; cf. 4 Ezra 5:43–44). Mankind is perfected by grace, not by nature (ibid.; cf. 4.7, ed. Harvey, 2:153–54: freedom). Therefore man must first follow his nature (4.63.3).

From the earliest traditions of Christianity, Irenaeus inherited the symbolic-typological reading of history.[39] Irenaeus enlarged it in his own way, and grafted the notion of divine accommodation onto it: "Fourfold is the form of animals, fourfold the Gospels, fourfold the disposition of the Lord. Therefore four testaments were given to the human race; one before the Fall under Adam; the second . . . under Noah; the third the law under Moses; the fourth, which renews man and recapitulates all in itself, is through the Gospel."[40] Indeed, Christ himself *recapitulated*, in his life, all the ages, all the history of mankind.[41]

If the continuity from the Old to the New Testament had to be upheld against heretics, discontinuity had to be stressed against the Jews or against those pagans who saw in Christianity nothing but a new version of the older and more authentic Judaism; and in Christians *homines rerum novarum cupidi*. By calling itself "new," Christianity introduced a powerful tension into the ancient and medieval mentality, which tended to venerate old institutions as truly authentic; an ambivalence not unlike the connotations of "revolution" in the American political vocabulary today. Tertullian showed how

---

God could set his example from the beginning; man needed time to emulate and adjust to it—he was yet a child (4.63.2, cf. 1 Cor. 3:2). The "image *and* likeness" of God is not a description of the primordial condition of mankind, but rather of its goal. The *image* is man's nature, the *likeness* his future goal (4.63.2, ed. Harvey, 2:296). Cf. Ladner, *Idea of Reform*, pp. 83ff. Irenaeus drew this view from Theophilus of Antiochia who argued that Adam in Paradise was not perfect, only perfectible; his progress had to be at first protected from outside influences; even after the Fall mankind continued its progress. Theophilus, *Ad Autolicum*, ed. Otto, pp. 124ff.; Loofs, *Theophilus von Antiochien "Adversus Marcionem" und die anderen theologischen Quellen bei Irenaeus*, p. 63 n. 3, p. 410 and *passim*; Kamlah, *Christentum*, pp. 112–14; Bengsch, *Heilsgeschichte*, pp. 189–91.

[39] Woolcombe, "Biblical Origins and Patristic Developments of Typology," in *Essays on Typology*, pp. 39–75; J. Danielou, "The New Testament and the Theology of History," pp. 25–34.

[40] "Qualis igitur dispositio Filii Dei, talis et animalium forma; et qualis animalium forma talis et character Evangelii. Quadriformia autem animalia et quadriforme Evangelium et quadriformis dispositio Domini. Et propter hoc quattuor data sunt testamenta humano generi; unum quidem ante cataclysmum sub Adam; secundum vero . . . sub Noe; tertium vero legislatio sub Moyse; quartum vero, quod renovat hominem et recapitulat in se omnia, quod est per Evangelium, elevans et pennigrans homines in caeleste regnum": *Adv. haer.* 3.11.11, ed. Harvey, 2:50. Cf. 4.34.5, ed. Harvey, 2:216 (ages). The vision of progress is again close to Theophilus (Loofs, *Theophilus*, p. 62).

[41] *Adv. haer.* 4.11.2, ed. Harvey, 2:158 (with reference to Justin Martyr); 4.62: Διὰ τοῦτο καὶ ὁ κύριος ἡμῶν ἐπ᾽ ἔσχατον τῶν καιρῶν ἀνακεφαλαιωσάμενος εἰς αὐτὸν τὰ πάντα, ἦλθε πρὸς ἡμᾶς; Loofs, *Theophilus*, pp. 357ff.; E. Scharl, *Recapitulatio mundi: Der Rekapitulationsbegriff des Hl. Irenaeus und seine Anwendung auf die Körperwelt* (Freiburg 1941); Bengsh, *Heilsgeschichte*, pp. 107, 120.

the Mosaic law was neither the first nor the last stage in the gradual perfection of the *disciplina* throughout history.[42] All laws made by and for man have to change *pro temporibus et causis et personis*;[43] the divine law also changed gradually (*per gradus temporum*), "because human mediocrity was incapable of accepting everything at once."[44] In his affections and acts, in his mercy, his wrath, as in his laws, God adapts himself to the human condition.[45] Innovations, if in the way of truth, should not be feared: Christ said of himself that he is truth, not custom.[46] Tertullian became a montanist. Even in the age of grace he expected a still more perfect revelation to which Christianity is a slow preparation.

3. Accommodation and Political Theology: Eusebius and Augustine

The age of Constantine called for a real political theology—in the ancient and modern sense of the term. Eusebius of Caesarea exposed the entire history of the world, sacred as well as profane, as a process guided by divine accommodation. He employed both the Stoic and Epicurean accounts of the gradual formation (and deformation) of culture in his description of the gradual restitution of hu-

---

[42] Tertullian, *Adversus Judaeus* 2.9, p. 1343; *De virginibus velandis* 1.4, pp. 1209, 1211. The gradual progress of justice: *De virg. vel.* 1.6; Otto, "*Natura*" und "*dispositio*": *Untersuchungen zum Naturbegriff und zur Denkforms Tertullians*, pp. 194ff. While it is true that Tertullian, unlike Irenaeus, does not see human progress as a *deificatio* (Otto, "*Natura*," pp. 56ff., 210ff.), he does see the history of justice (*aequitas*) as a process of natural adjustment (*adaequatio*) of humanity and God: *Adversus Marcionem* 2.27.7, p. 507; Eibl, *Augustin und die Patristik*, p. 160; Funkenstein, *Heilsplan*, pp. 25–26.

[43] Tertullian, *De praescriptione haereticorum* 24.2, p. 206; *Adv. Jud.* 2.10, p. 1343: "pro temporum conditione . . . in hominis salutem."

[44] "Ut quoniam humana mediocritas omni semel capere non poterat, paulatim dirigeretur et ordinaretur et ad perfectum produceretur disciplina ab illo vicario Domini Spiritu Sancto": *De virg. vel.* 1.4, p. 1209. Cf. *De patientia* 3.9, p. 325 (*mediocritas humana*); 6.1, p. 329 (*mediocritas nostra*) and the next note. Against Marcion's *subito Christus, subito Johannes*, Tertullian emphasizes the gradual shift from the Old to the New Testament (*Adv. Marc.* 3.2.3, p. 510): "Nil putem a Deo subitum, quia nihil a Deo nos dispositum." Cf. *Adv. Marc.* 4.2.4, p. 566; *De carne Christi* 2, p. 874.

[45] Tertullian, *Adv. Marc.* 2.27.1, p. 505 (quoted above p. 43 n. 3). Cf. *Adversus Praxen* 16, p. 1180; *De testimonio animae* 2, pp. 167–78; *De anima* 16, pp. 802–803. On the divine "condescension" Duchaltelez, "La 'condescendance' divine et l'histoire du salut," pp. 593–621.

[46] Tertullian, *De virg. vel.* 1.1, p. 1209. For the career of this quotation down to Gregory VII cf. Ladner, *The Idea of Reform*, p. 138 nn. 29–31. Tertullian, as Anselm of Havelberg in the twelfth century, insists that only with the aid of "innovation" can Satan, who invents new things daily, be fought off: *De virg. vel.* 1.4, 3.2, pp. 1209, 1211.

manity to its original "royal nature."[47] Originally man was, of course, a monotheist. Original monotheism is an old theologumenon: the Bible itself narrates that the first man conversed with God freely, though not always on friendly terms. In Paradise, humanity was still in its childhood phase, and meant to achieve perfection evenly. With the Fall man soon deteriorated into utter political anarchy and—with the exception of a few just men—forgetfulness of God; man lost his true nature to become a brute and grew wildly as a formless matter (ἄγρια ὕλη), "without society, without arts."[48] After the flood, mankind arose, in a slow progress, from polyarchy and polytheism to monarchy and monotheism: religious, cultural, and political levels are interdependent and one cannot lag much behind the other. Without a universal monarchy, the redeeming creed (σωτήριον δόγμα) of Christianity could not have spread all over the inhabited world; and conversely, the Christians, a Church from among all the nations, form a new nation (νέον . . . ἔθνος) without local allegiances or parochial, ethnic patriotism. They are the natural, real citizens of the Roman Empire.[49] In Eusebius's de-apocalypticized, world-oriented eschatology, the Roman Empire was predestined to evolve into a kingdom of God.

His was a much more effective defense of Christianity than older arguments for its tolerance on the basis of give and take (do ut des)— the claim, for example, that the prayer of Christians sustains the Empire;[50] but the closer, new linkage between "Christianity and

[47] Eusebius of Caesarea, In Psalmos 8.7–9, Migne, PG 23: 129. Further references: Cranz, "Kingdom and Polity in Eusebius of Caesarea," p. 51 (and n. 23). The "political theology" of Eusebius was first treated under this term by Peterson, "Der Monotheismus als politisches Problem: Ein Beitrag zur Geschichte der politischen Theologie im Imperium Romanum,"in Theologische Traktate, pp. 88ff.; the term "political theology," renewed in the thirties of this century by Carl Schmitt, has its origin in antiquity: Varro distinguished between mythical, political (civil) and natural theology, above I.A.n.1 (Jaeger).

[48] Eusebius, HE 1.2.19, ed. Schwartz, p. 8: ἀλλὰ καὶ ὅ τε πόλιν οὔτε πολιτείαν, οὐ τέχνας; οὐκ ἐπιστήμας ἐπὶ νοῦν ἐβάλλοντο, νόμων τε καὶ δικαιωμάτων καὶ προσέτι ἀρετῆς καὶ φιλοσοφίας οὐδὲ ὀνόματος μετεῖχον; νομάδες δὲ ἐπ' ἐρημίας οἷά τινες ἄγριοι καὶ ἀπηνεῖς διῆτον . . .See also Cranz, "Kingdom," p. 51 n. 27; Peterson, "Der Monotheismus," p. 89; Funkenstein, Heilsplan, pp. 32–34.

[49] HE 1.2.22ff. ("slow movement back to polity and civilization . . . through the world," Cranz, "Kingdom"); HE 1.4.2; 10.4.19 (universal monarchy and Christianity); cf. Zimmermann, Ecclesia als Objekt der Historiographie, pp. 21–23, 39. The Christians do not live in a "hidden corner of the earth" is probably directed against Celsus; above III.B.1.

[50] Such was the argument of Tertullian or Melito of Sardes. Cf. H. Berkoff, Die Theologie des Eusebius von Caesarea, pp. 53ff.; id., Kirche und Kaiser: Eine Untersuchung zur Ent-

culture," *Thron und Altar*, was not without its dangers. To the generations of Christians nourished with the Eusebian political theology, the crumbling Western Empire meant the end of Christianity or the end of the world.[51] Augustine assured them that it meant neither. His "major and arduous work," *De civitate Dei*, depicted a totally separate origin, progress, and goal to both cities—the city of God "wandering on earth" and the worldly city. The hiatus between them does not necessarily stem from a recognition of different gods, but even from their worship of the one God. It is the difference between *uti* and *frui*, using God and serving him.[52] The earthly city achieves at best an unstable earthly peace. Worldly polities owe their existence not to justice, but to a craving for power (*libido dominandi*) which is satiated only with the advent of a universal empire.[53] But power has its own dialectics: where it succeeds most it evokes the strongest resistance. The *pax terrena* achieved by the Roman Empire is a temporary condition only. Whether the Empire is, or is not, Christian is of no great significance from the perspec-

---

stehung der byzantinischen und theokratischen Staatsauffassung im 4. Jahrhundert, pp. 14ff., 31ff. (Tertullian), 85ff.; Mommsen, "St. Augustine and the Christian Idea of Progress," pp. 346ff., 359ff., 368. The expression "Christianity and culture" was taken from Overbeck's seminal book. On other authors who tied the fortunes of Christianity with those of Empire see Funkenstein, *Heilsplan*, pp. 35–36.

[51] Hieronymus, *Commentarium in Danielem*, CCSL 15A, pp. 794–95. On the four monarchies see Swain, "The Theory of Four Monarchies: Opposition History under the Roman Empire," pp. 1ff.; Hieronymus, *Epist.* 123.16.4, CSEL 56, p. 94: "Quis salvum est, si Roma perit?"; cf. J. Straub, "Christliche Geschichtsapologetik in der Krisis des römischen Reiches," pp. 52ff., 60ff.

[52] Augustine, *DcD* 15.7, pp. 460–61; "Et hoc est terrenae proprium civitatis, Deum vel deos colere, quibus adiuvantibus regnet in victoriis et pace terrena, non caritate consulendi, sed dominandi cupiditate. Boni quippe ad hoc utuntur mundo, ut fruantur Deo; mali autem contra, ut fruantur mundo uti volunt Deo; qui tamen eum vel esse vel res humanas curare iam credunt." The distinction between *uti* and *frui* goes back to Philo's distinction between κτῆσις and χρῆσις. Cf. Baer, *Yisrael ba'amim* [Israel among the Nations], p. 49.

[53] *DcD* 19.12, pp. 657–59: peace as a natural law, common to man and beast. *DcD* 19.13: even the earthly city strives for peace and order: "ordo est parium dispariumque rerum sua cuique loca tribuens dispositio." The expansion of the earthly city from families to cities, then to world power can, however, be attained only by force; and the larger the order, the less stable: "qui utique, velut aquarum congeries, quanto maior est, tanto periculis plenior" (*DcD* 19.7, p. 671), because men prefer to live "rather with their dog than with an alien." Nature, which implanted the striving for an earthly peace, also sets its limits. The order of the earthly city originated, and is maintained, by *libido dominandi*, which always meets with resistance: *DcD* 19.15–16, p. 682ff.; *DcD* 3.10, pp. 71–72; and Kamlah, *Christentum*, p. 321; Funkenstein, *Heilsplan*, pp. 46–49.

tive of the city of God.[54] And though the city of God does not impede the *pax terrena* and may even welcome it, it does not identify with it. It is, on this earth, an alien resident (*peregrinus*) and a wanderer. Its periods, events, and heroes, although they *correspond* to the events of a secular history, neither presuppose nor imply them. The parallelism of events, persons, and institutions in both sacred and profane history was taken by Eusebius as a sign of their interdependence; to Augustine, it manifests their difference.

A strong revival of apocalyptic motifs marks the *DcD*. The total dualism of society and history reminds us of the separation between "the city of the vain" ('*ir shav*) and "the holy community" ('*adat kodesh*): the latter was seen by the Qumran sect as the avant garde of the new αἰών in the midst of the old, the *verus Israel*.[55] But, unlike the apocalypticians, Augustine emphasized, against the Donatists, that one cannot know until the end of days who will be saved; the Church on earth is an inseparable mixture of "wheat" and "chaff" for the duration of its pilgrimage.[56] In what sense, then, is the *civitas Dei peregrinans in terris* one and the same city as the *civitas Dei coelestis*, with its stable population of angels and saints? Augustine warns us not to view them as *two* cities, even if it seems so; and it seemed so to some of his interpreters. Indeed, Augustine's language changed from a philosophical one in his youth to a juridical one in the *De civitate Dei*. Then he spoke of *duo genera hominum*; now he speaks of a city in the sense in which, say, Cicero defined a *res publica*. Its citizens share a *consensus iuris*, both those who will always remain in it and those who will lapse or be found wanting.[57]

---

[54] *DcD* 5.25, p. 119; cf. *DcD* 19.17, pp. 683–85. Fuchs, *Der geistige Widerstand gegen Rom in der antiken Welt*, pp. 23–24, 90ff.; less pronounced is the distinction between sacred and profane history. Wachtel, *Beiträge zur Geschichtstheologie des Aurelius Augustinus*, p. 68. Augustine counts the blessings that accrued to the Empire from Christianity; he also says that the Church uses the "earthly peace" (*DcD* 19.17, pp. 684–85); but he never indicates, as did Eusebius or Prudentius and even Orosius, that the Church needed a universal monarchy so as to expand and thus fulfill its mission.

[55] Flusser, "The Dead-Sea Sect and Pre-Pauline Christianity," pp. 220ff., 227–29; cf. 1 Q p Hab. 10.1.10, ed Heberman, p. 47 (*ir shav-adat sheker*). Cf. above n.13.

[56] Augustine, *Enchiridion* 118, ed. Scheel, p. 73; *DcD* 22.18, p. 837 (the growth of Christ's body). Augustine polemicizes against the Donatists' notion of a *corpus bipartitum* (Tychonius), and replaces it with his notion of a *corpus permixtum*. Cf. T. Hahn, *Tychonius-Studien: Ein Beitrag zur Kirchen- und Dogmengeschichte des vierten Jahrhunderts*, pp. 37, 57–60. P. Brown, *Augustine of Hippo: A Biography*, pp. 212–25.

[57] Augustine, *De vera religione* 27.50, p. 219 ("duo genera . . . quorum in uno est turba impiorum . . . in altero series populi uni Deo dedicati"). On the Pauline terminology.

Down to the first presence of Christ, both cities developed as one *civitas permixta* in analogy to an organism: the first five ages of the world (*aetates mundi*),[58] shared by both cities, were prefigured in the five days of creation and correspond to the first five ages (infancy, childhood, youth, adulthood, middle age) in the life of each individual. These ages have a canonical limit of years—not so the sixth age. In the history of the world, the last age began with Augustus; since then, the world simply "grows old" in slow decay. In the history of the chosen people, it began with the coming of Christ, the second Adam—just as the first Adam was created on the sixth day. The progress of the city of God from now on is not comparable anymore to biological processes: it runs through *spirituales aetates*, and is measurable "not by years, but with advancements" (*non annis, sed provectibus*).[59] Each of these periods of the world has its own beauty (*pulchritudo sua*); each had institutions and theophanies accommodated to the level of humanity at the time—I shall return to Augustine's discourse on that matter shortly. At the end of history, its entire course will reveal itself as a magnificent, harmonious melody—*velut magnum carmen*.[60]

Augustine's various divisions of history into periods—the three time spans (*ante legum, sub lege, sub gratia*),[61] the seven ages (*aetates*)—

F. E. Cranz, "The Development of Augustine's Ideas on Society before the Donatist Controversy," *Harvard Theol. Rev.* 58 (1954): 255ff. *DcD* 2.21, pp. 52ff.; 4.4, pp. 101ff.; 19.21, pp. 687ff. (Cicero, *De re publica* 1.25, ed. Ziegler, pp. 24-25); F. G. Maier, *Augustin und das antike Rom*, pp. 189-90. Augustine emphasizes that one can speak "only of two, not of four cities": *DcD* 12.1, p. 355. I elaborated the interpretation in *Heilsplan*, pp. 45-48: the *civitas Dei coelestis* and the *civitas Dei peregrinas* (and, symmetrically, the earthly city and the *civitas diaboli*) are each *one* in respect to the *will* of its members to belong to it (Cicero's definition of the state: *consensus iuris*). They are *two* only in respect to the *ability* of its citizens to remain there: those who are the citizens of the *Civitas Dei coelestis* (angels) will remain there forever; many of those who belong now to the *civitas Dei peregrinas* will change their citizenship. Another solution Kamlah, *Christentum*, pp. 133ff.

[58] Augustine, *De vera religione* 26.49, pp. 187-260, esp. p. 218; *De Genesi contra Manichaeos* c. 24, Migne, *PL* 34: 193-94; *De diversis quaestionibus* 44, Migne, *PL* 40: 28; *DcD* 16.3, 12, 43; 18.27, 45, pp. 503, 515, 548-50, 618, 641-43; *Retractationes* 1.25.44, p. 121; Boll, "Die Lebensalter: Ein Beitrag zur antiken Ethnologie und zur Geschichte der Zahlen," pp. 89-145.

[59] *De vera religione* 26.49, p. 218; cf. *De Gen. c. Man.* (last n.): the sixth age is confined *nullo annorum tempore*.

[60] Above IV.c.n.8 (*De div. quaest.* 44). Augustine, in answer to the question "quare non ante venit Christus," continues ibid.: "Nec oportuit venire divinitus magistrum, cuius imitatione in mores optimos formaretur, nisi tempore iuventutis."

[61] Cf. above n. 17.

surely originate in the apocalyptic traditions, Jewish and Christian. His elaborate paralleling of the days of creation and the ages of the world was one of the most influential figurative devices—at least in western medieval Europe—so influential that even Jews adopted it.[62] The image was not altogether new. The apocalyptic visionaries based their calculations of the end on the realization of the metaphor "For a thousand years in thy sight are but as yesterday when it past, or as a watch in the night" (Ps. 90:4). It found its way into the *Midrash* as well as into Christian literature before Augustine.[63] But Augustine also differs from this tradition. He deliberately sought to de-apocalypticize a hitherto apocalyptic image by shifting the significance of the analogy from the *duration* of the world (which nobody knows and nobody should presume to calculate)[64] to the *structure* of history. The days of creation correspond to the world's ages not in the number of years (one thousand for each period) but in the contents of each day of creation. Hence the great care that Augustine took in the elaboration of the figurative details.

4. Prediction without Divination: The Twelfth Century

Original speculations on the meaning of history did not emerge in western Europe after Augustine and Orosius until the eleventh century. Images and *topoi* inherited from Antiquity continued to serve a basically static sense of the insignificance of present events. The symbolical reading of the history remained as a merely exegetical device, a part of the transliteral understanding (*spiritualis intelligentia*) of the Scriptures. Recent history—the present—had no particular characterization: it remained as a part of an undifferentiated sixth age, a truly "middle age" (*medium aevum*) between the first and coming presence of Christ. All that could be said about it was that the world "grows older."[65] Such was certainly the attitude toward

[62] Abraham bar Hiyya, *Sefer megillat ha'megalle*, ed. Poznanski, pp. 14–47; Ramban, *Perush hatora*, ed. Chawel, to Genesis 2:3; Abarbanel, *Perush hatora*, p. 146. On Isaac ibn Latif see Heller-Willensky, "Isaac Ibn Latif," in *Jewish Medieval and Renaissance Studies*, ed. Altman, p. 218.

[63] Above n. 17.

[64] *DcD* 20.7, pp. 708ff.; cf. *DcD* 18.54, pp. 653ff. (against the calculation of 365 years allotted to the Church on earth; T. Hahn, *Tychonius*, p. 79; *Epist.* 199.12.46, p. 284 (Wachtel, *Geschichtstheologie*, pp. 86–87); *De div. quaest.* 58, Migne, *PL* 40: 42–44.

[65] The term "middle age" was sometimes used to designate the time *sub lege*, e.g., Irenaeus, *Adv. Haer.* 4.39, ed. Harvey, 2:233: "circumcisio et lex operationum media ob-

political events; although Church theoreticians did at times employ a historical perspective—even the principle of accommodation—to explain the variety of customs and liturgical forms in the one Church, but no more than that.

The first original conception of recent history in western Europe may have been the "Five Books of Contemporary History" written by Raoul Glaber, a monk of Dijon in the beginning of the eleventh century. He distinguished strictly among the *eventus, signa*, and *figurae* of history.[66] As a mere narrator, he recorded new events as if merely "continuing" the work of his predecessors. Beyond them, he looked for special events, for "signs" of divine intervention and of the coming end of history, particularly around the turn of the Millennium; connecting and interpreting, he looked for structures—*figurae*. Of the latter, he found particular delight in Augustine's six-days scheme and in the *quaternitas*—the correspondence between the paradisical rivers, the Platonic virtues, the Gospels, and periods of history. The fourth virtue, that is, justice, comprises all others and is their *fundamentum et finis*: history until his days is the history of the earthly implementation of Justice, culminating in the empire of Charlemagne. The structure is not new: we found it in Irenaeus; Ambrose applied it exegetically; Glaber at times repeats the latter verbatim.[67] But Glaber used the scheme to interpret the

---

tinuerunt tempora"; Randolfus Ardens, *Homiliae* 2.13, Migne, *PL* 1552042c: "Fuit autem lex ordinata, id est statuta medio tempore inter tempus legis naturalis et tempus gratiae." The secularized version of the term may be Bacon's source (above n. 17). On further connotations down to the humanists' usage Huizinga, *Zur Geschichte der Begriffes Mittelalter*, pp. 213ff., 226–27.

[66] Rudolfus Glaber, *Historiarum sui temporis libri quinque* praef., 1.1.4, in *Raoul Glaber, les cinq livres de ses Histoires*, ed. Prou, pp. 1, 5, and *passim* (*eventus, novitates*); cf. Glaber, *Vita sancti Guillelmi abbatis Divionensis*, Migne, *PL* 142: 718c; *Hist.* 4, praef., p. 90 (*signa*); cf. Vogelsang, "Rudolfus Glaber, Studien zum Problem der Cluniazensischen Geschichtsschreibung," pp. 57–60 (a summary of the dissertation appeared in *Studien und Mitteilungen zur Geschichte des Benediktinerordens*, pp. 25–38, 277–97); *Historia* praef., 1.1.2, p. 2 and *passim* (*figura*). On the structure of the book Sackur, "Studien über Rudolfus Glaber," pp. 377ff.; Funkenstein, *Heilsplan*, pp. 77–84.

[67] *Historiae* 1.1.2–3, ed. Prou, pp. 2ff.; Ambrosius of Milan, *De Paradiso* 3.19, ed. Schenkel, p. 277: "In his ergo fluminibus quattuor virtutes principales quattuor exprimuntur, quae veluti mundi istius incluserunt tempora," etc. Ambrose ties it also as once did Philo, *Legum allegoriae* 1.19ff. (63–67), pp. 187ff., to the cardinal virtues. In the same passage, Glaber refers also to the *sex aetates*, but not enough to make him, as Vogelsang ("Glaber," pp. 24ff.) wanted, into an Augustinian. On Glaber's symbolical propensity (without reference to Ambrose) Rousset, "Raoul Glaber," p. 12.

singularity of recent history; this was new, a premonition of the mighty intellectual trend that followed.

No century of medieval historical thought was as productive, as innovative as the twelfth century. Its enchantment with new and richer periodizations, with new and richer figurative perspectives, was not art for art's sake; nor did it merely add details to an existing framework. The war of investitures, the Reconquista, the Crusades, had awakened a sense of the specific significance of recent history, of the deeds and events of the *moderni* within the history of salvation. From Rupert of Deutz through Geroh of Reichersberg to Joachim of Fiore, the so called "symbolists" discovered that recent history was as worthy of exegesis as that of the Old and New Testament; and they dealt with the totality of history in a manner reserved hitherto to scriptural exegesis only. The symbolic-typological interpreter justified his knowledge with an oft-quoted verse (Dan. 12:4): "many shall pass, and knowledge will multiply";[68] he believed that the structure of history becomes gradually more transparent the closer we approach its end. He did not need new prophecies or visible signs, as earlier generations did, to foretell the course of the remaining future: in his symbolic inferences he discovered a method of *prediction without divination.*

Of all the new intellectual movements of the twelfth century, this "speculative biblicism" was the most conservative in method and the most revolutionary in its consequences. The method found its consummation in Joachim's *Concordia veteris ac novi testamenti.* To each person of the Trinity he allotted one period of history: the period of the Old Testament was the period of the Father; of the New Testament that of the Son; of the coming *evangelium eternum* that of the Holy Spirit. Each period prepares the next one, each period reveals the events and persons of the previous one on a higher level. His teachings were proclaimed by some radical Franciscans to be the very *evangelium eternum* that he anticipated.[69]

So intense was the fascination with the typological reading of history that it sometimes did not fail to impress Jewish exegetes—even

---

[68] Above n. 11.

[69] Benz, *Ecclesia spiritualis: Kirchenidee und Geschichtstheologie der franziskanischen Reformation*, pp. 244–55. On Joachim and the symbolistic tradition, Grundmann, *Studien*; Dempf, *Sacrum Imperium*, pp. 229–68 (he coined the term "symbolism of the 12th century"); M.-D. Chenu, *Nature, Man, and Society*, pp. 99–145.

though it was, of the four senses of the Scriptures, the one most recognizable as specifically Christian. Nachmanides, as Abraham bar Hiyya before him, borrowed the Augustinian parallelism between the days of creation and the periods of history, and added a host of additional figures to prove that in their deeds and events, the patriarchs prefigured Jewish history just as God, in the six days of creation, prefigured world history. Nachmanides even borrowed the technical term for such prefiguration from Christian exegesis: he calls them *dimyonot* or *tsiyure devarim*,[70] and ascribes to them a predetermining force stronger than I can find in any other text, Christian or Jewish. He did not cover the traces of origin of his method well; a later anti-Christian polemicist, Yair ben Shabetai of Corregio, put the words of Nachmanides into the mouth of the Christian adversary.

At times linked with the new typological interpretations and at times separate from them, the principle of accommodation was likewise employed to interpret anew the totality of history, to express the sense of new achievements and of the specific place of the present in light of historical retrospection. In the notion of accommodation, the twelfth century found a rational interpretation of the trends leading toward the present and beyond it. Two examples may suffice.

Anselm of Havelberg described the operation of God as *pedagogice et medicinaliter* in words taken from Gregory of Nazianz: between the two "revolutions of life" (*transpositiones famosae vitae*), dividing the periods *ante legem, sub lege, sub gratia* and coming together with cosmic changes (*cum attestatione terraemotus*), and again the third revolution expected to end history, humanity has been "slowly adjusted" (*paulatim usa est*) through additions, subtractions, and changes.[71] With this argument, Anselm justified to a point the di-

---

[70] Ramban, *Perush hatora*, ed. Chawel, 1:279 (preface to Exod.): the occurrence of the fathers, as prefigurations (*tsiyure dvarim*), are "of the category of creation"; 1:77 (to Gen. 12:6): "And know that every divine decision, whenever it turns from the potentiality of a decree to the actuality of similitude (*dimayon*, prefiguration), such a decree will be fulfilled under all circumstances." For further examples and detailed discussion see Funkenstein, "Nachmanides' Typological Reading of History," pp. 35–59; trans. in *Studies in Jewish Mysticism*, ed. Dan and Talmage, pp. 129–50.

[71] Anselm of Havelberg, *Dialogi* 1.5, Migne, *PL* 188: 1147; partly using verbatim Gregory of Nazianz, *Oratio* 5.25; cf. Ghellinck, *Le Mouvement théologique du 12ᵉ siècle*, p. 376 n. 8 and Funkenstein, *Heilsplan*, pp. 184–86 n. 75. Cf. also above IV.C.4.

versity of the eastern and western churches and suggested a slow period of preparation for a reunion. Anselm extended the argument later to justify the confusing multiplicity of orders, and the diversity of religious movements (*quare tot novitates in ecclesia hodie fiunt*). They are a sign of vigor and a necessity at this particular stage in the development of the Church. He distinguishes seven such smaller periods—later on, the Joachimites will call them *aetatunculae*[72]—corresponding to the seven seals of the Apocalypse; of these *status ecclesiae*, he sees the present Church in the fourth, characterized by the strategy of indirect approach of Satan—not through heathens and heretics, but through *falsi fratres*. The accommodated strategy of the Holy Spirit must consist in an intensification of the *varietas*, in ever new incentives and reforms, in ever new variants of *viri religiosi*, of new religious movements. "This variety was not made due to the mutability of God, but rather due to the changing weakness of the human race and its change from generation to generation."[73]

His contemporary, Otto of Freising, combined the opposites: the Eusebian "Imperial theology" with the Augustinian two cities. Augustine, we remember, used organological metaphors in order to separate radically the course of the *civitas Dei peregrinans in terris* from that of the *civitas terrena*; the *procursus* of both is independent even if synchronic. Otto of Freising accepts and deepens the Augustinian analysis of the origins and course of political power. The cycles of earthly politics, discernible in the four major *translationes imperii*,[74] are prompted by the very *libido propagandae dominationis*—

---

[72] Joachim of Fiore, *Enchiridion in Apocalypsim*, in *Joachim von Floris und die joachimitische Literatur*, ed. Huck, p. 290. On the history of the Apocalypse-exegesis, and on Anselm's place in it, see Kamlah, *Apokalypse und Geschichtstheologie*, pp. 66–70.

[73] "Facta est autem haec varietas non propter invariabilis Dei . . . mutabilitatem, sed propter humani generis variabilem infirmitatem et temporalem mutationem de generatione in generationem": *Dialogi* 1.13, Migne, *PL* 188: 1160a–b. Cf. Kamlah, *Apokalypse*, p. 69; Berges, "Anselm von Havelberg," p. 52. Anselm may have remembered the remarks on change and variety in the Church as an adjustment to the *fragilitas humana* in Walahfrid Strabo (above IV.C.1. and n. 10). Tertullian spoke of the *mediocritas humana*.

[74] Otto of Freising, *Chronicon*, ed. Hofmeister, 1 prol., p. 7; 3 prol., p. 134; 4.31, p. 223; 6.36, p. 305; 7 prol., p. 308. He speaks of further *translationes* within the Roman Empire (to the Greeks, Franks, Ottonians); Goez, *Translatio imperii: Ein Beitrag zur Geschichte des Geschichtsdenkens und per politischen Theorien im Mittelalter und in der frühen Neuzeit*, pp. 112ff.; Otto even speaks once of *retranslatio* (6.22, p. 285.15; unnoticed by Goez). The *translationes* are a movement from east to west (1 prol., p. 8; 5 prol., p. 227; 5.36, p. 260). Otto emphasizes, in each of them, the slow decline of the one in preparation of the ascent of the other.

Augustine spoke of the *libido dominandi*—to which civil society owes its existence, expansion, but also decay. Only *terrore* could men be forced from their brutal, primitive existence in solitude into societies;[75] without it, no crafts or sciences would evolve. Will for power generated the *pax terrena* down to the formation of world monarchies (since the greed for power is insatiable). It also generates equal antagonism which will dissolve them from within; the apex of might is the beginning of its slow decline. Note the principle of heterogeny of ends that already characterizes the origin and immanent mechanism of the *civitas terrena*. Aiming at one thing, another is achieved—even by the Church. These *corsi e ricorsi* of growth, maturity, and decay are as valid for the Carolingian and Salian Empire as they were true of the Persian and Roman. Note that the organological metaphors are less important than the almost mechanical chain of causes, which nevertheless serves a preordained goal; for the history of Israel (in the flesh, and later in the spirit), the history of revelation, and the history of the Chruch are accommodated to these immanent laws of political realities. Christianity needed for its mission a world empire (Eusebius), and had to immerse itself in the dialectics of worldly power (*dominium*),[76] which it had to acquire in ever greater measure since Constantine; each *translatio imperii* added to its might, and when it came to the summit of power (since the struggle over investitures) it entered also the

[75] *Chronicon* 3 prol., pp. 132–33, answering the question "quare non ante venit Christus," contains a long quotation from Eusebius (above IV.D.3 and n.48) concerning the gradual process of humanity from anarchy to monarchy. Describing the foundation of the first, Assyrian, Empire (1.6, p. 44) Otto departs from one of his main sources—Frutolf of Michelsberg's world chronicle (*Chronicon*, ed. Waitz, p. 34), which described a primitive *aurea aetas* prior to Ninus (following Frechulf of Lisieux); Landsberg, *Das Bild der alten Geschichte in mittelalterlichen Chroniken*, pp. 47ff.

[76] *Chronicon* 4 prol., p. 183: "Ego enim, ut de meo sensu loquar, utrum Deo magis placeat haec ecclesiae suae, quae nunc cernitur, exaltatio quam prior humiliatio, prorsus ignorare me profiteor. Videtur quidem status ille fuisse melior, iste felicior." This famous passage is an echo, perhaps, of Hieronymus, *Vita Malchi*, Migne, *PL* 23: 5: "quomodo et per quos Christi Ecclesia nata sit, et adulta, persecutionibus creverit et martyriis coronata sit: et postquam ad Christianos principes venerit, potentia quidem et divitiis maior, sed virtutibus minor facta sit." Otto, however, is convinced of the inevitability of the slow gain of power of the Church at the cost of the Empire: the war of investitures is just the final station in the process. I have argued elsewhere (*Heilspan*, pp. 98–109; differently from Spörl, *Grundformen mittelalterlicher Geschichtsschreibung*, p. 42) that Otto transformed Augustine's *civitas permixta* into a distinction between the *ecclesia permixta* (the Church on earth which is always infested by *reprobi*) and the *civitas permixta* (the Christian state in which regalia and sacerdotalia hold an unstable balance). The latter began with Constantine or Theodosius.

path of its own decline.[77] The accommodation of the Church to the world, albeit necessary, has its price yet to be paid.

5. From Organological to Corporative
Metaphors: Ockham and Dante

Interest in universal-historical speculations, symbolical or rational, abated in the academic environment of the later Middle Ages; history, after all, was not even an independent discipline within the *artes liberales*.[78] The level of reflective historical writing reached by Otto of Freising was not reached again until the seventeenth and eighteenth centuries. Later medieval thought adopted the principle of accommodation for other purposes. The long tradition of the idea was a source for important elements in the construction of legal and political theories.

That laws change according to time and place has been a classical maxim.[79] At times it was buttressed, as mentioned before, by a "political climatology": the temper of peoples changes according to the climate they inhabit, hence different constitutions are best for them. Rome's ideal, moderate climate made an ideal *regimen mixtum* possible.[80] Relativization of laws can also be ascribed to those Roman jurists who favored the distinction between *ius naturale* and *ius gentium* (and not all of them did): they were willing to limit the validity of the universal institution of slavery to the "law of nations" only. In the eleventh century, Gregory VII could surprise his adversary with the unheard-of claim that the pope can institute *new* laws (*novas leges condere*);[81] in the thirteenth century, the existence of a flexible, alterable *lex positiva* was a truism. Laws, we hear from Marsi-

---

[77] *Chronicon* 7 prol., pp. 308–309, esp. p. 309.10: "Verum quia regno decrescente, ecclesia. . . . in presenti quoque in magnum montem crescens in magna auctoritate stare coepit." The Church, during the investiture struggle, combated the Empire "non eius tantum, id est spiritali, sed suo proprio, materiali scilicet [gladio]" (p. 309.2). By "spiritualis gladium" he means the excommunication, by "materialis" the deposition of the king. Both are unheard of (*Chronicon* 6.35, pp. 304.21ff.; cf. Sigebert of Gembloux, *Epistula adversus Paschalem papam*, p. 463). But the Church only hastened an inevitable process (7 prol., pp. 308.ff), at the end of which stands the respiritualization of the Church. Here as elsewhere, Otto constructs teleological as causal explanation: the causal chains amount to a divine plan. Cf. Brezzi, "Ottone di Frisinga," pp. 129ff., esp. p. 160.

[78] Wolters, "Geschichtliche Bildung im Rahmen der Artes Liberales," in *Artes liberales: Studien und Texte zur Geschichte des Mittelalters*, ed. Koch, pp. 160ff., 74ff (Otto of Freising as an exception).

[79] Kelley, "Klio and the Lawyers."

[80] Above n. 29.

[81] Caspar, ed., *Das Register Gregors VII*, pp. 202–203.

lius of Padua, can be added, subtracted or altogether changed not only from one age to another, but even between periods of the same age.[82] Eventually, such scattered insights culminated in Bodin's systematic, historical comparison of all available constitutions,[83] a veritable *historical* account of the *ius gentium*—as Vico would also try to establish later, better equipped to do so.

Moreover, theoreticians since the thirteenth century had outgrown the uncritical use of organological metaphors—whether historical or political. Canonists and legists exchanged the metaphor of the state as a body and its hierarchy as limbs for the definition of the state as an abstract—though no less valid—form of unity. The state (or the Church) is a *universitas*, a corporation; in the new vocabulary, it was a *persona ficta*, a representative entity only, though capable of legal action.[84] Constitutions became, so to say, personality features rather than organic properties. And instead of reference to the biological cycle of groups—the ages of the world, of the Church, of a nation—more and more attention was given to the concrete historical circumstances in which political institutions operated, to the concrete historical differences between periods. Compared to earlier historical reflections which were reviewed, much was gained and some was lost by this shift to the concrete. Gained was an increased sense of the actual historical interdependence of institutions and the events within a period. Lost was the evolutionary sense of an almost necessary sequence of periods.

In both the organological and symbolical periodizations mentioned before, the nerve of the argument was the congruence of events and of sequences of events—be it on teleological, and analogical, or even almost mechanical-causal grounds. But relations, the Terminists argued time and again, should never be hypostatized. Just as there is no absolutely necessary reference point in nature—the center of the universe could be reduplicated, if God so wanted,

---

[82] Marsilius of Padua, *Defensor pacis* 1.11.3, ed. Scholz, pp. 54–56, esp. pp. 54.21ff.: "Quod quidem videre sat est experiencia nota per addicionem et substraccionem ac totaliter in contrarium mutacionem quandoque factam in legibus, secundum diversas etates et secundum diversa tempora eiusdem etatis."

[83] Skinner, *The Foundations of Modern Political Thought*, 2:284–301; Franklin, *Jean Bodin*, esp. pp. 59–79.

[84] Gierke, *Das deutsche Genossenschaftsrecht*, III: *Die Staats- und Korporationslehre des Altertums und des Mittelalters und ihre Aufnahme in Deutschland*, esp. pp. 426–36; Kantorowicz, *The King's Two Bodies*, pp. 302–13. The actual legal employment of the theory lagged considerably behind the theory.

in many worlds—there is no absolutely necessary reference point in history either. We remember—had God wanted it, the savior of the world could have been a stone or a donkey.[85] "Periods," to Ockham, are contingent constellations of conditions and things alterable at any point in time. Whether Ockham analyzes the history of property (*dominium*) or the history of the papacy,[86] he insists on a sharp distinction of periods without any attempt to describe a necessary evolution of periods. But he does seek the *congruentia temporum*.[87] Humanity was permitted to acquire dominion after the Fall— to compensate for the loss of the natural dominion over things Adam had in Paradise; this is a statement of fact, not a necessity. The very freedom of humans is a statement of fact, not a necessity; but due to this fact, no definite political order, no "ideal" constitution— whether temporal or ecclesiastic—can be called "necessary." The Church's duration is guaranteed, should all of its members have lapsed but one remaining old, faithful woman.

Dante's apotheosis of the Roman Empire transformed another strand of accommodation arguments and elevated them to a height never reached before. The "political theology" of Eusebius, repudiated by Augustine, reappropriated by Orosius and, later, by all quarters, saw the congruence of Augustus and Christ; the Empire and Christianity are interdependent. Dante went several steps further. Christ, the embodiment of all human beings, could have been punished in order to expiate the sin of all humanity (embodied in Adam) only by a judge who represented a truly *universal* government.[88] The *felix culpa* of which Ambrose spoke,[89] it seems, has now been transported from the beginning of history to its middle: Au-

---

[85] Above II.D.1. and n. 3. Even though the *Centiloquium* was not written by Ockham, he, like Scotus, agreed, e.g., that God, *de potentia eius absoluta*, could yet save Judas Iscariot—*etsi damnatus*.

[86] Ockham, *Opus nonaginta dierum* cc. 14, 88, in *Opera politica*, ed. Offler, 1:439, 662–63. Cf. Miethke, *Ockhams Weg*, pp. 467–77; Dempf, *Sacrum Imperium*, pp. 512–14.

[87] Ockham, *Octo quaestiones de potestate papae* 3.11, in *Opera politica*, ed. Offler, 1:113: "unde et de voluntate Dei, qui secundum congruentiam temporum cuncta disponit, regnum Filiorum Israel, quod primo fuit unum, postea fuit in duo regna divisum." At question are possible and necessary exceptions to the *optimus principatus*. For a similar expression (*qualitas temporum*) above IV.A.n.25.

[88] Dante Alighieri, *De monarchia* 2.13(11), in *Opere*, ed. Moore and Toynbee, p. 363: "Si ergo sub ordinario iudice Christus passus non fuisset, illa poena punitio non fuisset: et iudex ordinarius esse non poterat, nisi supra totum humanum genus iurisdictionem habens, quum totum humanum genus in carne illa Christi portantis dolores nostros . . . puniretur."

[89] Above IV.A.n.6.

gustus, in whose name Christ was crucified, was nonetheless an ideal earthly monarch. In their penetrating studies, F. Kern and E. Kantorowicz[90] have shown to what extent Dante separated humanity and Christianity, not only in the individual, but also and particularly in the body politic. Augustine also distinguished sharply between the "earthly city," whose natural goal is the "earthly peace," on the one hand, and the "heavenly city," with its peace on the other: unstable and engulfed in violence the one, eternal the other. Dante reversed only the appraisal: an "earthly paradise" by human natural faculties alone is attainable. Humanity is a corporation (*universitas, civitas*). The actualization of all human potentialities is never possible through an individual or a group, but only by all of humanity organized in one body politic.[91] It can be ruled by a monarch who, since he possesses all, needs nothing—and hence never acts out of self-interest. A most remarkable, and little noted, facet of Dante's argument is the attempt to prove that the people of Rome are born to world-rule because, as a corporate body, as a people, they embody the monarchical virtues: the Romans conquered the world not out of self-interest, but in the interest of justice and peace.[92] Dante conveniently overlooks the ancient accusation and the many well-known occasions when they created a desert and called it peace. Rome's history is a truly providential history.[93] Another noteworthy facet of his argument is the emphasis on monarchy as a genuine unity within a variety. Human groups need different governments according to time and climate; the monarchy is the only constitution that permits both unity and variety.[94] More than anyone before him, Dante stressed the autonomy of human history as the endeavor to attain human perfection in a

---

[90] Kantorowicz, *Two Bodies*, pp. 451–95; Kern, *Humana Civilitas*, esp. pp. 7–40.

[91] Dante, *De monarchia* 1.3(4), p. 342; 1.7(9), p. 344.

[92] Dante, *De monarchia* 2.6, p. 356; 2.10, p. 360: "Quod si formalia duelli serrata sunt (aliter enim duellum non esset), iustitiae necessitate de communi adsensu congregati propter zelum iustitiae, nonne in nomine Dei congregati sunt? . . . nonne de iure adquiritur quod per duellum adquiritur?" A duel was conducted without self-interest. The monarch, too, is without self-interests (1.12[14], p. 347). We obtain, then, a symbolic affinity of Christ, the ideal monarch, and Rome as the ideal monarchy: they act for the good of humanity, not their own. Both were saviors.

[93] *De monarchia* 2.8–12, pp. 358–62. The outcome of an (ideal) duel, indeed, manifests God's judgement.

[94] *De monarchia* 1.14(16), p. 349: "quum dicitur, humanum genus potest regi per unum supremum Principem, non sic intelligendum est, ut minima iudicia cuiuscumque mu-

human-made polity. His sacramentalization of the human polity and human history, apart from the fact that he was a layman, makes his thought a veritable, and perhaps the earliest, example of a secular theology.

My few remarks on the later medieval reception and transformation of accommodational traditions are not intended to exhaust the topic, and are largely based on the research of others. I merely meant to stress two features that both link and separate the early medieval from the early modern versions of special, historical (or political) providence. The early Middle Ages thought of both the state and of history in organological terms. Early modern theoreticians often preferred mechanical-physical terms. Between both, later medieval thought preferred corporational terms to discuss politics and the examination of circumstances to discuss historical institutions. Secondly, and more importantly; between the relative, vague autonomy ascribed to the human collective endeavor in the early Middle Ages and the almost absolute autonomy that became the presupposition of early modern political and historical theories, stands, again as a link between both but distinct from both, the serious attempt to define the precise scope of that autonomy, of "the dignity of man." It was the true "Entdeckung des Menschen" (Burckhardt).[95]

E. HISTORY, COUNTER-HISTORY, AND SECULARIZATION

1. The Disdain of History: Sebastian Franck

All accommodational interpretations of history exuded a measure of optimism, of trust in the basic success of the Church, of truth, of spirituality. Decay pertained only to the penumbra of history, to the

---

nicipii ab illo uno immediate prodire possint. . . . Habent namque nationes, regna et civitates inter se proprietates, quas legibus differentibus regulari oportet. . . . Aliter quippe regulari oportet Scythas, qui extra septimum clima viventes . . . et aliter Garamantes, qui sub aequinoctiali habitantes." Dante's respect for national peculiarities corresponds to, and is ultimately grounded upon, his infinite respect for the individual; *De vulgari eloquentia* 1.3, in *Opere*, p. 380: "Quum igitur homo non naturae instinctu sed ratione moveatur; et ipsa ratio vel circa discretionem, vel circa iudicium, vel circa electionem diversificetur in singulis, adeo ut fere quilibet sua propria specie videatur gaudere." While this may not be a precise notion of individual personality (such may not have existed before Romanticism), it is the closest approximation to it I know of in the Middle Ages. But cf. Augustine, *DcD* 21.12.

[95] Burckhardt, *Die Kultur der Reniassance in Italien, Ein Versuch*, in *Werke*, 3:206–41.

affairs of the *gentes* or even of the states in general. In the views of
even the more pessimistic among medieval authors, the Church
could have suffered, and was expected to suffer again, only short in-
tervals of lapses or retrogressions. The reformation movements
changed this perspective radically. Whether the decay of the Church
started, as Luther thought, about the time of Augustine or, as others
did, much earlier, it was a long time indeed: from the vantage point
of any Protestant church or sect the majority of Christians, Catholic
or Protestant, had lived and continued to live in a near total eclipse
of truth. A history of Christianity that, granted that the world nears
its end, consists on balance of more stretches of vice and error than
of piety, can hardly be seen as sacred history. The vestiges of God
and Christ, the life and efficacy of Christianity, are to be sought not
in the public-historical domain, but rather in the private realm of
the single conscience. With the individualization of Christianity and
its eschatology—their *Verinnerlichung*—came a disdain of history.

It found its expression in the historical and other writings of Sebastian Franck.[1] All that matters in the unpredictable, stormy passage of history is the single human being, his struggles, emotions,
motives. If Irenaeus saw the life of Christ recapitulating all the history of mankind,[2] Franck stated it of every single person. "Whoever
sees a natural person, sees all of them. All persons are one person."[3]
All history, even the life and death of Christ, is an allegory of that
which happens to each of us. No person, not even Christ, can suffer
for another, die for another, redeem another. Christ redeems only
in that he is born in each of one of us, "lives, dies, and resurrects."[4]

[1] On him see Hegler, *Geist und Schrift bei Sebastian Franck: Eine Studie zur Geschichte des
Spiritualismus in der Reformationszeit*—still indispensable; Dilthey, "Auffassung und
Analyse des Menschen im 15. und 16. Jahrhundert," in *Gesammelte Schriften*, 2:81–89;
E. Seeberg, *Gottfried Arnold: Die Wissenschaft und die Mystik seiner Zeit*, pp. 516–34; De-
Jung, *Wahrheit und Häresie: Untersuchungen zur Geschichtsphilosophie bei Sebastian Franck*.
On his background see Peuckert, *Die grosse Wende: Das apokalyptische Saeculum und Lu-
ther*, pp. 178–83, 527–29. "Counter-history" is a term used by Biale, *Gerschom Scholem:
Kabbala and Counter-History*; pp. 199–201: Arnold.

[2] Above IV.D.2.

[3] Sebastian Franck, *Paradoxa* nos. 92–93, ed. Wollgast, p. 162: "Deshalb wer einen na-
türlichen Menschen sieht, der sieht sie alle. Alle Menschen Ein Mensch. Es ist alles
Adam. Wer in der Stadt ist, der ist in der ganzen Welt. . . . Die alten haben eben dasselbe
mit ihren spitzen Schuken gemeint, was wir jetzt mit unseren zweischräteligen meinen."

[4] Frank, *Paradoxa* nos. 109–14, pp. 180–92, esp. pp. 187ff.; E. Seeberg, *Gottfried Arnold*, p. 524.

Nicolaus Cusanus had already individualized the theme of divine accommodation. To each of us, he said in *De visione Dei*, God appears in our image and likeness: to the young as a young man, to the old as old.[5] Sebastian Franck's theology can be read as an expanded commentary of this theme. In all climates and religions there were ceremonies, petrified dogmas, a *pontifex maximus*; in all climates and religions there can be found true worshippers, be they "heathen, Turks, or Jews."[6] The truth is expressed not in opinions or ceremonies but in our individual perceptions, sensitivity, life, and acts.

2. Gottfried Arnold as a Counter-Historian

Not everywhere and always was the disdain of history articulated so strongly. On the whole, though, it is safe to say that the Protestant occupation with history tended at first to be critical-polemical rather than systematic-speculative. Both stances, the disdain of history and the sharpening of critical faculties, culminated toward the end of the seventeenth century in the pietistic counter-history of Gottfried Arnold's *Unparteyische Kirchen und Ketzerhistorie*. The term calls for a justification. Counter-histories form a specific genre of history written since Antiquity—it is curious that it has not been identified sooner. It consists of the systematic exploitation of the adversary's most trusted sources against their overt intent: in the fortunate phrase of Walter Benjamin, counter-histories "comb the sources against the grain," as Marxist historiography indeed does to reconstruct the history of the victim rather than that of the victors.

A counter-history of sorts was one Manetho's hostile account of Jewish history.[7] It was based mainly on an inverted reading of biblical passages. Does not the Bible admit that the people of Israel

---

[5] Nicolaus Cusanus, *De visione Dei* 6, in *Werke*, 1:300(215): "Sic si leo faciem tibi attribueret, non nisi leoninam iudicaret, et bos bovinam: et aquila aquilinam. O domine quam admirabilis est facies tua quam si iuvenis concipere vellet: iuvenilem fingeret, et vir virilem, et senex senilem." It is the inversion of the most common *topos* in the critique of religion since Xenophanes, which Judaism and Christianity continued as a critique of idolatry. Above III.B.n.1.

[6] E. Seeberg, *Gottfried Arnold*, p. 527; Weigelt, *Sebastian Franck und die lutherische Reformation*, pp. 44–46. On the comparison of the biblical and Egyptian ceremonies above IV.C.5.

[7] Stern, ed., *Greek and Latin Authors on Jews and Judaism*, 1: *From Herodotus to Plutarch*, pp. 62–86 (Manetho), 389–416 (Apion). Cf. Heinemann, "Antisemitismus," Pauly-Wissowa, *RE*, Suppl. 5.3–43; Levy, *Olamot nifgashim* [Studies in Jewish Hellenism], pp. 60–196.

lived as outcasts in seclusion in the Egyptian province of Goshen; that Moses grew up as an Egyptian; that a riffraff (*asafsuf*)—a mixed multitude (*erev rav*)—accompanied the Hebrews on their way out of Egypt, and that they conquered Canaan by force and drove out its indigenous inhabitants? Indeed, for the Hebrews are neither a venerable nation nor is their constitution authentic and worth preserving. Rather, they started as an Egyptian leper colony, secluded and despised, until they called to their aid the Semitic tribes of the Hyksos, and established a reign of terror for over a century (reminiscent, perhaps, of Joseph). Expelled by Iachmes I, the Hyksos, together with the outcasts, left Egypt, led by a renegade Egyptian priest named Osarsiph (Moses). He gave them a constitution that was, in all respects, a plagiarized, inverted mirror-image of Egyptian mores. Or, as Tacitus was later to say: "Moses . . . introduced new laws contrary to those of the rest of mankind. Whatever is sacred to us, is profane to them; and what they concede, we regard as sacrilege."[8] They conquered Canaan by force and established a commonwealth worthy of outcasts—secluded and disguised by a sense of election—calculated to perpetuate their rebellious spirit and their hatred of the human race (*misanthropia; odium humani generis*). It was an ingenious propaganda. Indeed, Manetho's description of the way in which outcasts preserve their sense of value by constructing a counter-ideology in which their discrimination is interpreted as a sign of special election is strongly reminiscent of what modern sociologists of knowledge describe as the formation of a "counter-identity."[9]

A counter-history was Augustine's account of Roman history in the *De civitate Dei*, which I discussed earlier.[10] Cicero had written his *De re publica* with the intent to show that the history of Rome is the unfolding of *iustitia*: Augustine uses the same Roman sources to show that it is only the history of greed and lust for power. Roman history shows that, "remota iustitia, quid sunt imperia nisi magna latrocinia?"[11] And (to mention but one more interesting example) the Jewish "Narratives of the History of Jesus" (*Sefer toledot Jeshu*),

---

[8] Tacitus, *Hist.* 5.4.

[9] Berger and Luckmann, *The Social Construction of Reality*, pp. 166–67. It is a curious coincidence that the authors chose as their example a putative leper colony.

[10] Above IV.D.3.

[11] Augustine, *DcD* 2.21, p. 52; Fuchs, *Der geistige Widerstand*, pp. 1ff.

written in the seventh century,[12] is likewise a counter-history in this precise sense. It employed the Gospels in order to turn Christian history on its head. Jesus was the son of an illicit affair; he became a magician, a powerful seducer of the masses. The Jewish legal establishment, at the end of its wits, knew no better remedy than to have a man of its own ranks (Judas Iscariot) volunteer to infiltrate the movement in disguise and destroy it. The hero of the Gospels turned into a villain, the villain into a hero.[13]

These few examples should suffice to show that counter-histories form a distinct genre. Earlier I remarked that, in the classical and medieval historical method, the historical fact was viewed as immediately given and understandable, and therefore the eyewitness, if truthful, was seen as the best historian.[14] There was no middle ground between the truth and the falsehood of a historical account, a realm of involuntary distortion. The exception proves the rule. The construction of a counter-history assumes that some truth can be elicited even from a falsified document.

Protestant historiography was driven, from the outset, toward the writing of a counter-history of the Church; unlike earlier exercises in polemical historiography, a new art of historical-philological criticism, already cultivated by generations of humanists, was now well established and available. Gottfried Arnold's "impartial" history of Christianity was indeed a *critical* counter-history. There is hardly anything I could add to its exhaustive and insightful study by Erich Seeberg, except to connect it to my present theme, conceptions of providence and accommodation.

"Heresies are necessary"—*oportet et haereses esse*.[15] These Pauline words acquired, in the Middle Ages, a historical-providential connotation: heresies were the providential challenge to which the development of dogma, and even the rejuvenation of the Church

[12] Kraus, *Das Leben Jesu nach jüdischen Quellen*; Dan, *Hasipur ha'ivri biyme habenayyim* [The Hebrew Story in the Middle Ages], pp. 122–32.

[13] In its second, later part, the *Sefer toldot Jeshu* narrates the early history of Christianity in a similar vein. Peter, a Jewish sage sent to infiltrate and impede the movement from within, succeeds in separating the Church and its customs from the main body of Judaism.

[14] Above IV.A.2.

[15] Grundmann, "Oportet et haereses esse"; *id.*, *Ketzergeschichte des Mittelalters*, in *Die Kirche in ihrer Geschichte*, ed. Schmitt and Wolf, pp. 1–2 (reevaluation of heresy in the Reformation and post-Reformation).

through new religious orders, was the response.[16] Heresies are an incentive through the negation of truth. Arnold reversed the evaluation. Sectarians and so-called heretics were the only historical vestiges of Christianity in the time of its decay.[17] Going "back to the sources," in a religious as well as historiographic sense, he could show that whenever the corrupt establishment defined a movement as heretical, it did so because it abhorred being reminded of the true, spiritual, nondogmatic, and nonceremonial origins of Christianity; that Christianity is incomprehensible and apolitical by the very "scandalous" example of its founder.[18] Gottfried Arnold did not seek reason in history; he put his trust rather in the subterranean, though continuous, instances of defiance of worldly—or accepted—wisdom which are the trademark of martyrs and sectarians alike.[19] The "true" history of Christianity was a secret, private history; not even the appearance of Protestantism changed this basic diagnosis. It may be named the secularization of history in a similar sense to Augustine's demand, against the Eusebian political theory, to sever the history of the world from the history of Christianity. Except that Arnold included the history of the Church in decay— the history of the *saeculum*.

It may be that Franck and Arnold do not represent the middle road of Protestant readings of history, but they did draw radical consequences from a tendency innate in most reform movements to secularize history—the history of empires, laws, and ceremonies. The handiest response that Catholic authors could offer was to choose, from the vast collection of versions of providence in history, those which could most easily be clothed in a more modern and palatable attire. The fiercest attack against the secularization of history and the denial of its providential course was made, perhaps, by Bossuet.

3. Bossuet and La Peyrère

Earlier I discussed the dialectial theodicy of the later prophets: the very weakness of Israel, they claimed, is a sign of God's power

---

[16] Above IV.C.4, D.4 (Anselm of Havelberg).

[17] E. Seeberg, *Gottfried Arnold*, pp. 183–97 (Scripture and early Christianity).

[18] Gottfried Arnold, *Unparteyische Kirchen und Ketzerhistorie*, 1 §9, p. 24; E. Seeberg, *Gottfried Arnold*, pp. 66–68, 150–63.

[19] Jesus himself stood trial as a heretic: E. Seeberg, *Gottfried Arnold*, p. 224; cf. pp. 219ff. (heresy); 176ff. (invisible Church).

rather than of his inability to protect his chosen people. For indeed it is a mighty God who uses as instruments to punish a small nation vast empires—while "they do not know it."[20] Whether or not there is merit to my suggestion that there we ought to look for the ultimate source of all versions of the "cunning of God" or the "cunning of reason" in history, it is evidently so in the case of Bossuet: "there is no human power which does not serve in spite of itself [malgré elle] other designs than its own"; conquerors are but "instruments of divine vengeance." Only superficially, only if one considers small segments of time or "particular causes" is the course of history "surprising."[21] For God does not always intervene in history directly; divine providence consists of the employment of human passions, self-interests, and motives for its own designs. Bossuet goes on to revive, in an idiom enriched by the humanistic historical scholarship of his time, the Eusebian political theology with its strong sense of the accommodation and correspondence between sacred and profane history; down to the apotheosis of the monarchical principle which Bossuet, as is well known, held in its purest absolutist form. To deny providence in history—sacred or profane—is to deny order, authority, morality, to deny the divine itself.[22]

A far more original—and bizarre—response to the secularization of history was developed by a Calvinist of Marrano origins who later converted to Catholicism. The world of letters was unanimous

[20] Above IV.D.I.

[21] J.-B. Bossuet, *Discours sur l'histoire universelle* 2.5; 3.8, ed. Truchet, pp. 198–99 (divine vengeance), 428–29: "Quand les césars flattaient les soldats, ils n'avaient pas dessein de donner des maîtres à leurs successeurs et à l'empire. En un mot, il n'y a point de puissance humaine qui ne serve malgré elle à d'autres desseins que les siens. Dieu seul sait tout reduire à sa volonté. C'est pourquoi tout est surprenant, à ne regarder que les causes particulières, et néanmoins tout s'avance avec une suite réglée." Throughout the discourse, Bossuet stresses first and foremost the divine pedagogy; only the divine pedagogy raised the Jews, a stubborn and not particularly gifted nation, above the level of the gentiles. History is, because of its character as divine pedagogy, a veritable, living *speculum regale*—the *Discours* was written *ad usum delphini*—for it teaches the king the ways of pedagogy.

[22] J.-B. Bossuet, *Politique tirée des propres paroles de l'Écriture sainte* 4.1, ed. Le Brun, p. 178; Keohane, *Philosophy and the State in France: The Renaissance to the Enlightenment*, pp. 251–58, esp. pp. 256f. (self-interest propelling public good). On Bossuet's continuous fight to suppress Simon's biblical criticism see Hazard, *La Crise de la conscience européenne*, pp. 203ff.

in the condemnation of Isaac la Peyrère's *Pre-Adamitae*.[23] When Augustine once spoke of *duo genera hominum* he meant it metaphorically;[24] la Peyrère meant it literally. He based the separation of profane, gentile history and Jewish, sacred, providential history on biological grounds. Adam and his descendants were latecomers to history: long before them existed the race of the pre-Adamites from which most of the gentiles descended. The theory accounted for the discrepancies of the biblical account, whose Mosaic authorship and absolute authenticity la Peyrère was among the first to doubt, again in a systematic critical vein.[25] The theory accounted for the narrow geographical-ethnic horizon of the Bible, for the existence of nations with certified historical traditions long predating the putative biblical chronology *ab mundi conditione*, for the existence of inhabited continents to which the Gospels could not have come. In fact, la Peyrère also constructed a counter-history of sorts: he employed the Bible to prove, *malgré lui*, the vestiges of a history of which the biblical narrator was conspicuously silent. And finally, the theory could, though seldom did until much later, be combined with a better explanation of the presence of evident fossils than the usual reference to them as "sports of nature," *lusus naturae*.[26] Only Jewish history, and the history of gentiles grafted onto the Jewish stock, is true providential history; it will close with the *Repell de Juifs*, their return to the grace of God.

I mention la Peyrère's theory because it leads us back to our starting point, Vico's synthesis of historical knowledge. Vico accepted a mild version of the polygenetic theory—the "world of nations

---

[23] Popkin, *The History of Scepticism from Erasmus to Spinoza*, pp. 214–28. There may be a hidden irony to the name *Pre-Adamitae*. The Adamites were a sect accused of licentious practices (naked ceremonies). La Peyrère could have thought of the Pre-Adamites as their prefiguration; but he does not say so.

[24] Above IV.D.3 and n.57.

[25] Strauss, *Spinoza's Critique*, pp. 64–85. Strauss pays less attention to earlier, reformatory sources of biblical criticism, especially among anti-Lutheran spiritualists. Agrippa von Nettesheim compiled a long list of lost biblical books which Sebastian Franck copied so as to show that the Bible could not have been intended to be the only and ultimate "God's word"; otherwise no part of it would have been lost, for "Gots wort bleibt ewig, is alweg gewesen und wird alweg sein"; *Paradoxa* nos. 47–50; Weigelt, *Franck*, p. 50.

[26] Haber, *The Age of the World: Moses to Darwin*, pp. 277ff., mentions la Peyrère and the palingenetic theory only in passing. It was easier, it seems, to separate the age of the world from the age of mankind (e.g., Buffon, Cuvier: ibid., pp. 124ff., 196ff.) in order to account for geological phenomena. Palingenetic theories were rather motivated by ethnographic considerations.

(*gentes*)" descends from those "giants" who, the Bible mentions, sometimes mixed with the "daughters of men" (Gen. 6:4).[27] More important, Vico sought after a better way to save the "history of the *gentes*" from being altogether devoid of divine providence. Direct providence, he maintained, governs the history of the chosen people, Jews and then Christians; but the indirect providence governs the affairs of nations by the very laws that govern the enfoldment of human societies. Our sketchy review of the many turns and twists of the idea of accommodation in its exegetical and historical implications should equip us for a better understanding of both the originality of Vico's thought and its debt to a long tradition.

F. VICO'S SECULARIZED PROVIDENCE AND HIS "NEW SCIENCE"

1. Vico between Realism and Utopianism

With his "system of the natural law of the *gentes*" Vico attempted to mediate between contradictory political theories. On the one hand he sympathized with the intentions underlying the long tradition of natural law and even the search for ideal states, but on the other hand he recognized that purely theological justifications of social orders rest on wishful thinking. "Philosophy considers man as he should be and so can be of service to but very few, those who wish to live in the Republic of Plato."[1] Far from ignoring the persuasive strength of mechanistic-egotistic accounts of the origin and growth of social orders, Vico maintained as much as Hobbes, Mandeville, or Spinoza that neither social instinct nor the urge for a perfect society are immediate facts of man's original nature. Public benefits, we recall, arise at best out of private vices: "Legislation considers man as he is in order to turn him to good uses in human society. Out of ferocity, avarice, and ambition . . . it makes civil happiness."[2]

---

[27] Vico, *SN* §§ 61 (Giants), 126 (separation of sacred and gentile history), 171, 369–73; Vico also makes it clear that all gentile chronology starts *after* the flood—the discrepancy between biblical and other chronologies was one of the strongest undermining forces to the authority of the Bible. Bossuet, la Peyrère, and Vico fought a losing battle to save it, each of them in his way.

[1] *SN* § 131.

[2] *SN* § 132. Cf. also *Autobiography*, trans. Fish–Bergin, p. 138: "Up to this time Vico had admired two only above all other learned men: Plato and Tacitus; for with an incomparable metaphysical mind Tacitus contemplates man as he is, Plato as he should be."

Society, in other words, is not a natural product, but an artifact; Vico shared with Hobbes the fundamental principle that "we make the commonwealth ourselves."[3] But if it is not a product of human nature, must the social orders and institutions be seen as mere conventions or impositions? Vico denied this Hobbesian consequence and developed instead his doctrine of man's collective, acquired, or "historical nature" as mediating between nature and society.

2. The Dialectics of "Nature" and "Ideal History"

Vico's links to Hobbes and the Hobbesian tradition are manifold. He, too, stresses the epistemological primacy of matters political over the physical sciences: *verum et factum convertuntur*. Civil society is for him as it was for Hobbes a human artifact. Social structures are neither a product of innate social inclinations nor part of the state of nature. In a way, Vico even radicalizes Hobbes's initial polarization of the natural and civil state. Hobbes could not but endow all men, even in *status naturalis*, with a modicum of reason. Indeed, only the faculty of "foresight" accounts for the boundless antagonism within the stateless society. Animals, far from a *bellum omnium contra omnes*, may form natural societies.[4] Antagonism springs from man's knowledge of his vulnerability and of the finitude of natural resources. Foresight breeds acquisitiveness and the urge for protection. But foresight also enables, for the very same reasons, the transition from the natural to the civil state. Man's fear in the state of nature is rational. Against Hobbes, Vico returns to the conceptions of primitive man as a sheer brute: "From these first men, stupid, insensate and horrible beasts [*stupidi, insensati ed orribili bestioni*] all the philosophers and philologists should have begun their investigations of the wisdom of the ancient gentiles; that is, from the giants in the proper sense."[5] The very humanity of man is an artifact.

---

[3] Thomas Hobbes, *Six Lessons*, in *The English Works of Thomas Hobbes*, ed. Molesworth, p. 184.

[4] Hobbes, *Leviathan* 2.17, ed. Macpherson, p. 225.

[5] *SN§* 374. Vico also characterizes the dominant passion of his primordial man as fear—but fear of gods rather than of his fellow men: "Of such natures must have been the first founders of gentile humanity when at last the sky fearfully rolled with thunder and flashed with lightning. . . . Thereupon a few giants, who must have been the most robust, and who were dispersed through the forests on the mountain heights where the strongest beasts have their dens, were frightened and astonished by the great effect . . ." (*NS* §377).

Is this to say that being within civil society means *contra naturam vivere* and that natural law, even in the minimal sense of "dictates of reason" which Hobbes preserved, is a vacuous phrase? Hobbes, I shall try to show later, mediated between nature and society by reference to the natural aetiology of the state. Its origin is dictated by the sense of self-preservation. Spinoza, on the other hand, altogether relativized the distinction between bodies natural and artificial; the state and its institutions, much as any physical compound, are nothing but a balance of forces. Vico sought the mediation between nature and society along a different path. He reinterpreted nature to stand for the very *process* through which man acquires a *second*, social nature. Consequently, he claimed that topical, or historical reasoning is the only instrument with which one can grasp the phases of this process and its driving moments.

"This New Science or metaphysics," Vico announced his program, "studying the common nature of nations [*la commune natura delle nazioni*] in the light of divine providence, discovers the origins of divine and human institutions [*divine ed umane cose*], among the gentile nations, and thereby establishes a system of the natural law of the *gentes*, which proceeds with the greatest equality and constancy [*che procede con somma egualitá e constanza*] through the three ages. . . . The age of the gods . . . The age of the heroes . . . The age of men, in which all men recognized themselves as equal in human nature."[6] The term "common nature" is ambiguous. It stands both for a regular process of development *and* for any of its states, the last of which being the age, if a Marxist term may be used, of "social emancipation" or "true equality." "Natural law" is thus founded neither on social instincts nor on a computation of enlightened interests. It is rather the very immanent, regular, "ideal" process through which civilization emerges time and again as man's acquired, collective "second nature." "The nature of peoples is first crude, then severe, then benign, then delicate, finally dissolute."[7]

Vico names this process of transformation an "ideal eternal history [*storia ideal eterna*] traversed in time by every nation in its rise, development, maturity, decline, and fall."[8] The ambiguous use of

---

[6] *SN*§ 31; on the history of the *topos* of the three ages above IV.A.n.11, D.n.17. (Censorinus, Christian tradition).

[7] *SN*§ 242.

[8] *SN*§ 245.

the term "ideal" corresponds to the ambiguous use of the term "natural." The latter had a formal and material connotation: it stood for
a process as well as for the distinctive features of each of the phases
of the process, predominantly the original phase. The former
stands, again, both for the process and its goal. More precisely, the
"ideal history" does not describe the actual historical course of a definite social entity; rather it measures the actual history against the
(methodological) norm of the regular sequence of the periods of social creativity. These periods and their immanent sequence are truly
ideal types. They stand for the course of a "nation" if imagined in and
by itself, isolated from outer influences.⁹ The "ideal history" is a
limiting case or a necessary fiction, much like Hobbes's state of nature.

"Nature" is not only a given state but also a "process," and what
is more, it is not a process reducible to external "necessities." The
sequence of human social institutions, though initiated by man's environment, is not governed by them throughout. Of itself, although "occasioned" by outer necessities,¹⁰ the human spirit, which
is one in all of its manifestations, emanated all human institutions in
regular phases. The order of man's collective institutions determines the order of his ideas. Here, as so often, Vico extrapolates key
concepts of heterogeneous philosophical systems from their ontological context to use them in the realm of history alone. His theory
of occasioned spontaneity bears reminiscences of the Occasionalists; the principle of *l'ordine dell'idee dee procedere secondo l'ordine delle
cose* is a polemical limitation of Spinoza's axiom.¹¹ The emphasis on
human history as a history of man's spontaneous creativity allows
Vico to challenge the mechanistic-deterministic interpretation of
the foundation (and progress) of civil society. In direct polemics

⁹ The appearance of theologians precipitated, in twelfth-century France, the "prema-
ture passage from barbarism to the subtlest sciences," just as the appearance of the philosophers in ancient Greece (ibid., §§ 158–59); the course of Roman history was partly
determined through the circumstance that it was founded in the proximity of well-de-
veloped cities (ibid., § 160).

¹⁰ *De uno principio*, in *Opere*, 2:55 (quoted above p. 203 n. 3; esp. *sed occasio fuit* etc.).
Cf. also next two notes. Vico's view that only God knows those things which he made,
wherefore our knowledge of physics is necessarily fragmentary, was likewise shared by
the Occasionalists. Cf. below v.A.

¹¹ *SN*§ 238; Spinoza, *Ethica* 2 prop. 7; Vico limits the validity of the assertion to the
cose umane. Cf. Nicolini, *Commento storico alla seconda Scienza nuova*, 1.94.

against Hobbes, he intends to show how "divine providence initiated the process by which the fierce and violent are brought from their outlaw state to humanity. . . . It did so by awakening in them [*con risvegliar in essi*] a confused idea of divinity, which they in their ignorance attributed to that to which it did not belong. Thus through the terror of this imagined divinity, they began to put themselves in some order. . . . This principle of institutions Thomas Hobbes failed to see among his own 'fierce and violent men' . . . and fell into error with the 'chance' of his Epicurus."[12] Between the challenge and the human response Vico inserts the mediation of imagination.

3. Imagination in Construction and Reconstruction

The mediatory function of imagination, immanently developing into "reason," prohibits a purely mechanistic reconstruction of the origins and advancement of "nations." Vico set out to prove that the method of historical inquiry is throughout different from the method of reconstructing physical laws. Assuming the intellect to be uniform in all its "guises," introspection, that is, the imagination turned inwards, became for Vico the main instrument of historical understanding. Vico seems to see in the Prothagorean *homo mensuram omnium* principle a natural mode of cognition, serving the construction of society (imagination) as well as its reconstruction (the understanding of its origin). The principle in this, its second, reflective sense, is misleading if used statically. "It is another property of the human mind that whenever men can form no idea of distant and unknown things, they judge them by what is familiar and at hand."[13] The uncritical, or static projection of one's own image on alien societies or remote phases in *their* own society engenders the "conceit of nations" as well as the "conceit of scholars." But the same principle turns into a constructive tool of understanding if used as a dynamic principle, as a method to reconstruct the phases of society out of the analogy to our own development. Vico thus gave new foundation to the speculative *topos*, discussed above, that the *aetates hominis* are a recapitulation of the *aetates mundi*.[14]

[12] *SN*§§ 178–79; cf. n. 10 above. Vico also gives a historical meaning to Hobbes's *conatus*: §§ 340, 504.

[13] *SN*§ 122; the full quotation above IV.A.2 (and n. 27).

[14] Above IV.D.3 and n. 58 (Augustine).

Employed with discipline, the imagination (or introspection) of the historian is not to be understood as the rational isolation of causes or narrative revivification of the past. It rather means the capacity to reconstruct the mentality, that is, the network of images and corresponding institutions constituting each phase within the "ideal" process of socialization. Imagination was the driving force of cultural transformation, and it is the driving force of the interpretive endeavor to reconstruct them. This interpretive reconstruction, aided with a knowledge of the universal structures of imagination acquired through introspection, has three distinct scopes. (i) Vico's "poetic logic" is, much as primitive thought for Lévi-Strauss, a "logic of the contingent" or of the a posteriori.[15] Imagination responds spontaneously to similar circumstances with the same structuring categories (tropes). Their knowledge enables the reconstruction of the original circumstances—the "occasions"—of images (topoi) and their corresponding social institutions; the fear of the gods, induced by natural phenomena, acts as the "occasion" for the emergence of the family. Such original images are constructive errors, permitted or even evoked by the cunning of providence.[16] (ii) The interpreter will not succeed in rendering meaning to discrete, disparate images, institutions, or cultural instances unless he uncovers the patterns of cross-references underlying them, that is, their context. Imagination allows the interpreter to find, for example, in Zeus the symbol of class structures. Vico assumes that all spheres of human activity in one age express the same mental configuration. (iii) This done, the interpreter will be able to uncover the immanent logic in the transformation of one set of images into another, of one cultural matrix to the next in the line of growth.

Again, as we saw in the case of "ideal" and "natural," Vico seems to profit from the difference and ultimate identity of two connotations of "imagination"—the imagination of the historian and the collective imagination of the society that the historian investigates; because they are all of the same stock, because our mental disposition has not shed all traces of the past (which each of us recapitulates in childhood), we are also equipped to extricate ourselves, in the

---

[15] Lévi-Strauss, La pensée sauvage, ch. 2; cf. also Leach, "Vico and Lévi-Strauss on the Origins of Humanity," pp. 309–17.

[16] Cf. Faj, "Vico as a Philosopher of Metabasis," in Giambattista Vico's Science of Humanity, ed. Tagliacozzo and Verene, pp. 87–109.

imagination, from the present, and place ourselves in an alien, primitive mentality. In this way Vico leads us toward a thorough *historical* meaning of his celebrated identification of *verum* and *factum*. It, too, has two interdependent connotations, an objective and a subjective one. The science closest to us is the science of humanity because "we made this world of nations" ourselves. But in each period *of* history, that which society constructed—gods, laws, institutions—was its truth, was true for its members—as absolutely as is our science to us.[17] There is no other truth than that constructed by man—except, of course, revelation.[18]

4. Common Sense and Providence

In the beginning of this chapter I claimed that the novel, revolutionary, methodological concept that underlies Vico's historical as well as political reasoning is the concept of historical-social *contexts*: each society in each of its successive (ideal) "times" can be determined through some internal, integrating principle rather than, as hitherto, in contraposition to other segments of the historical time. Helped by his imagination, the historian uncovers the "harmony" of an age, the "correspondence" or "accommodation" to each other of all of *cose uomani* in a given period.[19] Historical interpretation is an exercise in contextual reasoning. Yet, Vico's idea of sociohistorical contexts expresses, beyond the insight in the necessity of immanent contextual interpretation, an almost aesthetic category; they constitute an *Einheit in der Manningfaltigkeit*. All manifestations of an age are facets of one and the same mental configuration. Vico uses, in a sense peculiar to him, the term "common sense" to name

[17] *SN*§ 376: "In this mode the first men of gentile nations . . . created things by means of their own ideas, but by a creation which differs infinitely from that of God. For God, in his purest understanding, knows things and creates them in knowing them; [whereas], because of their powerful ignorance, men create by dint of highly corporeal imagination and, this being so, they created with such wonderful sublimity . . . that it perturbed to excess the very men who, by imagining things, created them, whence they are called 'poets' which, in Greek, means 'creators' . . ." (he later quotes Tacitus: "fingunt simul creduntque"). On knowledge by doing cf. below ch. v. On mythmaking Mali, "Harehabilitatsia" pp. 125–92, 236–85.

[18] And with it the history of the chosen people; above IV.E.3.

[19] *SN*§ 32 ("Convenevolmente a tali tre sorte di natura e governi, si parlarono tre spezie di lingue . . ."); § 311 ("tra loro conformi"); § 348; §979 ("mode of the time"). In the following pages I rely on my article "Periodization and Self-Understanding in the Middle Ages and Early Modern Times," pp. 3–23.

this definite mental configuration of each age, the harmonic principle of each period.[20] This, in essence, is what Karl Mannheim was later to call *der totale Ideologiebegriff,* based on the "collective experience" of groups.[21]

In the emphasis on the spontaneous and immanent harmony as the frame of intelligibility, we detect Vico's affinity with Leibniz, even if we discard Vico's adherence to the doctrine of "metaphysical points" as insignificant. Indeed, with his distinction between possibility and compossibility,[22] Leibniz had already developed the logical foundation of the concept of contextual harmony and made it a cornerstone of his metaphysics. If the predicate-in-notion principle (*praedicatum inest subiecto*) is to be taken literally, the monads are contexts of attributes. Monads cluster into "possible worlds," inasmuch as they are, beyond their logical possibility, compossible on the grounds of the principle of sufficient reason.[23] They are a logical, epistemological, but also an aesthetic category of contextual unity.[24]

Vico's "collective mentality of an age" thus has two complementary aspects. On the one hand, it secures the anonymity of historical processes: from now on, heroes, cultural or political, are dethroned from the place they had occupied in genetic accounts. *Zeitgeister* of all kinds are imagined as force fields mightier than the mightiest individual. On the other hand, we remember that Vico stressed time

---

[20] *SN*§ 142: "Common sense is judgment without reflection shared by an entire class, an entire people, an entire nation, or the entire human race."

[21] Mannheim, *Ideologie und Utopie,* pp. 60–64, 154–55. The difference between Vico's *sensus communis* and Bacon's *idola fori* lies, to speak Mannheim's language, exactly in that Bacon's idols have the function of uncovering (collective) errors, that they are not universal or *wertfrei,* that they belong to the *partikulärer Ideologiebegriff* (p. 58). He does not mention Vico in his historical survey (Vico, however, mentions the idols theory of Bacon as a forerunner of his own inquiries). Cf. also Stark, "Giambattista Vico's Sociology of Knowledge" in *Giambattista Vico: An International Symposium,* ed. Tagliacozzo, pp. 297–307 (without discussion of the "common sense"). It is also interesting to note the similarities and differences between Vico's "common sense" and Ibn Khaldùn's *Asabiyah* (*The Muqaddimah,* ed. Rosenthal, 1:lxxvii; 261–65, and *passim*). Both concepts occupy approximately the same positional value in their systems, but the latter indicates rather an emotional configuration, the former an intellectual or mental one. See also Mali, "Harehabilitatsia," pp. 344–71.

[22] Above II.H.3.

[23] Above III.E.2–3.

[24] On the various aesthetic presumptions and influences of Leibniz see, for example, Baeumler, *Das Irrationalitätsproblem,* pp. 38–60 and *passim.*

and again the spontaneity of this collective mentality of an age. By itself, though "occasioned" by outer necessities, the human spirit emanates all human ideas and institutions in regular phases. Because of this universal though ideal regularity,[25] introspection (imagination), became for Vico the main instrument of historical understanding. Again we detect the affinity with Leibniz, whose monads represent, as do Vico's societies, a genuine unity only because they produce their perceptions and apperceptions of themselves: they "have no windows"; they are spontaneous. We remember that Leibniz also gave the most radical epistemological meaning to the logical postulate *praedicatum inest subiecto.*

With his own conception of spontaneity of the collective imagination (or *sensus communis*), or rather in his theory of occasioned spontaneity, Vico believes he has mediated between the reality of man's original, brutal nature and the ideal of eternal law, between Hobbes and Grotius. Society emerged neither by nature nor by convention but by both, since man could and did transform his brutal nature by himself; he acquired a "historical nature." Vico insisted that "natural law" was not based on social instincts or on deliberate reasoning or on necessities (or norms) but on the very immanent, regular, "ideal" *process* through which civilization emerges time and again as man's acquired nature. This is Vico's version of the *List der Vernunft*, of how private vices transform into public benefits.[26] He calls it "providence," a term also standing for the immanent dynamics of the regular transformation of one phase into another. The final phase of human equalization, while envisaged as necessarily monarchical, bears nonetheless all the marks of a successful state in

[25] The lack of a sense for "individuality" in Vico was argued by Meinecke, *Die Entstehung des Historismus*, in *Werke*, ed. Hofer, 2:63–69. Meinecke's sense of individuality consists not merely in insistence on singularity, but involved, at least in the view of those who tried to find a systematic formula for "historicism," an *epistemic resignation*. The concept demands historical constructs to contain "einen starken unvertilgten Rest von Anschauung" which makes them "individuelle Totalitätsbegriffe" (Troeltch, *Der Historismus und seine Probleme*, p. 36). Meinecke recognizes a few "forerunners"; cf. also his "Klassizismus, Romantizismus und historisches Denken im 18. Jh.," in *Werke*, ed. Kessel, 4:264. Vico, although he recognized the different "guises" of social expression and the singularity of each historical phase (for example, *SN*§ 148), could not argue their total individuality precisely because he insisted on the basic similarity of the human collective imagination and the phases of its productivity in every society.

[26] Above IV.A.I.

Spinoza's terms. It is a balance of conscious, enlightened self-interest. But it also reminds us of Dante's vision of a *humana civilitas*.

Thus, in a paradoxical turn of expressions, "providence" came to signify man's emancipation from nature or even from God, the spontaneity of his social endeavors. Here Vico expresses, beyond any particular correspondence we might find between his thinking and this or that contemporary theory, the sense and self-understanding of modern thought which stressed, in endless variation, the autonomy and "dignity of man." Vico's aim is to demonstrate the gradual growth of human independence, that is, the rational determination of man's collective fate. In this sense Vico introduced historical philology, or contextual reasoning, as a means by which to mediate between the mechanical and teleological interpretations, between nature and reason through the concept of history. With the help of this version of a *List der Vernunft*, Vico can reintroduce providence into history and thus resume, on a richer base, a tradition of Christian philosophy of history going back to Irenaeus of Lyons, seeking to establish the correspondence between the divine plan of salvation and the immanent nature of man.

5. Human Autonomy and Spontaneity

Hobbes, Spinoza, and Vico sought to defend and to define existing political institutions on the basis of a realistic assessment of "man as he is, not as he should be." They represent basic prototypes of anti-utopian thinking grown out of the experience of the modern state. Their defense of political realism is based on a radical interpretation of society or the state as a product of human efforts, or labor. Yet, an ultimate difference separates Vico from both Hobbes and Spinoza. Hobbes and Spinoza believed society to be an outcome of rational design, though Hobbes believed only one design could ensure the durability of the state. Vico, on the other hand, believed it to be a product of a long evolution in which the individual or a group could not, or should not, interfere. The "mechanization" of political theory or its rejection thus helped to formulate two basic patterns in the varieties of the conservative ideologies to come. We can call them, in anticipation, the positivistic and the evolutionary defense of existing orders. Hobbes and Spinoza believed that the state needs the mature participation of each of its members; every order, inasmuch as it is genuine order, is worth conservation, yet conserving it is an incessant and conscious task for all. This type of conservatism shares its premises with the utopian or revolutionary ideologies it detests, namely the recognition of the state as a product of design; it merely negates the wisdom of radical changes. Unwillingly, one could say, Hobbes and Spinoza even prepared the rational idiom of revolutions, while Vico, well aware of the dangers implied in regarding the state as a design always in need of deliberate adjustments to new realities, insisted on the anonymous, almost instinctive amelioration of the human condition. In it he saw the working of providence, the "invisible hand" of God or nature.

The medieval principle of divine accommodation was now transformed into a principle of human creativity. Nowhere is it more evident than in the new attitude toward the very notion of God. Cusanus epitomized the medieval tradition (and transformed it) when he spoke of God as appearing to each individual in his own personal human image.[27] Vico did not simply warm up Xenophanes' criticism of Greek religion; he invested it with a profound historical meaning. Our capacity to imagine gods, our constructive imagination, is the only driving force of history, a *fact* that, like society itself, is beyond truth or error. Lucretius, to whose account of the origins of society Vico owed much, wanted to impress on us the debilitating effect of the fear of the gods: "tantum religio potuit suadere malorum." Vico, whose thought was enriched by centuries of accommodational interpretations of history, relativized both accounts. All religions but one may be an error; but they are a constructive error, the motive force behind the amelioration of the human condition, and therefore they are of divine origin.

[27] Above IV.E.n.5.

V

DIVINE AND HUMAN KNOWLEDGE:
KNOWING BY DOING

A. A NEW IDEAL OF KNOWING

1. Contemplative and Ergetic Knowledge

When we see things or know them, says Malebranche, we see and know them *in* God. God's knowledge is the sum total of his always "clear and distinct" ideas.[1] Some of these ideas of singulars God also endows with existence (Malebranche, like most rationalists and empiricists of his century, was a Nominalist).[2] Inasmuch as we possess clear and distinct ideas—of mathematics, for example—and even when we intuit objects *extra animam*, we participate in God's intelligence; no complicated psycho-physical mechanism is needed to mediate between our knowledge and its objects. Berkeley adopted a similar solution to guarantee the validity of our empirical knowledge, that is, of that being (*esse*) which (unlike Malebranche's affirmation of an external world) is nothing but "perceiving" (*percipi*).[3] No less radical a case for the identity of God's mental acts and ours was made by Spinoza: a clear and distinct concept of the self and its mental acts involves the awareness that they are but so many modifications of one divine attribute, cogitation; and that they stand in a one-to-one correspondence with physical states, which are but so many modifications of the only other known attribute, namely extension.[4]

---

[1] Malebranche, *Recherche* 3.2.6, *OC*, 1:437–47 *Eclaircissements* 10, *OC*, 3:121 ("que nous voyons toutes choses en Dieu"). "Quae utique vetus est sententia, et si sano sensu intelligatur, non omnino spernenda": Leibniz, GP, 4:426.

[2] *Nicholas Malebranche: The Search After Truth*, trans. Lennon, pp. 759–861 (philosophical commentary), esp. p. 761, 813–15. By "Nominalist," to repeat, I do not mean a thinker who denies the validity of relations or other universals, but rather a thinker who maintains that only singulars *exist*.

[3] McCracken, *Malebranche and British Philosophy*, pp. 205–53; 208 n. 15 (literature since Luce, *Berkeley and Malebranche*). The link was stated early, not only polemically: Burthogge, *An Essay Upon Reason*, p. 109. It is not surprising that Leibniz should say of Berkeley, as he did of Malebranche (above n. 1), "multa hic recte et ad sensum meum": Kabitz, "Leibniz und Berkeley," p. 636.

[4] Above II.F.1.

For all of them, it seems, the difference between divine and human knowledge became quantitative rather than qualitative. God may possess infinite knowledge, ours is finite; God knows all at once intuitively, our thinking processes are discursive. Yet what we know, we know exactly as God knows it—it is, in fact, the same *act* of knowledge by which we and God know something. Even Descartes, who conceded to God the right and might to invalidate mathematical truths, stated that he preferred the term *idea* over others because "it was the term commonly used by philosophers for the forms of perception of the divine mind."[5] This divine form of perception now became ours. Divine knowledge, in the Thomistic tradition, was above all introspective. By knowing itself, the divine intellect knows ipso facto everything other than itself; its knowledge is the simple unity of the "knower and the known."[6] To Descartes this mode of knowledge became also the characteristic mode of human knowledge, inasmuch as it contains clear and distinct ideas; these are innate, we know them by introspection. And if, as Descartes believed, some of our ideas about God are also "clear and distinct," it would follow that we know God in the same way as he knows himself—at least partially. Descartes did not say so, but Malebranche, who understood this implication, tried to avoid it by disclaiming a "clear and distinct" notion of the self.[7] Malebranche held, however, that we know God's infinity clearly and distinctly; in the Scotistic tradition, he let this knowledge be established by the

[5] Above II.A.n.12.

[6] Thomas Aquinas, *Summa Theol.* 1 q.14 a.6: "Sic igitur dicendum est quod Deus seipsum videt in seipso, quia seipsum videt per essentiam suam. Alia autem a se videt non in ipsis, sed in seipso, inquantum essentia sua continet similitudinem aliorum ab ipso." Cf. ibid. a.1: "species cogniti est in cognoscente" (also *In Ar. de anima* 3.8 lect. 13b). Malebranche quotes 1 q.15 a.2; 5 q.14 a.6; *Recherche* 4.11.3, *OC*, 2:97. Maimonides, *Guide* 1.68 (*shehu hasechel hamaskil vehamuskal*; Buxtdorf, *Doctor perplexorum*, p. 121: "Deum esse intellectum, intelligens et intelligibile"). On the Aristotelian principle of knowledge of same by same (ὅμοιον τῷ ὁμοίῳ) see Schneider, *Der Gedanke der Erkenntnis des Gleichen durch Gleiches in antiker und patristischer Zeit*, pp. 65–76 (Thomas).

[7] Malebranche, *Recherche* 3.2.7 §4, *OC* 1:451–53; *Eclaircissements* 11, *OC*, 3:163—71. "Consciousness" is here not only in a middle status between sensations and ideas. It is the precondition of sensations, accompanies them, we know it through them. This is an interesting predecessor of Kant's notions of consciousness, as the reply, "which always accompanies my perceptions (*Vorstellungen*)." But Malebranche does not employ it to guarantee the intelligibility of the world. To the contrary: consciousness is the paradigm and source of sensations and *error*.

ontological proof.[8] Earlier I discussed the transparency of God in the seventeenth century; here we find its ultimate expression—let me call it the *contemplative rationale* for the (partial) identity of human and divine knowledge.

Of Guelincx and Malebranche—but not of Spinoza—the opposite can equally be said: that they did as much to deepen the gap between the human and divine intellect as they did to narrow it. In contrast to our passive knowledge, they claimed, God's knowledge is in some respects knowledge through and by *doing*. This is true not only of entities created *ex nihilo*, but also, and especially, of causes and forces. There are no other causes or forces in nature except for God's always active will.[9] Every single instance of an impact of one physical body on another, every instance of interaction between soul and body are instances of immediate divine causation, including our knowledge of it. Divine causation replaced both mechanical and psycho-physical causation altogether. Rather than being actors in the world, human beings and angels turned into mere "spectators."[10] Let me call it the *pragmatic*, or *ergetic rationale* for the *difference* between the divine and human knowledge.

These two models of knowledge—passive the one, active the other—pertain to altogether different domains. Contemplative knowledge is knowledge of ideas and their necessary relations.

---

[8] *Recherche* 4.11.2, *OC*, 2:95: "Or, cette idée simple et naturelle de l'être ou de l'infini renferme l'existence nécessaire: car il est évident que l'être (je ne dit pas un *tel être*) a son existence par lui-même. . . . Il se peut faire que les corps ne soient pas, parce que les corps sont de *tels êtres*, qui participent de l'être, et qui en dépendent." This is both his answer to the Gaunilo-style objection that one can prove, *eodem modo* the existence of an "infinitely perfect body": and his correction to Descartes's version of the ontological proof: necessary being can be only attributed to a perfect being, or being-as-such. Malebranche also says (3.2.6) that infinity can be known, but not comprehended by the soul; which only means that though we have a clear idea of infinity, we do not have such an idea of its content. Henrich, *Der ontologische Gottesbeweis*, pp. 23–28, ranks him as a successor of Descartes in that he commences with the necessary (rather than most perfect) being (above II.A.2 and n. 10). But inasmuch as Malebranche recognized a deficiency in the proof and corrects it the way he does, he comes closer to the Cambridge Platonists or Leibniz.

[9] *Recherche* 3.2.5 (existence); indeed, Malebranche can explicitly do away with forces (which Descartes only implicitly may have done, above II.E.1): "All natural forces are therefore nothing but the will of God" (6.2.3); to opine differently leads to "the most dangerous error of the ancients," namely, polytheism. Cf. also Cassirer, *Das Erkenntnisproblem*, 1:559–67.

[10] Arnold Guelincx, Γνῶθι σεαυτόν sive *Ethica* 1.2.2.8; 14. Ibid. 1.2.2.4: "Qua fronte dicam, id me facere, quod quomodo fiat nescio?" That is, doing is knowing, there can be no act without knowledge.

Contrary to Descartes, Malebranche scorns the assumption that God could have "created" eternal truths: God is the sum total of them. Knowledge by doing, that which we named factual or ergetic knowledge, pertains to the domain of actual beings and their inter-action. It is the domain of the contingent. The distinction between *potentia Dei ordinata* and *absoluta*, it seems, acquired an entirely new meaning: in the domain of thought (or ideas), there was no room for any exception to "order."[11] In the domain of reality, there was no other cause than God's absolute, incessant *actus voluntatis*, in which regular constellations function merely as an "occasion."[12] Human knowledge is knowledge of the former only; the latter, namely actual beings and their interaction, can only be known to the creator in the manner of their creation. We do not know things, not even ourselves; we know ideas alone, among which only one idea involves its being of necessity.

2. Malebranche's Sources

Were both rationales, both modes of knowledge such a break with the medieval tradition? Malebranche was knowingly indebted to the Augustinian tradition, to the conviction that all true knowledge has its source in divine illumination. In the presence of mathematical, immutable truths in our mind, Augustine saw an eminent proof for the vestiges of the divine intellect in ours.[13] But illumination can at best account for just the positive half of Malebranche's epistemol-ogy. As to the distinction between knowledge of ideas and knowl-edge of beings, it bears a superficial resemblance to the Thomistic synthesis between the Augustinian/Neoplatonic and the Aristote-

---

[11] Malebranche, *Eclaircissements* 10, *OC*, 3:198: "Les vérités sont immutables et néces-saires, aussi bien que les idées. Il a toujours été voui que 2 fois 2 sont 4, et il est impossible que cela devienne faux"—impossible even to God; the order of ideas is "coëternelle" and "nécessaire" with God (*OC*, 3:86). Malebranche polemicizes here and in other similar instances against Descartes's notion of *created* truth. Cf. Reiter, *System und Praxis: Zur kritischen Analyse der Denkformen neuzeitlicher Metaphysik im Werk von Malebranche*, pp. 124ff., esp. 134–38.

[12] *Recherche* 6.2.3, *OC*, 2:312. For the origins of the term (and other terms for indirect causation) cf. Specht, *Commercium*, pp. 29–56 (Scholasticism, Descartes); 144–45 (Cor-demoy); 162–75 (Guelincx, Malebranche).

[13] E.g., Augustine, *De libero arbitrio* 2.8; 12. Malebranche recognized his indebtedness and insisted on the difference: e.g., *Recherche* 3.2.6, *OC*, 1:443 (quotes *De trinitate* 14.15). See also Moraux, "Saint Augustin et Malebranche," in *La Philosophie et ses problèmes: Ou-vrage en homage à Mgr. Jolivet*, pp. 109–36.

lian account of knowledge. Thomas also distinguished between
God's knowledge of forms and of beings (singulars). The forms are
represented in God's mind as eternal, divine ideas: to know them,
God needs only to reflect upon himself. We recall also how Thomas
secured God's knowledge of singulars: it is the most immediate
mode of knowledge, namely knowledge by doing, by giving being
(*esse dare*); it needs neither the senses nor another medium.[14] God
imparts it on separate intelligences, which cannot even infer the existence of singulars, as we do, from sense data, through the mediation of sensible species. It follows that God, the only being capable
of giving being, is also the only being to have an immediate knowledge of singulars qua singulars.

Further possible medieval links to the Occasionalists come to
mind. In the last chapter we noted some of the similarities and differences between Descartes's notion of intuition and the Scotistic-
Terministic tradition. Since Duns Scotus the human intellect was
made to share with the divine the same kind of immediate existential knowledge. The *cognitio intuitiva intellectiva* secures knowledge
of singulars (at least qua existents) without the mediation of species.
We may possess, according to Duns Scotus, such knowledge already here on earth; some of us will certainly possess it in the beatific vision promised thereafter.[15] Ockham went further: every cognition of singulars is, of necessity, intuitive. In exceptional cases he
admitted *de potentia Dei absoluta* the possibility of an intuitive cognition of non-existent things—although he refused to call it an evident cognition.[16] It is a cognition of a most peculiar structure, not
always grasped by commentators. Whenever (if ever) it takes place,
God functions both as its cause and as its object. Perhaps Ockham
meant to say that, if I have such an intuitive cognition, say, of a non-
existent chair, I would be seeing (in a Wittgensteinian idiom) God
"as" a chair. Now Ockham's exceptional case became the ordinary
human mode of cognition to Malebranche. There is no way to ascertain whether the chair I see matches the existing one "out there."

---

[14] Above III.B.3 and n. 37. Cf. Connell, *The Vision in God*, pp. 73–91 (angelic knowledge in Thomas); 91ff. (Scotus); 146–48 (Augustine). Connell, to put it in my terms,
looks only for the origin of the contemplative similarity—not for that of the ergetic dis-
similarity (knowledge by doing). And he emphasizes traditions concerning abstractive
rather than intuitive knowledge.

[15] Above III.B.3 and n. 42.

[16] Above III.B.3 and n. 56.

The chair I see is in part an idea of God—inasmuch as I can see it as a modification of the *étendue idéal*.[17] In part this chair has predicates and properties that are not at all divine—a projection of sensations, associations, and linguistic habits. Whatever the chair I conceive has in excess of spatio-temporal modifications is a mere image. Again we are reminded of Aureoli's *esse apparens* and its problems.[18] In short, many aspects of Malebranche's contemplative rationale for the identity of human and divine knowledge seem to have been anticipated in the fourteenth century, the first time in which a sustained epistemological debate took place in the Middle Ages in a well-defined terminological horizon.

Similarities should also be noted between Spinoza's or Malebranche's epistemology and the Averroist doctrine of *unitas intellectus*, condemned in the Middle Ages and revived in some Renaissance philosophies; the Averroists drew the radical consequences from the Aristotelian identification of knowledge, the known, and the act of knowing.[19] The reduction of all causality to God's incessant activity had indeed its precursor in the extreme voluntarism of the *Ishari'a*, transmitted to the West through Averroës and Maimonides. Many learned contemporaries of Guelincx and Malebranche reminded them of that dubious pedigree.

These various traditions are not irrelevant to the understanding of Guelincx and Malebranche. Retracing the different roots of their surprising and extreme positions has many merits, not the least of which is that it makes us aware of other sources than the Cartesian mind–body problem, even though the latter was the soil in which the terminology of occasions first grew.[20] Yet, the hunt after similar traditions in the past can easily obscure the novelty of the problems that Guelincx and Malebranche confronted, and the boldness of their solution. All the medieval traditions I mentioned are concerned with the knowledge of entities and their properties. The Occasionalists, notably Malebranche, were much more concerned with the knowledge of *relations*: the relations of ideas to ideas, of ideas to things, of things themselves. Truth is nothing but the "rap-

[17] Above pp. 88–89.

[18] Tachau, "Vision," pp. 60–78.

[19] Above n. 6. Already Renan, *Averroës et l'Averroisme*, pp. 125–26, compared the Averroists to Malebranche in respect to the *unitas intellectus*.

[20] Specht, *Commercium*, above n. 12.

port" between ideas.[21] Malebranche's epistemology addressed a totally different ideal of *knowing* than the Middle Ages; a new vision of a universal method which is mathematically oriented and overarches all disciplines, and results in a "mechanical" explanation of all phenomena. The Occasionalists, I shall argue, sought to preserve this idea even though they recognized its theological dangers. They set out to save it by extricating its thorns.

3. The New Method as Ergetic Knowledge

Descartes captures the imagination of his and later generations with a vision of a new method of scientific inquiry. It was, and remained, much harder to characterize than to praise. Several catch phrases are easy to hand, but they seem vague and slippery. The new method purported to be universal, a canon of principles and procedures overarching all disciplines. Descartes claimed that it was a tool for the discovery of new truths, not only for the exposition of old ones; a genuine *ars inveniendi* whose practitioners, unlike the Lullian art, know what they talk about because they start with simple intuitions and combine them according to infallible rules.[22] The method proceeded by resolution and composition. It succeeded in uniting algebra and geometry, in the mathematization of mechanical laws and the mechanization of natural phenomena. It had built philosophy, physiology, and psychology on a new foundation. Most "mechanical philosophers" of the seventeenth century embraced many of these hopes and claims; but what, precisely, did they mean? How did mathematics become the paradigm and language of science? What were the connotations of "machine" and "mechanical"?

To begin with, the application of mathematics to other disciplines always depended on the willingness to overcome the classical, Aristotelian injunction against the mixing of entities of different genera.[23] The vice of *metabasis*—the transportation of methods from one discipline to another—became, in the seventeenth century, a virtue. The success of the "new science" of mechanics led to its em-

---

[21] *Recherche* 1.2.2, *OC*, 1:52–53 (truth as the rapport of two entities among themselves); 6.1.5, *OC*, 2:286–87: "Il y a des rapports ou des vérités de trois sortes. Il y en a entre les idées, entre les choses et leurs idées, et entre les choses seulement"; cf. Reiter, *System und Praxis*, pp. 206–209.

[22] Below n. 24.

[23] Above II.B.2; below V.B.2 (metabasis).

ulation in other disciplines. Hobbes and Spinoza envisaged even a mechanistic theory of social order. But the success of mechanics, Galileo's analysis of terrestrial motions, and Newton's synthesis of those and Kepler's law of planetary orbits, would have been impossible without a slow and gradual change of mathematics itself. The classical Greek view of mathematics as an inventory of (ideal) mathematical entities and their absolute properties—a perception that forbade the representation of one kind of mathematical entities by another—eventually receded before a new perception of mathematics as a science of relations and structures:

> But for all that I had no intention of trying to master all those particular sciences that receive in common the name of Mathematics; but observing that, although their objects are different, they do not fail to agree in this, that they take nothing in consideration but the various relationships or proportions which are present in these objects, I thought that it would be better if I only examined those proportions in their general aspect.[24]

Because mathematics turned into a formal language of relations, not only could numbers be represented by figures and vice versa, but also nonmathematical relations—motions, forces, intensities—could be expressed in a mathematical language.

These were, in part, medieval developments; I shall try to assess their role and impact. Yet deeper still, we detect in both Descartes's vision of a new method and in others' of the seventeenth century an entirely new ideal of knowing—of acquiring knowledge—into which all the ideals of knowledge I discussed earlier seem to merge. This new ideal, somewhat elusive yet powerful down to our own days, was the ideal of knowing through *doing* or knowing by *construction*. Francis Bacon, who still believed in the task of science to discover the "forms" of things, also believed that to discover a form is the same as being able to produce the thing in question.[25] This is

---

[24] Descartes, *Discours* 2, AT, 6:19–20; *Philosophical Works*, ed. Haldane and Ross, 1:93 (translation). Rule 1 of the *Regulae ad directionem ingenii*—probably his earliest work—already implies the advocation of *metabasis*: the sciences, unlike the arts, should not be studied "in isolation from each other": all sciences are interconnected. This was clearly recognized by Ortega y Gasset, *Der Prinzipienbegriff bei Leibniz und die Entwicklung der Deduktionstheorie* § 22.

[25] F. Bacon, *Novum Organon* 2.1, in *The Works of Francis Bacon*, ed. Ellis et al. 1: "Super datum corpus, novam naturam sive novas naturas generare et superinducere, opus et intentio est humanae potentiae. Datae enim naturae formam, sive differentiam veram, sive

the reason why "science is power." Descartes, whose geometrical interpretation of matter-in-motion was so radical that it barred him from the understanding of forces, nonetheless believed that he could reconstruct the making of the universe, as it is, by merely combining the "clear and distinct" ideas of an original mass and the laws of motion. He could succeed where Plato's construction of the universe failed, because his mathematics (so he believed) could deal with change; and it could do so because it was another mathematics, constructive in a much broader sense than ancient geometry. Glanvill, invoking the Platomic image of the Geometer-God, added that "the *Universe* must be *known* by the *Art* whereby it was *made*."[26] Hobbes thought that the task of science is first to destroy the world and assume the existence of the self only with its "phantasms," and then reconstruct the world systematically with the help of an arbitrary, unequivocal system of signs. For the same reason he believed that the science of politics is closer to our understanding than the science of nature: we, who "made the commonwealth ourselves," can more easily reconstruct its making mentally. Giambattista Vico epitomized the ergetic ideal of knowledge in his famous phrase "verum et factum convertuntur." We know for certain only those things that we have constructed ourselves; Vico, who reacted against the claims of the "mechanical philosophers" of nature, insisted that we only have such a knowledge by construction of a society; and that only God knows nature in the manner of knowledge by doing.

This new, *ergetic* ideal of knowing stood squarely against the old, *contemplative* ideal. Common to most ancient and medieval epistemologies was their receptive character: whether we gain knowledge by abstraction from sense impressions, or by illumination, or again by introspection, knowledge or truth is found, not constructed. Implicitly or explicitly, most "new sciences" of the seventeenth cen-

---

naturans naturantem . . . invenire, opus et intentio est humanae scientiae." Cf. ibid. 5; 2.41 (only genetic knowledge is true knowledge). Cassirer, *Das Erkenntnisproblem*, 2:11–28; Rossi, *Francis Bacon: From Magic to Science*, pp. 14–16 (the alchemical pedigree of this terminology).

[26] J. Glanvill, *Plus Ultra or the Progress and Advancement of Knowledge*, p. 25. This can be read as a mere repetition of an old *topos*, namely that God made everything *mensura et numero et pondero* (*Sapientia Salom.* 11:21): cf. Curtius, *Europäische Literatur*, pp. 493–94, 527–29 ("Gott als Bildner"). Or it may also reflect the new ideal of knowledge-by-construction.

tury assumed a constructive theory of knowledge.[27] Guelincx and Malebranche, I believe, rebelled first and foremost against the implicit dangers of this new constructive ideal of knowledge by confining it to the realm of ideas and their combinations. For the mechanical interpretation of nature could easily lead to the presumption that we know the making of the universe in the manner of the creator. The Occasionalist reserved knowledge by doing for God alone, and he did so much more radically than any medieval author because they, too, shared the admiration of the mathematical science of nature. Ancient and medieval science confined the operation of machines to artifacts, and conceded to us knowledge by doing at least of the latter. This distinction collapsed with the "mechanical philosophy" of the seventeenth century, and the image of the machinelike universe in its modern guise threatened to erode the wall between human and divine knowledge much more thoroughly than any contemplative concept of knowledge could. The Occasionalists faced the danger by conceding that, indeed, all knowledge of *reality* is through acting, and by boldly asserting that, therefore, all knowledge of reality is solely God's.

B. CONSTRUCTION AND METABASIS, MATHEMATIZATION AND MECHANIZATION

1. Construction and Motion in Ancient Greece

In reviewing the various possible antecedents of "knowledge by doing," it seems as if I have omitted the oldest and most immediate paradigm. "To do," "to produce," "to construct," "to generate" were terms used for constructions in Greek geometry.[1] Constructions were a distinctive feature of geometry from its beginnings that were not forgotten in the Middle Ages. But their means and role became gradually restricted. Proclus Diadochus, in his commentary on the first book of Euclid, tried to systematize and reconcile various attitudes toward the status of constructions. The opinions he quotes go back six or even eight hundred years before his time.

---

[27] H. Arendt, *The Human Condition*, esp. pp. 294–304, developed several lines of thought close to those in this chapter—except that I do not venture a diagnosis of "modern man." She also refers to the *verum-factum* principle, and to Hobbes, and so does Löwith, "Vicos Grundsatz" (above IV.A.I).

[1] *The Works of Archimedes*, trans. and ed. Heath, pp. clxxiv f.

Some, he says, wanted to reduce all "problems" (of construction) to "theorems," because geometrical entities cannot suffer generation or corruption (Speusippus). Others (Menaechmus) wanted to reduce all theorems to problems, because some constructions provide "that which is sought for," others make us see the qualities of a mathematical object. Both, he concludes, are right:

> The followers of Speusippus . . . because the problems of geometry are of a different sort than those of mechanics, since the latter are concerned with perceptible objects that come-to-be and undergo all kinds of change (αἰσθητὰ γὰρ ταῦτα καὶ γένεσιν ἔχοντα καὶ παντοίαν μεταβολήν). The followers of Menaechmus are also right because the discovery of theorems does not occur without stepping down (προόδου) into matter, I mean intelligible matter (ὕλην . . . νοητήν). In going forth into matter and shaping it, our ideas are correctly said to resemble acts of production (γενέσεσιν). For the movement of our thought in projecting its own idea is a production . . . of the figures in our imagination and their properties. But it is in the imagination that the constructions, sectionings, positions, comparisons, additions, and subtractions take place, whereas the contents of our understanding (διανοίας) remain fixed, without any generation or change.[2]

Problems and theorems differ nonetheless. If we view both as adding a predicate to a subject, then constructions add only possible predicates (an equilateral triangle can, but need not, be inscribed into a circle), while theorems add necessary predicates (the basis angles of such a triangle are necessarily equal). Proclus also refers to a tradition that goes back to Oenopides through Zenodotus and Poseidonius: constructions only prove *existence*, theorems prove properties.

Constructions actually served as existence-proofs in Greek geometry, though only in geometry and even there not solely.[3]

---

[2] *Procli Diadochi in primum Euclidis Elementorum librum commentarii*, ed. Friedlein, pp. 78–79. I followed (with few changes) the translation of Morrow, *Proclus: A Commentary of the First Book of Euclid's Elements*, p. 64.

[3] Zeuthen, "Die geometrische Konstruktion als Existenzbeweis," pp. 222–28. That this is true of geometry only has been rightly emphasized by Szabó, *The Beginning of Greek Mathematics*, pp. 317–22: in arithmetics, Euclid proves the existence of a prime greater than any given prime (*Elements* 9.20) without being able to construct it ($p_n! + 1$ is not necessarily a prime). Becker, *Grundlagen der Mathematik in geschichtlicher Entwicklung*, pp. 94–95, shows that even in geometry a distinction between abstract and constructive existence-proof has later been drawn (based on Philoponos, *Aristotelis Physicorum libros . . . commentaria*, p. 112.27–29).

Whether or not Oenopides was the first to use them as existence-proofs, it seems that he was the first Greek mathematician we know of to have separated geometrical from mechanical problems. He was credited by Proclus with the solution of two fundamental construction problems (Euclid, *Elements* 1.12: to draw a perpendicular to a line from a given point which is not on it; *Elements* 1.23: to construct a rectilinear angle on a given point on a straight line equal to a given rectilinear angle). On the face of it, these problems seem too simple for the level of constructions already reached by the end of the fifth century. His achievement, it has been conjectured, was that he demanded of constructions to employ a circle and straightedge only and that he showed how it can be done.[4] It is more or less taken for granted that the restriction to lines and circles reflects their status as "the most beautiful and perfect" figures.[5] This may be so, but it is insufficient as an explanation why constructions with circles and straightedges help to separate geometry from mechanics—if indeed this was the purpose of Oenopides. It must be shown why circles and lines and all figures that can be generated from them could be held to be "nonmechanical" in a specific sense, whereas others could not.

There existed, at least since the *quadratrix* of Hippias of Ellis, a family of figures composed of "mixed lines." Sometimes they were simply called "generated by two motions."[6] The motion involved in the generation of the *quadratrix* (to which later mathematics added the cochloid and the spiral) is *not* mere displacement, as is the motion involved in the generation of a circle by a rotating line or later in the generation of "revolving bodies." Such motions do not

---

[4] Proclus, *In primum Euclidis*, ed., Friedlein, pp. 283, 333 (in the name of Eudemus). Heath, *History of Greek Mathematics*, 1:175; Fritz, *Schriften zur griechischen Logik*, 2:154–61; Szabó, *The Beginning*, pp. 273–79 (identity with Euclid's first three postulates).

[5] Fritz, ibid., 2:156: "Vielmehr können Kreis und Gerade nur deshalb gewählt worden sein, weil sie als die 'schönsten und vollkommensten' Kurven betrachtet wurden." My argument is that, though simpler, I do not see why compass and straightedge are less "mechanical" devices than those one has to use to solve problems, say, of "verging" (νεῦσις), unless one takes into consideration not just "motion" (displacement), but also the rate of motion (motion in time). Moreover: the straight line, by definition, cannot be "perfect" since it is unlimited.

[6] Simplicius, *In Arist. physic. comment.*, ed. Diels, p. 60.7–18; according to Iamblichus, he says, Carpus actually called his curve "generated by double motion." Proclus, *In primum Euclidis*, pp. 104–106, speaks of "mixed" motion. In the discussion of the cylindrical helix he actually speaks of two different, non-uniform motions.

take *time*, and could in fact be imagined as instantaneous. If motion is required to explain the generation of straight lines and circles (but many Greek mathematicians did not think so), then it is—to use again the language of Proclus—an "imaginary motion" only κίνησις φανταστική), a mere displacement.[7] The higher curves, however, are generated by *motion-in-time*, that is, by genuine κίνησις. The *quadratrix* is generated by two adjacent sides of a square *ABCD*, one moving from *B* to *A* parallel to the basis *AD*, the other rotating *with equal speed* from *AB* to *AD*. The point *P* at the intersection of both lines describes the *quadratrix*. Perhaps other constructions that used "kinematic means" already existed in the time of Oenopides.[8] Perhaps the restriction to line and straightedge was meant to bar motion-in-time from geometry. But whether or not such considerations capture the reasons behind the initial moves to separate geometry from mechanics, they were prominent later. Curves generated by motion-in-time remained second-class citizens. They were not regarded as legitimate means to solve the problems they were capable of solving, for example, the trisection of an angle. Physical motion—motion-in-time—had no place in geometry; the "intelligible matter" of which Proclus spoke cannot be conceived as generated in time. The term is not even restricted to the Platonic-Neo-

---

[7] Proclus, *In primum Euclidis*, p. 185.8ff. But Proclus does not explicitly say that the imaginary motion does not involve time. He explicates, at this place, Euclid's first three postulates; Euclid, of course, refrained from the language of motion when defining circles and lines (Szabó, *The Beginning*, p. 277). So did Aristotle, e.g., *Metaphysics* B2.194a3–7 (straight or curved independent of motion). But Heron did not: Heath, *Mathematics in Aristotle*, p. 93 (with comparison to Gauss). That Proclus nonetheless meant motion without time may be inferred from an analogous case. Creation-*in*-time, he says of Plato's demiurge (below V.B.n.66), serves only to distinguish theoretically parts of a composite timeless whole. Construction as imaginary motion nonetheless emphasizes the *activity* and spontaneity of mind; knowledge, in the Neoplatonic tradition, was active in the sense that it was a kind of intrinsic motion.

[8] Proclus, *In primum Euclidis*, pp. 272.7 (Hippias and Nicomedes), and 356.11, only attributes him with describing the properties of the *quadratrix*, not using it to trisect the angle (note: ἕτεροι δὲ ἐκ τῶν Ἱππαίου etc.). Pappus does not even mention Hippias: *Pappi Alexandri Collectionis quae supersunt* 4.30–32, ed. Hultsch, pp. 250.33–258.19; "Kinematic means": Becker, *Mathematische Existenz: Untersuchungen zur Logik und Ontologie mathematischer Phänomene*, p. 250 (both νεῦσις and the *quadratrix*). Whether to attribute the strict rules governing construction to a Platonic reform (as Becker wants) or to a slow recovery from the Parmenidean-Zenonian challenge I am not competent to decide. None of these authors distinguishes between motion and motion-in-time. On other construction see Thomas, *Selections Illustrating the History of Greek Mathematics*, 1: *From Thales to Euclid*, pp. 257–363, esp. pp. 263–67, 335–47.

platonic tradition. Aristotle, too, uses it to distinguish between the substrate of mathematical and physical objects.[9] He insists, though, that they have no separate existence. Mathematical properties are won by abstracting from bodies their physical properties qua matter that undergo change, that is, motion-in-time. It may well be the source of Malebranche's *étendue intelligible*.[10] The cardinal difference between his and the Greek tradition is that his "ideal extension" encompasses, of necessity, motion-in-time. It was construed to sanction the reunification of geometry and mechanics in the seventeenth century rather than to separate them. In the balance of this section I want to show, in very rough outlines, how the ideals of mathematization and mechanization converged into a new methodical ideal of knowing by construction that the Greeks rejected almost from the outset.

2. The Prohibition of METABASIS

Not only was time, at least in many quarters, excluded from the definition of "mathematical objects"; these were also barred from representing or explaining genuine temporal processes. In the beginning of the second chapter I discussed briefly the early failure of the ideal of the mathematization of nature in Antiquity.[11] Plato's geometrical-mechanical construction of the cosmos provided no more than a metaphor; he admitted himself that the imperfect and changeable material entities can be represented by mathematical entities only imperfectly. Aristotle added that, since mathematical entities abstract from the main physical property—change—there is little use of mathematics in physics. The usefulness of mathematics for physics became confined to static structures: bodies at rest (balance) or periodic motion that is, in a sense, "both rest and motion"—the regular, uniform, simple celestial motions. Aristotle further barred any analogy between celestial and terrestrial motions on mathematical grounds. The uniform motion of a point along a curve is incomparable with its motion along a line, straight or bro-

---

[9] Aristotle, *Metaphysics* Z10.1036a10; H6.1045a34. Proclus, *In primum Euclidis*, p. 52, objects to the Aristotelian location of the "imagination (with mathematical entities) in the passive intellect. Cf. Mueller, "Aristotle on Geometrical Objects," pp. 156–71.

[10] Above, II.F.2; V.A.1–2.

[11] Above, II.B.1.

ken.[12] The various constrictions that Aristotle imposed on a mathematical conceptualization of change are just different versions of his strict injunctions against transitions from one genus to another—μετάβασις εἰς ἄλλο γένος.[13] Just as in nature a specific difference can appear only in one genus, so also in the sciences arguments and methods of one should not be "carried over" into another (which is not subaltern to it). To try and explain the causes for change mathematically is like trying to prove a mathematical theorem from the immortality of the soul. Aristotle, unlike Plato, did not believe in an overarching scientific method common to all sciences.[14] Genuine disciplines are autonomous and require different principles (ἀρχή). For the very same reasons, arithmetical propositions themselves cannot be explained geometrically; because the one deals with numbers, the other with magnitudes.[15] More yet: even with geometry, only figures "of the same kind" can be compared to each other—lines to lines, curves to curves, areas to areas.[16] So deep-seated is Aristotle's fear of mixing genera that one wonders whether its roots are not deeper than the ontological commitment to a rational, unique, non-arbitrary classification of the world. A recent anthropological theory stressed the fear of mixed, not clearly definable objects sensed by many cultures and expressed by prohibition of access or usage.[17]

The theory of proportionality, about which Aristotle makes some interesting historical remarks,[18] does not transgress this prohibition—though it may seem so. Proportions are not mathematical objects (i.e., numbers or magnitudes) and not even absolute properties of mathematical objects; they are relations only, and relations can be compared across separate genera or even domains of inquiry.

---

[12] Aristotle, *Physics* x4.248a10–b7. On the one hand it is absurd not to assume that a body can move with the same motion on a circle and a straight line. On the other hand we must accept this conclusion because the line and the circle are incommensurable (οὐ σύμβλητα, ἀσύμβλητα). Aristotle even denies that we could say of a curve that it is longer than a line!

[13] Above, II.B.2 and nn. 10–13.

[14] Wieland, *Die aristotelische Physik*, pp. 187–202, 202–30.

[15] Aristotle, *Posterior anal.* A7.75a38–b8.

[16] *Metaphysics* I 1.1053a24–30: ἀεὶ δὲ συνεχὲς τὸ μέτρον etc.

[17] Douglas, *Purity and Danger, passim*; Leach, "Anthropological Aspects of Language: Animal Categories and Verbal Abuse," in *Mythology*, ed. Maranda, pp. 39–67; 47: "taboo inhibits the recognition of those parts of the continuum which separate the things."

[18] Below n. 21.

To say that a given area is "bigger than" a given line is meaningless; to say that two areas have the same ratio as two lines is not. Aristotle even seems to regard the proposition that a curve is longer than its subtended line objectionable,[19] which Euclid does not: it is the foundation of the method of exhaustion. If so, then Aristotle will have distinguished between the incommensurable (ἀσυμμετρία) and the incomparable (ἀσύμβλητα).[20] But he stresses the proportionality of incomparables: previous generations, he says, did not recognize the "commensurable generality" of proportions and permitted only proportionalities "of lines and lines, numbers and numbers, areas and areas."[21] It may be that out of the mathematical theory of proportionality Aristotle later developed a theory of "focal meanings" attributed to him by Owens, Patzig, and others[22]—a theory that enabled him (if indeed he held it) to develop metaphysics, or the science of being qua being, as a universal science. At times he even speaks of a universal mathematics,[23] and I believe that it is not a science with definite objects and properties but a science of relations or perhaps even universal-formal algorithms of the kind employed, for example, in his theory of mixtures.[24]

Mathematics—as Aristotle recognizes—is, of course, a language. A good part of geometry consists of the representation of figures "of the same kind." By analogy, however, some properties of figures can stand for a few well-defined, nonfigural properties. Lines do, in the tenth book of Euclid's *Elements*, represent numbers because both share the property of iteration. Aristotle would presumably not object; he even permits lines to represent *time* because both share the character of a continuum. Circles and lines can also, of course, represent the path of bodies in motion. But "motion-in-

---

[19] Above n. 12.

[20] *Metaphysics* A1.983a16; *Physics* H4.248a18; above n. 12 and III.c.n.12.

[21] Aristotle, *Posterior anal.* A5.74a16–b4, esp. a19–25; cf. the comments of Heath, *Mathematics in Aristotle*, pp. 41–44 and Livesey, "Metabasis," pp. 47–48.

[22] Above II.B.n.11.

[23] *Metaphysics* EL.1026a23–7; M2.1077a9–10; Heath, *Mathematics in Aristotle*, pp. 223–24, thinks of a kind of algebra.

[24] *De gener. et corrupt.* B5.332b6–333a15; or *Ethica Nicom.* E3–5.1131a10–1134a15; or again in the logical writings Aristotle uses letter schemata with strict rules of operation. The transformation of elements in the *Meteorology* is actually a kind of Abelian group. It may be that Aristotle refers to both mathematical and nonmathematical formalism, inasmuch as it is a formalism, by the name of "catholic mathematics."

time" cannot be represented in *one* figure or symbol. A line may represent *either* the length and shape of a distance traversed *or* the time, but not both. Our expression of velocity as distance over time is, indeed, seen from the vantage point of Aristotle, a *metabasis*, a mixing of categories. Much less could the distribution or intensity of a quality be represented: quality and quantity are different categories. And finally, the rate of change—acceleration or retardation—is an altogether impossible notion, "there cannot be motion of motion or becoming of becoming or in general change of change."[25]

Had Aristotle admitted geometrical figures generated by motion-in-time, they could have represented physical motions-in-time, perhaps even accelerations. There existed, as we saw, a tradition of figures generated by combined angular and linear uniform motions: Aristotle, and even Euclid, were conspicuously silent about them. The so-called revolving bodies of Euclid, I argued, are no exception: their motion means displacement only, of the same character involved in demonstrating the congruency of discrete figures; it could, for that matter, be thought of as instantaneous. Aristotle, following Eudoxus, represented the apparent retrograde motion of the planets with a *hippopede*. But the *hippopede* is not a figure generated *necessarily* by motion (as two concentric spheres around different axes). It can also be interpreted as the line described on the surface of a sphere by a cylinder internal to it.[26] Such were the constrictions of *metabasis*.

A century and a half later, Archimedes freed himself from some of these constrictions—but not, it seems, from all. He already felt secure enough with the method of exhaustion to have curves be represented by lines. He employed mechanical considerations in the finding—though not in the proof—of theorems.[27] The spiral of Archimedes, moreover, is not only generated by motion-in-time; it is perhaps the first curve to be generated by two *different* uniform motions, namely by the combined angular and radial motion of a point

---

[25] *Physics* E2.225b15: ὅτι οὐκ ἔστι κινήσεως κίνησις οὐδὲ γενέσεως γένεσις, οὐδ' ὅλως μεταβολῆς μεταβολή. Aristotle may be thinking of the fallacy of self-reference.

[26] Neugebaur, *A History of Ancient Mathematical Astronomy*, 2:677–80; Thomas, *Greek Mathematical Texts*, 1:14 (from Simplicius's commentary on *De caelo*).

[27] *Archimedes*, ed. Heath, supplement, pp. 7, 13–14 ("to investigate problems in mathematics by use of mechanics"). Archimedes distinguishes between investigation and proof, but he believed that the former is of universal value worthy of publication.

(a line rotating uniformly around one of its term points and a point uniformly moving outwards along that line).[28] The resulting motion amounted to a uniformly accelerated motion which the schoolmen of the fourteenth century were to call *motus uniformis difformis*. Nicole Oresme recognized this kinship.[29] Yet Archimedes did not explore any further the road that led to an adequate representation of non-uniform motion.

3. Aristotle's Prohibition Eroded; the Intension and Remission of Forms

Held firmly in early Middle Ages, the injunction against crossing methods eroded in the fourteenth century both in theory and in fact. The doctrine was incompatible with the Terministic theories of science. Ockham refused to derive the unity of a discipline from the alleged unity of its subject matter. Science is an arrangement of propositions that answer a given canon of questions, an aggregate of *habitus* brought under one aspect. The same proposition may enter different proof-schemata in different "sciences." There was no need to challenge Aristotle's injunction head on: more and more disciplines could simply be declared to be *scientiae mediae*.[30]

Among the instances of actual, constructive transference of methods, the medieval science of motion is of particular relevance to our discussion. Much has been written on the medieval interest in the quantification of qualities, or the "intension and remission of forms," which led also to new ways of representing motion. It was a home-grown interest, nourished by many sources: theological (the infusion of *caritas*), philosophical (what does a change in quality mean), medical-pharmaceutical (the compounding of drug effects), optical (the intensification of light), and methodological-mathe-

[28] Ibid., pp. 151–88, esp. p. 154. Only in the seventeenth century was circular motion, even if uniform, recognized as acceleration.

[29] Oresme, *Tractatus de configurationibus qualitatum et motuum* 1.21, in *Nicole Oresme and the Medieval Geometry of Qualities and Motions*, ed. Clagett, pp. 225f. Oresme speaks alternately of uniformly difform, difformly difform curves and motions. Clagett believes (p. 450) that the knowledge of Archimedes' spiral is derived from a medieval compilation. On Descartes and spirals below v.B.5.

[30] Miethke, *Ockhams Weg*, pp. 245–60; Livesey, "Metabasis," pp. 333–57. Livesey also shows how, in the fourteenth century, more and more sciences fall under the category of *scientiae mediae*.

matical (how to represent change).[31] Analogies were freely exchanged from one domain to another. I shall attend to only one aspect of this immensely rich subject: the growing employment of mathematics as a language, its implications and the awareness of it.

The philosophical problem, briefly stated, was simply that a quality (form) as such is equal to itself in all of its instantiations;[32] yet experience tells us that some qualities are present in different degrees in different subjects or even change degree within the same subject, as when something gets hotter and colder. To say that a subject "participates" more or less in a quality is to assume the ontic status of qualities; to say that the quality is present in a subject in different states of actualization is to say that no subject is really hot unless hottest. But grace is present in the begraced before it may be increased. It could, of course, also be said that some qualities include, in their very essence, a possible range of change—an answer leading to an infinite regress not unlike the third man's paradox.[33]

Of particular importance was the solution of Duns Scotus and his followers. As a common property, he held, a form is always the same and equal to itself. But contracted into an individual being, it is represented (or instantiated) as a concrete instantiation that was later called a *formalitas*. If I may be allowed a simile, the *formalitates* relate to their form as shadows of a figure to that figure—with a definite or indefinite range—except that, in this case, only the shadows "exist." The *formalitas* is a projection of a form into the individual, and thus an accidental property of that form. The range of a form in its individual instantiations could now be interpreted, as Franciscan theologians traditionally did even prior to the theoretical justification, as the addition of parts of that form—qua formality—which

---

[31] On the ancient antecedents (the ἁπλός of the Stoics; Galen, Philoponos) Sambursky, *Das physikalische Weltbild*, pp. 423–30; medical sources: McVaugh, "The Medieval Theory of Compound Medicines;" id., "Arnald of Villanova and Bradwardine's Law," pp. 56–64; other sources: A. Maier, "Die Struktur der materiellen Substanz," *Studien*, 3:3–35; id., "Die Calculationes des 14. Jahrhunderts," ibid. p. 263; E. Sylla, "Medieval Quantifications of Qualities: The 'Merton School,' " pp. 9–39, esp. pp. 12–24.

[32] Gregory of Rimini, *Lectura* 1 d.17 q.2, ed. Trapp, 2:321: "secundum proprietatem sermonis loquendo nulla forma augmentatur nec intensive nec extensive, sicut satis bene probat ratio, nec etiam suscipit magis et minus . . . quoniam forma, verbi gratia albedo, non fit magis albedo quam fuit prius" etc. He sums up, in this question, the current solutions (succession theory, addition theory) in great detail.

[33] A. Maier, "Das Problem der intensiven Grösse," in *Zwei Grundprobleme*, pp. 3–43.

coalesce *ad unum*.[34] This particular account of intensification lent itself almost naturally to a geometrical or arithmetical representation.

The very notion of "form" underwent significant changes during
the fourteenth century. Forms ceased to be the ontological backbone of the world and the sole instruments for its cognition. Scotus
joined those who dissolved the unity of substantial forms; instead,
many forms were said to coalesce in one substance—down to the
individual form. After Ockham and largely due to his influence,
epistemological discussions shifted ground from an *assimilatory* to a
causal account of cognition: the act of cognition ceased to be seen as
an identity with or a becoming one of the forms of things with the
intellect, a process mediated by sensible and intelligible species (*adequatio rei ad intellectum*). Rather, objects were now supposed to cause
in us intuitive and abstractive notions (*notitiae*), which function
as terms (*incomplexa*) of propositions. Only the latter can be true or
false, and hence the object of science: be they mental propositions
(Ockham), actual written or spoken ones (Holcot), or propositional
objects (*complexe significabilia*: Rimini, Pierre d'Ailly). In the Terministic discourse, forms were reduced to either qualities or relations of singulars.

It is, then, clear why the attention of fourteenth and fifteenth century schoolmen could have shifted from philosophical questions of
justification to the logico-mathematical questions of representation. The Mertonian tradition, though representing intensities by
extension—lines, areas—seldom mixed the categories, that is, seldom let a figure represent combined intensive and extensive properties.[35] Nicole Oresme's method of "configuration of forms" did
both in the framework of a general theory of representation. Assuming that the extension of quality be represented (or imagined)
by a line, and the various intensities of that quality as the different
heights to every point on the line, then the figure bounded by the
line and the curve linking all heights represents the precise distribution of that quality in the line (such will be the figure representing

---

[34] Duhem, *Système* 7:462ff.; Maier, ibid., pp. 44–58, denies, against Duhem, all links of
the problems of *intensio et remissio* and the *latitudo formarum*. It ought to be emphasized
that the so-called addition theory did not require a Scotistic commitment; e.g., Gregory
of Rimini, above n. 32.

[35] Sylla, "Medieval Concepts of the Latitude of Forms: The Oxford Calculators," pp.
223–83, esp. p. 278; *id*, "Medieval Quantifications" (above n. 31).

accelerated motion). Assuming that a quality is extended over an area, then the qualitative configuration will be represented, in the same way, by the volume bounded by the surface of a solid. As to the distribution of a quality extended over a three-dimensional figure, Oresme remarks with a sigh that we unfortunately lack a fourth dimension by which it could be represented separately; but it can again be mapped into space, and the subject thus may be said to have "two bodies"—a doctrine not unknown to medieval theologians and jurists.[36]

Here, in the theory of Oresme, the symbolic character of geometry becomes more pronounced—beyond the already considerable flexibility of the *Calculatores*. It is just a matter of perspective, or usefulness, whether a figure (say, an area) represents an area or something else (e.g., *velocitas totalis*). If the latter, then, to stay with the same example, time and distance are truly united in *one* symbol. Oresme's part in the later emergence of analytic geometry is contestable. But his part in the transformation of geometry into a formal language, capable of describing changing variables, is not.

What place did the "Nominalism" of the fourteenth century have in these developments? We recall that, for Ockham, mathematical notions were altogether connotative. Extension, number, time, degree were, altogether, concepts addressing relations between singulars rather than naming singular objects or absolute properties of them. Such connotative notions, if construed without redundancies, are not without a *fundamentum in re*; but they should not be hypostatized either. The Terminists ceased to view mathematics as an inventory of mathematical objects and their absolute properties; this was a necessary, but not sufficient, condition for the growing awareness of the formal nature of mathematical expressions.

To Ockham, as well as to some of the *Calculatores*, it may have seemed as if only such mathematical arguments that have a meaning in the interpretation of nature were worthy of consideration.

---

[36] Oresme, *De configurationibus* 1.4, ed. Clagett, p. 176: "Et quamvis qualitas superficialis ymaginetur per corpus, et non contingat esse vel ymaginari quartam dimensionem, tamen qualitas corporalis ymaginatur habere duplicem corporeitatem: unam veram ad extensionem subiecti secundum omnem dimensionem, aliam vero solum ymaginatam ab intensione ipsius qualitatis infinities replicabilem secundum multitudinem superficierum subiecti." The "body" so integrated over the *superficies* of all bodies will have, of course, a *volumen*; but will it have a definite *figura*? On the "two bodies" in medieval political theory above IV.D.5 and n. 90.

Others, however, admitted willingly that their models were mental constructs only. Various modes of representation were applied for different problems; more often than not they amounted to no more than a device for visualization. But again they prove how far Aristotle's prohibition had eroded. The contiguity of two qualities, each of which still occupies a range from zero to infinity, was represented by the horned angle.[37] It was, perhaps, the most elegant way to visualize the Scotistic coalescence of *formalitates* in an individual subject. Concentric circles, or a triangle, or even a simple line could represent the order of perfections in respect to God, the *ens perfectissimum*. Even Kant, we shall see, constructed simple, nongraded perfections as so many degrees of an overarching property, namely "reality"; the *ens realissimum* thus was synonymous with the *ens perfectissimum*.[38]

Not all the kinematic speculations of the fourteenth century were introduced as purely imaginary constructs. The study of fractional and irrational exponents was promoted by Bradwardine's so-called dynamical "rule," one of the few attempts of fourteenth-century schoolmen to apply the new mathematical-logical techniques to a real physical question. The Peripatetic theory of motion postulated a direct proportion between force and velocity, and an inverse proportion between the velocity and the resistance of the medium. If so, then any force, however small, acting on any resistant object, however large, must move it some distance, for F/R is always a positive magnitude. Even a fly must be capable of moving an elephant. Thomas Bradwardine solved the problem by assuming that increments of velocity depend on an exponential increase of the ratio of force to resistance.[39] The rule, which enjoyed wide acceptance, is of course useless to a classical physicist. Yet, an important change is

[37] Murdoch, "*Mathesis*," pp. 238–46, esp. pp. 242ff.

[38] Johannes de Ripa, *Questio de gradu supremo*, ed. Combes and Vignaux, pp. 143–222 (and above II.D.2); and Murdoch, "*Mathesis*." On Kant below VI.A.2; A. Maier's dissertation on Kant's category of reality and the problem of intensive magnitudes may have led to her later interest in the Scholastic origins of this notion and to medieval science.

[39] Crosby, *Thomas Bradwardine, His Tractatus de Proportionibus: Its Significance for the Development of Mathematics*, p. 112; A. Maier, *Die Vorläufer Galileis*, pp. 86–100; Murdoch, "*Mathesis*," pp. 225–33; Clagett, *Mechanics*, pp. 421–503. It is true that "Given the medieval convention of expression, Bradwardine's solution appears supremely simple and straightforward: double the velocity, double the force-resistance proportion" (Murdoch and Sylla, "The Science of Motion," in *Science in the Middle Ages*, ed. Lindberg, p. 233). On the other hand, it is the proportion of a proportion rather than a simple proportion.

indicated by it. Hitherto, forces could be represented as velocities or "motions" directly, and an increase of force could be represented by a proportional addition or subtraction of lines representing motion. Bradwardine's law permits force to be represented only indirectly: increase of force is measured by the *rate* of increase of velocity—very similar to Oresme's interpretation of *impetus* discussed earlier, and perhaps its source of inspiration.

4. Mathematics, Reality, and Harmony

From the vantage point of modern physics it may seem as if the mathematization of physics in the seventeenth century involved, in one respect, also its de-mathematization. Hitherto mathematics (or better, geometry) dictated to physics which formulae and figures ought to be looked for in nature: simple proportions, perfect geometrical figures. In the seventeenth century, it seems, natural philosophy emancipated itself from the tyranny of mathematics—and was, therefore, capable of *using* much more mathematics, or even generating new mathematics, taking its cue from problems of physics. Instead of being told by mathematics which constructs it ought to use, physics now turned to mathematics with concrete problems to be solved for *a* formula or figure—not necessarily the simplest or even a perfect one in mathematical terms. If circles do not describe celestial orbits, then the ellipse, a less perfect but more general figure, does. Again, the physicist of the seventeenth century employs mathematics rather as a language than as an inventory of real entities. And it seems further that some of this shift in the relationship between mathematics and physics is already noticeable in Bradwardine's rule. It substituted Aristotle's simple proportion with a proportion of a proportion. But this account is far too simplistic, and calls for many qualifications.

The *Calculatores* introduced a new style of mathematical kinematic reasoning. But they and their followers did not analyze the free fall of bodies with their new tools, or the real motion of projectiles. They may have felt that most real motions are too complex to describe kinematically. They certainly did not abandon the Aristotelian tradition that assigned to natural motions a geometrical representation by the "perfect" or "simple" figures of circle and straight line, and restricted the intelligible real motions to those constructable by the aid of compass and straightedge. The perfection of the

universe, its harmony, was still conceived as a static, geometric consonance; it consisted of a maximum of geometrical symmetry. At
best one can attribute to some theoreticians of the fourteenth century an awareness of the discrepancy between the simple geometric
representation of simple motion and the complexity of actual mathematical-kinematic analyses, not unlike the discrepancy between
the geometric description of celestial motions and their exact physical interpretation.

Nicolaus Cusanus expressed this resignation more strongly in
that he refused to attribute even to natural motions a perfect geometrical shape. He also stated quite clearly that mathematics is an
artificially constructed language, that mathematical entities are *entia
rationis* generated by us. As the ultimate conceptual abstraction,
mathematics is our best tool for understanding nature; it also forces
us to realize why, in the domain of the infinite, all our categories
collapse: an infinite circle is *eo ipso* also a straight line. Both the success and failure of the mathematical conceptualization are an image
of God's ideas—of the world he created "by measure and weight,"
and of himself as a *coincidentia oppositorum*. The descent from the
paradoxical mathematics of the infinite to the domain of finite magnitudes that are distinct and particularized because they obey the
principle of noncontradiction is analogous to God's contraction
into the creation.[40] Yet, Cusanus also epitomized the medieval reluctance to employ measurements in nature *because* they can never
be precise. Together with the ideal of absolute rigor, the seventeenth century also gave up the ideal of absolute exactness of measurement—only in such a way, as Anneliese Maier observed, were
the exact sciences made possible.[41]

The analysis of real motions, in kinematic or dynamic terms, had
to wait until physical thought was emancipated from the tyranny of
geometry and of geometrically perfect constructions. The parabola
that describes the motion of projectiles, the ellipse that describes
planetary orbits, are neither perfect figures nor even constructable
from perfect figures without "mechanical" means. The seventeenth
century did not abandon the notion of perfection, or harmony, of
the cosmos; it replaced the geometric-statical symmetry of the Pla

---

[40] Nicolaus Cusanus, *De docta ign.* 2.4; 6, ed. Wilpert, 1:44–46, 48–50 (34–38); above
II.D.3.

[41] A. Maier, *Metaphysische Hintergründe*, pp. 308–402, esp. 402.

tonic and Peripatetic tradition with a notion of dynamic consonance. With the growing insight into the symbolic-formal character of mathematics, "simplicity" came to mean generality rather than absolute symmetry.

Some of the more general preconditions for the transition to a dynamic notion of universal consonance were mentioned in the second chapter. Here I wish merely to draw attention to the important role of the abandonment of the obsession with constructable figures both in the detailed examination of concrete motions and in the general conceptions of universal "perfection." The gradual, by no means complete, loosening of this obsession is well illustrated by the changes in Kepler's cosmological-physical views from the *Mysterium cosmographicum* to the *Harmonices mundi*. The *Mysterium*, as Kepler remarked later, confined its search after harmony to the "matter," that is, shapes and sizes of the orbits so as to fit into the hierarchy of the regular bodies. But these, he says in the *Harmonices*, are merely the "building blocks," not the form and life of the cosmos that was built—he alludes to the Timaeus—"after the well articulated image of a living body." The harmonic relations are the "forms," corresponding to the relations between constructable polygons—constructable with compass and straightedge. These accommodate the actual motions of the planets, whose orbits are not "constructable," whose regularities must be explained by the mechanical, physical "forces" of attraction and repulsion.[42] The new "harmony" is constructed from motions-in-time.

None of this was anticipated by the natural philosophers of the fourteenth century. But they had a role in shifting mathematics from an inventory of ideal entities into a symbolic formalism capable of many interpretations. Jakob Klein described this shift for the

---

[42] Kepler, *Harmonices Mundi Libri V*, 5.9 prop. 49, in *Gesammelte Werke*, ed. Caspar, 6:360–63. Cf. Koyré, *The Astronomical Revolution*, pp. 256ff. (physical model), 326–43 (*Harmonices*); Koestler, *The Sleepwalkers*, pp. 388–98. Note that Kepler, *Harmonices* I prop. 1–49, pp. 20ff., uses Euclid's tenth book to distinguish between degrees of "knowing" (*scire*) of geometrical figures which is equivalent to their commensibililty or constructability with compass and straightedge. The essence of harmonies (expressed by rational proportions between polygons), as against the perfect bodies, is that they express motion-in-time (music). Kepler still viewed the ellipse as a less perfect figure than a circle. Yet it is worth noting that Baroque architecture showed a predilection for elliptic figures as no period before did. Wölflin, *Renaissance und Barock*, pp. 45–52. It expresses a new sense of dynamic harmony or "unity within variety."

sixteenth and seventeenth centuries:[43] it was not completed even
then. Upon this transformation depended the ability of mathematics to describe processes with the aid of function. The *Calculatores*
and their followers had in some instances a beginning notion of
both.

5. *Mathematics as Language*

The use of mathematics in the actual interpretation of natural processes and its growing formal-symbolic character went hand in
hand, even if mathematicians in the seventeenth century sometimes
emphasized the one aspect, sometimes the other. The program of
Descartes's analytical geometry consisted of the translation of algebraic into geometric terms (*introduire . . . termes d'Arithmétique en
la Géométrie*) or "expressing" (*exprime*) geometrical figures algebra-
ically.[44] The injunction against *metabasis* within mathematics was
openly violated; the barrier that seemed most fundamental to the
Greeks, and was not even disputed in the Middle Ages—the separation of numbers and continuous magnitudes—was removed.
More yet, Jules Vuillemin has shown that Descartes augmented the
distinction between "geometric" and "mechanical" curves with a
distinction between mechanical and "graphic" curves; the first two,
being solutions to algebraic equations to the second respective nth
degree, he accepted as legitimate; the latter, discontinuous curves he
rejected. Amongst the second category were the spiral and the *quadratrix*; he was often induced to attend to them by problems of motions, forces, and intensities.[45] In short, analytical geometry was, in

---

[43] Klein, *Greek Mathematical Thought*, pp. 150ff. See also Bochner, "The Emergence of
Analysis in the Renaissance and After," pp. 11–56, esp. pp. 22–25 (operations on real
numbers).

[44] Descartes, *La Géométrie* I, AT 6:370; Klein, *Greek Mathematical Thought*, pp. 197–
211, shows that the geometrical "imagination" corresponds to the actual shape of bodies
impressing our senses (esp. p. 210). "Geometry" to Descartes is first and foremost constructive geometry. See also Mahoney, *The Mathematical Career of Pierre de Fermat (1601–
1665)*, p. 44. On Descartes' limitations see Belaval, *Leibniz*, pp. 291ff.

[45] Vuillemin, *Mathématiques et métaphysique chez Descartes*, pp. 79–98 (classification,
construction) 35–55 (spirals). Spirals were already dealt with, *in extenso*, by Toricelli
and others. On the liberalization of means of construction in the seventeenth century
(Viéle, Kepler) and on Descartes's classification see also Bos, "Arguments on Motivation
in the Rise and Decline of a Mathematical Theory: The 'Construction of Equations,'
1637–ca. 1750," pp. 331–80 (reviewing the Greek sources he also fails, however, to distinguish between motion and motion-in-time).

Descartes's eyes, only an example of the heuristic force of his new universal method, valid in geometry as well as in optics, in physics as well as in psychology. What to the ancients was a cardinal vice, and to the Middle Áges a lesser one, now became a virtue: the transportation of models and arguments from one discipline to another.

Hobbes, though not himself a subtle practitioner of the art, viewed mathematics as a purely mental construct, an artificial language throughout. The artificial nature of mathematical concepts guarantees their absolute unequivocation. Mathematics is the paradigm for all other sciences because we created it ourselves out of nothingness: its veracity is entirely convertible into its construction. This, Hobbes believed, is true of every genuine science; in mathematics, however, it is more clearly apparent: truth lies in the consistency of our arbitrary construction. Mathematics, like thought in general, is nothing but computation.

No one in the seventeenth century came closer to the understanding of the formal-symbolic properties of mathematical reasoning than Leibniz. In arithmetic and geometry he saw instantiations—we would say models—of a *scientia generalis de relationibus* which he set out to construct.[46] Relations, as we recall, were in his eyes mere abstractions, without an ontic status, though indispensable for the ordering of phenomena. In other words, formal algorithms won priority over mathematical entities or properties. His version of the differential and integral calculus was, indeed, such a general-formal algorithm which he did not always know how to interpret.[47] Even more fundamental was Leibniz's *characteristica universalis* that was meant to serve as an *ars inveniendi* for all sciences. Simple ideas, he once hoped, could be assigned prime numbers; their combination would yield all compound ideas that there could be.[48] In his mature system Leibniz abandoned the hope of identifying simple properties;[49] but he did not abandon the hope of identifying formal algo-

[46] G. Martin, *Leibniz*, pp. 57–65.

[47] Boyer, *The History of the Calculus and Its Conceptual Development*, p. 212.

[48] *Dissertatio de arte combinatoria* (1666), GP, 4:27ff., esp. 43ff.; cf. GP, 7:187; Kneale, *Logic*, pp. 325–27; Schmidt-Biggemann, *Topica universalis: Eine Modellgeschichte humanistischer und barocker Wissenschaft*, pp. 186–211.

[49] Leibniz to De Volder (1703), GP, 2:249: "Doctrinam de Attributis quam hodie sibi formant (presumably Descartes, Spinoza, Malebranche) non admodum probo: quasi unum aliquod simplex praedicatum absolutum, quod Attributum vocant, substantiam constituat: neque enim ulla ego in notionibus invenio praedicata plane absoluta, aut quae connexionem cum aliis non involvant." Cf. above II.H.4.

rithms that would serve all possible inference-schemata. God, who already possessed a *scientia generalis*, created the world by "calculating his ideas."[50] The formalization of mathematics was, in a sense, its mechanization. Pascal and Leïbniz actually constructed a computation machine; the latter saw it as a crude model of an algorithm for a general science.[51]

6. The Three Meanings of Mechanization

Whether or not God's mind (or thought in general) is an ideal calculating machine, the physical universe was certainly viewed as one. It was an ideal clock—whether, as Newton thought, a clock that needed periodical rewinding or, as Leibniz insisted, a clock that runs perpetually with equal precision. The clock metaphor, however, now entailed altogether different features of the universe than its medieval or ancient predecessors.

The comparison of the universe to a great machine is an old metaphor that referred to the regular circular motion of the heavens.[52] In Antiquity and during the Middle Ages it was often much more than a metaphor. Complicated astronomical clocks were designed and built that visualized, represented, and facilitated the computation of celestial orbits. The Platonic academy seems to have possessed an armillary sphere: it underlies the creation of the universe

[50] GP, 7:191: "Cum DEUS calculat et cogitationem exercet, fit mundus." It is a marginal note to the following exchange in a *Dialogus* (written 1677): "B. Quid tum? cogitationes fieri possunt sine vocabulis. At non sine aliis signis. Tenta quaeso an ullum Arithmeticum calculum instituere possis sine signis numeralibus."

[51] Cf. Leibniz to Arnauld (n.d.; long before the sustained correspondence), GP, 1:81: "Habeo Machinas duas designatas, alteram Arithmeticae, alteram Geometriae provehendae. . . . Hoc si ad omnes figuras cogitabiles transtulerimus, non video, quid possit ad usum desiderari." Pascal's claims were much more moderate: *La Machine arithmétique*, in *Oeuvres complètes*, pp. 349–58. Cf. Goldmann, *Le dieu caché*, pp. 251–57.

[52] In Latin, *machina* can mean any artifact from the simple wheel to the system of the universe. Chalcidius translated *Timaeus* 32c (τοῦ κόσμου ξύστασις), 41d (ξυστήσας δὲ τὸ πᾶν) as "istam machinam visibilem"; "coagmentataque mox universae rei machina"; ed. Waszink and Jansen, in *Plato latinus*, ed. Klibansky, 4:25.7; 36.18; and the commentary p. 301.19 (*mundi machinam*). It may account for the fact that the metaphor was particularly favored by the school of Chartres; cf. Stock, *Myth and Science in the Twelfth Century: A Study of Bernard Silvester*, pp. 74 (*machina Fortune*—meaning the same as *rota fortunae*, the wheel of fortune), 199 (*machine corpore*), 208. Later usages: e.g., R. Bacon, *Questiones supra libros quatuor physicorum*, ed. Delmore, in *Opera hactenus inedita Rogeri Baconi*, 8:201.2–3: "ordinatio corporum universi et mundi machine congruentia, scilicet ne sit vacuum."

by the demiurge in the *Timaeus*.[53] A gradual, incessant technological improvement marks the history of these engineering devices down to the Middle Ages; they preceded the time clock and were its paradigm. From the fourteenth century onwards, the mechanical clock—now made much more precise through the use of the escapement—became the most admired man-made artifice, the paradigm of a perfect machine.[54]

In fact, the celestial motions of Aristotle, or even Ptolemy, could be represented by a mechanical device with reasonable accuracy. Kinematically, Aristotle allows the resolution of circular motions into circular motions only, or linear into linear. The apparent complexity of celestial motions needs, and must, be resolved into circular motions. Moreover, in both Eudoxian and Ptolemaic astronomy each planetary orbit is calculated and represented independently of the others. The system that combines them has only to provide a mechanism such that the motion of one planet not interfere—or influence—the motion of the others. Dynamically no forces need to be represented in the mechanical device because the spheres are moved by an agent separate from them—souls or the later separate intelligences. In short, ancient and medieval cosmology was susceptible to a mechanization in a literal sense of the word. Early modern cosmologies were not. The irony of the attribution of a "mechanization of the world picture" to the seventeenth century—and it is by no means wrong—lies in the simple, often overlooked circumstance that cosmologies based on the "new science" of mechanics were incapable of being represented by actual mechanical devices. Kinematically they resolved the circular or elliptical orbits into rectilinear components. Dynamically they assumed forces intrinsic to the system, and the system was conceived as a balance of motions and

---

[53] Cornford, *Plato's Cosmology*, p. 74 (to *Timaeus* 36c; cf. 40c and *Ep.* 2.312d). Farrington, *Greek Science*, pp. 40–41 has argued that Presocratic cosmologies—notably Anaxagoras'—were conceived "under the influence of techniques" watched in workshops. Then, perhaps, they were a kind of "knowledge by doing" in an even stronger sense than Plato's—but still not consciously so.

[54] Solla Price, "Clockwork before the Clock," pp. 81off. *Id.*, *Science since Babylon*, pp. 49–70. Against his thesis (the escapement as a perfecting of the *astronomical* clock; the need for timekeeping as a result, rather than a consequence, of the mechanical clock) Landes, *Revolution in Time: Clocks and the Making of the Modern World*, pp. 54–66. The importance of timekeeping as the expression of a sense of historicity unique to western Europe was already extolled by Spengler, *Untergang des Abendlandes*, 1:19, 171–75.

forces. It may be that today we could simulate the planetary system using electromagnetic fields; but it could not be done with the technological means of the seventeenth century. Needless to say, the machinelike working of an organism was likewise impossible to represent by a clock *en détail*.

Should we say, therefore, that the "mechanical philosophers" actually de-mechanized their universe, that the clocklike universe became for them a mere metaphor? Not at all. Ultimately, the mechanical, man-made devices had to remain, in the ancient and medieval perception, a metaphor only, because they represented *natural* through coerced, *artificial* motions. This distinction may have shifted in the later Middle Ages from the realm of immediate experience to the realm of reasoning; but it remained fundamental. It was abolished in early modern physics—hesitantly at first, radically thereafter. The mechanization of nature became neither a reality nor a metaphor, but a *model* and a *paradigm*.

But a model or a program for what? In a minimal sense "mechanical causes" stood for the desire to eliminate all but efficient causes from the interpretation of nature; to interpret all physical phenomena as "matter-in-motion." In this sense, the seventeenth century's ideal of mechanization was rooted in classical Atomistic thought. Now, it may seem as if the physics of Democritus was quantitative, while Aristotle's was qualitative. This is only true in respect to unrealized potentialities: the ancient Atomists never cared for a mathematical analysis of motion. The "shape, size, and order" of atoms moving in the void sufficed to explain why some of them joined together and others did not. The *clinamen* of the Epicureans intended to explain why atoms collide and form *vortices*. No further analysis was called for. To the contrary, from Atomistic quarters came the strongest attack on the very foundations of mathematics and apparently also on the foundations underlying Greek mathematical astronomy.[55]

While early medieval authors lost all sense of difference between Atomistic and Aristotelian physics—Isidore of Seville embraces both atoms and elements in peaceful coexistence[56]—Scholastics re-

---

[55] Above II.B. nn. 1, 7.

[56] Isidore of Seville, *Etymologiae* 13.2–3 (Lindsay). It is interesting to see how, without intending to do so, Isidore blurs the differences between the quantitative theory of matter of the Atomists and the qualitative elements of Aristotle. What the Atomists regarded as

stored some of the original argumentative context of the doctrine, though most of them were not interested in material invisibles. They tended, though, to restrict final causation to conscious acts only; and some of them warned, as also once Maimonides, against the presumption that mankind is the ultimate goal of creation.[57] The revival of Atomistic or corpuscular doctrines in the seventeenth century owes much less to Atomist doctrines than to a new, more specific sense of "mechanical."

7. Mechanization as Constructive Knowledge

In this, the narrower sense of the term, "mechanical" stood for the explanation of a given set of phenomena as a closed, semi-autarkic, balanced system of motions and/or forces; a system that sustains itself, at least for a while. Whether the elements of such a system were corpuscles or atoms was, in this respect, of secondary importance. The *balance* and interaction of motions and forces had to be constructively proven: that such-and-such putative motions coalesce in one rather than another way had to be shown through a mathematical analysis and synthesis, resolution and composition.[58] I do not think that Buridan's expansion of impetus mechanics to the heavenly motions can be seen as an early suggestion of such a system.[59] It allows planets to follow their orbits automatically. But their motions are still conceived as independent of each other; indeed, lack of friction and of any other influence or interaction allows the impetus given to each heavenly body to actualize the nature of impetus as a *res natura permanens*. Buridan's cosmology was "mechanistic" in the first rather than in the second meaning of the word.

---

an indirect proof for the existence of atoms—the random movements of dust particles that are seen if a ray of light penetrates a room through a window—Isidore takes to be a metaphor only. What they took as metaphor (letters as elements) he takes literally. Both theories of matter do not stand as different theories. Both seemed, perhaps, true to him because the words existed, and with them the *vis verbi* (*Etym.* 1.29.1).

[57] Maimonides, *Guide* 3.13; cf. above III.B.n.25 (Thomas); Descartes, *Principia phil.* 3.3, AT, 8.1, p. 81.

[58] "Eine Maschine," as defined by C. Wolff, *Deutsche Metaphysik* §557, p. 337, "ist ein zusammengesetztes Werck, dessen Veränderungen in der Art der Zusammensetzung gegründet sind . . . und dem nach ist die Welt eine Maschine"; quoted by Schmitt-Biggermann, *Maschine und Teufel: Jean Pauls Jugendsatiren nach ihrer Modellgeschichte*, p. 69.

[59] Above III.C.2. Oresme may have alluded to this (not his) theory when he compared the motions of the spheres to that of a clock with escapement: Oresme, *Le Livre du ciel* 2.2, ed. Menut and Denomy, p. 288; *Oresme*, ed. Clagett, p. 6 and n. 10; White, *Medieval Technology and Social Change*, pp. 125, 174.

Medieval cosmology never involved the search after clues for the making or breaking of the universe. Aristotle's doctrine of the eternity of the world—not in some form or forms, but with all of its forms—was an integral part of his physics and metaphysics. It is true that medieval schoolmen learnt how to preserve the Aristotelian cosmology and yet argue that it is compatible with creation out of nothingness.[60] Yet no medieval author had to reconstruct the early history of the universe in order to comprehend its present structure. The first chapters in Genesis had to be reconciled with cosmological theories, but added little to their understanding. That God chose to create the universe in six days rather than momentarily, *in ictu tempore*, had symbolical rather than cosmogonical reasons. Nachmanides even goes so far as claiming that the story of creation elucidates the miraculous, unnatural order of the elements now.[61] Early modern cosmologies, by contrast, involved cosmological speculations almost inevitably. They had to account for the rational construction of the complex balance of the motion that the universe presents now, and they did so from Descartes to Kant and Laplace.

Descartes, as once Plato in the *Timaeus*, calls his reconstruction of the mechanical processes governing the primordial random motion of particles a "plausible narrative." And elsewhere he says:

> I resolved . . . to speak only of what would happen in a new world if God now created, somewhere in an imaginary space, matter sufficient wherewith to form it, and if He agitated in diverse ways, and without any order, the diverse portions of that matter. . . . Further I pointed out what are the laws of Nature, and . . . tried to show that they are of such a nature that even if God had created other worlds, He could not have created any in which these laws would fail to be observed. After that, I showed how the greatest part of matter of which this chaos is constituted, must[!], in accordance with these laws, dispose and arrange itself in such a fashion as to render it similar to our heavens; and how meantime some of its parts must form an earth, some planets and comets, and some others a sun and fixed stars. . . . I did not at the same time wish to infer from all these facts[!] that this world has been created in the manner which I described; for it is much more probable that at the beginning God made it such as it was to be. But it

---

[60] Above III.D.n.18.

[61] Nachmanides, *Perush hatora* to Gen. 1:9, ed. Chawel, 1:14: air, not fire, is the most subtle of elements; its proper place ought to be above fire; divine decree keeps it under fire. Cf. Funkenstein, "Nachmanides," p. 45 (hidden miracles).

is certain, and it is an opinion commonly received by the theologians, that the action by which He now preserves it is just the same as that by which He at first created it. . . . [The] nature [of all things] is much easier to understand when we see them coming to pass little by little, than were we to consider them all complete to begin with.[62]

In other words, if left to itself, matter, endowed with a constant "quantity of motion," can be shown to have formed our universe inevitably. Contrary to Newton, Descartes cannot even conceive of another world obeying the same laws of nature as ours. The "imaginary space" is truly imaginary—while Newton, we recall, needed an infinite space so as to permit God, if he so wishes, to create other worlds with or without the same laws that govern our universe.[63] And Descartes can be caught red-handed calling his reconstruction "facts." Throughout this passage, one senses the echo of Aristotle's distinction between poetry and history: the former is more valuable, because it constructs events as they could always be.[64] Most important, however, is Descartes's assertion—not at all accepted "by all theologians"—that the logic of creation is the same as the logic of the preservation of the order of the universe; that there can only be one way of its construction and, therefore, of its reconstruction. Later, in the *Principia*, Descartes compared his hypothetical world and the God-made actual world to two clocks (*horologiae*) which are identical on the outside but have a different mechanism inside.[65] This can be read in two different ways. It can mean that construction of the world could be explained by yet another mechanism—this, however, he ruled out in the *Discours*. It could also mean that the world created by the same principles of mechanics in one act of God and the slowly constructed world of his hypothesis should be compared to two watches; but if so, Descartes's analogy is misleading; because the difference in this case is just a difference in the rate of assembling the parts of the construct. This is the sense in which some Greek commentators in Antiquity understood Plato's account of the construction of the world. "The cosmos itself," Proclus

---

[62] Descartes, *Discours* 5, AT, 6:42–44; cf. *Principia* 3.46, AT, 8.1, pp. 101ff. On the Cartesian theory of vortices, cf. Aiton, *The Vortex Theory*, pp. 30–58. As to its later career, Aiton shows that it was not refuted; it just faded away as scientists lost confidence in it.

[63] Above III.E.1.

[64] Aristotle, *Poetics* 1451b1–10.

[65] Descartes, *Principia* 4.204, AT, 8.1, p. 327.

quoted them, "exists everlastingly; but the discourse distinguishes
that which becomes from its maker and introduces in temporal order things that coexist simultaneously, because whatever is generated is composite."[66] The "temporal order" of the whole universe
was really an accidental property in any Greek cosmology (save,
perhaps, the Atomistic). The structure of the world was eternal.
Not so in Descartes's cosmology. Change, motion-in-time, was
not only the property of "matter" as against the structure of the universe: it was a structural property of the universe as a whole. We
may imagine this "time" as short as we wish, even condensed to an
instant—it must still be there. Descartes, if consistent, cannot interpret his cosmology the way Proclus interpreted Plato's: motion-in-
time underlies the very constructing *principles* of the world. This,
too, is captured by the simile of the clock. The simile was to have
important—and sometimes different—usages later. Already here,
in spite or even because of its ambiguity, it testifies to the extent of
the claims of a mechanical explanation of nature. Even those "mechanical philosophers" of the seventeenth century who shunned
such extreme claims retained some of them. Medieval schoolmen
believed that we know something about the static structure of the
"fabric of the world" together with many facts about it; but that
only God has the ultimate kowledge *propter quid*, that only God
knows the universe in the way it was made. The mechanical philosophers of the seventeenth century came close to believing that, even
if we can never hope to know all the facts about the universe, we
know nonetheless enough of its dynamic principles to reconstruct
its making in the way that God does.

The mechanical clock, in whose perfection the scientists and
craftsmen of the seventeenth century invested so much energy, was
the most suitable analogue to natural, mental, and social processes
for more than one reason. A clock is a machine that, once wound,
works of itself. Its work is not a work on something—pulling,
pushing, or lifting another object. Its work is performed by the very
regularity of its motions. So also does the universe—and the universe, moreover, *is*, by definition, the most precise time-telling device. Organisms are clocks too: a healthy body has a regular heart-

[66] Proclus, *In Timaeum* 1.382 (quoting Porphyry and Iamblichus against Plutarch and
Atticus). Cf. also above n. 7 (imaginary motion), n. 9 (abstraction).

beat, and the circulation of the blood had recently been made susceptible to a mechanical explanation. The *train of thoughts* or associations of ideas could be shown to be an inevitable process: perhaps like motion, perhaps themselves motion. Both connotations, the mental and the physical, were captured by the simile of two synchronized clocks. Guelincx introduced it into the mind–body debate of the seventeenth century,[67] and it became the most popular symbol and commonplace of this debate.

Moreover, the mechanical clock, which became the handiest example of the superiority of European culture over others, was also the supreme example of knowledge-by-construction. The knowledge of its construction is identical with the knowledge of its reconstruction. The Chinese, who lacked this knowledge, were incapable of repairing the European clocks of which they were so fond and which they imported in ever-growing numbers in the seventeenth century.[68] This may have led the authors of the *Logique du Porte Royale* to another famous simile. A Chinese Aristotelian, given a clock, will attribute its regular beat to its "sonorific quality."[69] We who know how clocks are made can dispose with the "obscure qualities" of the school—both in the understanding of the mechanics of clocks and in the understanding of the mechanics of nature.

8. Mechanization and Cosmology

Medieval cosmologies, such was the conclusion of the section before last, never involved of necessity a rational account of the formation of the universe, because their knowledge of the universe was not knowledge-by-construction. I am, of course, aware of the long tradition of philosophical reflections on the first chapters of the Book of Genesis, in commentaries on the *Hexaemeron* and outside them. Most of them drew a sharp line between creation and conservation, *opus conditionis* and *opus restaurationis*:[70] the universe was created out of nothingness, its order established during the six days of creation for all time. This seemed to be the plain sense of the Scrip-

---

[67] Guelincx, *Annotationes ad Ethicam* tr. 1 s.2 §2, p. 33 n. 19, in *Opera Philosophica*, ed. Land 3:211–12; Specht, *Commercium*, pp. 173-74 n. 97.

[68] Landes, *Revolution in Time*, pp. 39–44.

[69] Arnauld, *La Logique ou l'art de penser* 1.9, trans. Dickoff and James, p. 69.

[70] Hugh of St. Victor, *De sacramentis*, prol. 2.3, Migne, *PL* 176: 183–84; *De scripturis* 2, Migne, *PL* 175:11.

tures, unto which the philosopher might graft the cosmology of Aristotle. Two reasons, then, coalesced to render the question of the *formation* of the universe minor, if not redundant: (i) Aristotle's cosmology envisages an eternal universe—in its whole and all of its essential parts; and though medieval schoolmen corrected Aristotle as to the eternity of the universe, many of them retained enough of his cosmology so that their conceptual framework did not encourage the vision of a developing universe; (ii) the Scriptures themselves concentrate the work of creation into the first six days only, and even in those, it was stressed time and again, the order of creation did not follow from any intrinsic necessity: God could have created all at once. That he did not do so may have had pedagogical or other reasons.[71]

Assuming that—as I argued further—Descartes's rational account of the formation of the world was an archetype of knowledge-through-construction. Was it also a portent of physics to come? Did not, for example, Newton reject the Cartesian vortices because he did not wish "to feign hypotheses"? At least in the English horizon, it seems, knowledge-by-construction was confined to the experimental tradition (and may have been prepared by alchemical and magical practices); cosmogony was not perceived as a necessary part of astrophysics. And yet, Newtonian physics led—inevitably, I believe—to the Kant-Laplace hypothesis, a resuscitated version of Descartes's vortices with the Newtonian vocabulary. The mechanical account of the present state of the solar system (let alone the universe at large) assumed *eo ipso* a potential history of the cosmos. Even if one assumes, as Newton did, that all the planets and stars were created simultaneously and simultaneously set in motion, to keep the solar system in a perpetual balance, God has, as Newton concluded, to undo from time to time the effects of history—to restore the balance between gravitational and inertial forces which otherwise slowly tilt in favor of the former to cause the collapse of all matter into itself.[72]

In other words, the problem posed by the mechanical, construc-

---

[71] So already Philo, *De opificio mundi* 3.13-14, ed. Cohen and Wendland. That the world was created all at once was held, in the ancient Jewish exegesis, by R. Neḥemya (*Genesis Rabba* 12.3, ed. Theodor and Albeck). Only the derivatives (*toladot*) appeared successively during six days.

[72] Above II.G.2.

tive cosmologies of the seventeenth century was not—as was the case in the Middle Ages—to account rationally for the universe having a beginning in time. Indeed, we encounter outside the horizon of Scholasticism in the seventeenth century only few and uninteresting discussions of that matter. The eighteenth century's materialists were to revise it (and plead for the eternity of the world); Kant later placed the dispute among his antinomies. But if *creatio ex nihilo* ceased to be a problem, its problematics reappeared in a new guise, namely, how to account for the *conservation* of the universe that—at least in principle—was capable of intrinsic changes.

Descartes drew his rational account of the history of the universe so as to permit, at some phase, a stable balance of motions. Malebranche accepted the Cartesian account—but only, as Proclus once accepted Plato, as an account of possible constellations, not as a description of reality. Only God knows how he made the universe. Newton accepted the potential historicity of the universe, but bothered the Almighty to intervene and undo its effects. His universe, his physics, knew no conservation of force.[73] His fierce rejection of the vortex theory had good physical reasons, which he lays out in the *Principia*; it was also rooted in the wish to have God intervene unpredictably in the course of the world, and therefore, perhaps, also by the wish to pose a limit to our knowledge-through-construction. Leibniz remained, *more suo*, ambiguous. He rejected furiously the image of the universe as a mundane clock in need of rewinding. He also rejected, *in physicis*, perpetual motion, and did not deny that the universe has an intrinsic history. To the contrary, every possible world contains its history in its very concept, as much so as the monads that make it. But is such a possible world—like its monads—also eternal, governed by its own "principle of activity," read: conservation of forces? We look for an answer in vain.[74]

---

[73] Ibid.

[74] Monads are eternal by nature, i.e., suffer no process of coming-to-be (generation) or disintegration (corruption). God alone can create them out of nothingness or annihilate them. Possible worlds must, by definition, have the same property. If one of its monads is taken away, a possible world does not change: it turns ipso facto into another world. "Creation in time" is as meaningless as "annihilation" if time is merely a reference structure between monads or between predicates within monads (above II.H.4). Leibniz held, therefore, to his own version of a *creatio continua*. He spoke of "Fulgations continuelles de la Divinité de moment à moment" (*Monadology* §46, GP, 6:614) as the source of

In conclusion: applying knowledge-through-construction to the whole world was as inevitable as it was dangerous. It was dangerous because it makes mankind be "like God, knowing good and evil." Many seventeenth-century philosophers shunned its inevitable consequences; but only the Occasionalists had the courage to deny categorically that this kind of knowledge reveals reality.

C. THE CONSTRUCTION OF NATURE AND THE CONSTRUCTION OF SOCIETY

1. The Dialogue with Hobbes's Social Theory

The ideals of a mathematical and mechanical explanation of nature converged, in the thought of the "mechanical philosophers," into an ideal of knowing through construction. The vice of applying methods and procedures of one science to another was transformed into a virtue. The science of mechanics became a paradigm for a new psychology, a new medicine, and a new social theory.[1] Hobbes praised himself as the founder of the latter. No other thinker of the seventeenth century argued as consistently as he did for the constructive character of all human manifestations—language, science, political order. No one stressed more forcefully that all knowledge is knowledge by doing.

The core of his political theory lies in the novel insight that neither a social instinct (*inclinatio ad societatem*) nor indeed an urge for perfection, social or otherwise, is part of the basic endowment of human nature. Social organization of human beings—unlike some beasts—is not a natural product, but rather altogether an artifact. The continuous dialogue with Hobbes is, I believe, the distinguishing mark of modern political theories. The most important political thinkers since the seventeenth century did not reject him outright even if they were profoundly irritated by his claims. Instead, they absorbed the full force of his arguments before transforming them into a different, sometimes even a contrary, theory. Giambattista

---

beings, or sometimes of emanation (*Ecclaircissement des difficultés* etc. GP, 4:553) as the mode of perpetual creation. Medieval Aristotelians also maintained that the world is eternal by nature, even if God created it; some even ascribed this view to Aristotle himself.

[1] Schofield, *Mechanism and Materialism: British Natural Philosophy in the Age of Reason*, esp. pp. 40–87 (iatromechanics).

Vico is a most revealing example. He refused to accept the paradigmatic role of mechanics precisely because he endorsed the principle that truth and what is made are identical, *verum et factum convertuntur*.[2] Since we did not make nature, we cannot hope to understand it properly, either; but the science of humanity is entirely open to our investigation because—here Vico agrees entirely with Hobbes—society is a human artifact, because "we made the commonwealth ourselves." Our second, historical nature is entirely our own making. Another example for the fundamental importance of Hobbes's social theory may stand here for many others. When Marx denied that exchanging wares is a perpetual manifestation of a natural "propensity to barter" that elevates man above beast,[3] he did to economic theories what Hobbes did to political ones. Economic order became altogether a human, albeit necessary, artifact.

2. Atomism, Gassendi, and Hobbes

Hobbes's insistence on the thoroughly positive, thetic character of laws was not an altogether new position. Vico, among others, recognized that the modern brand of political realism stood within a long tradition.[4] The contraposition of φύσις and νόμος was introduced by the Sophists and adopted by the Atomists.[5] Strangely enough, few of Hobbes's modern interpreters have commented on the obvious link between the revival of the classical corpuscular-mechanistic cosmology in the seventeenth century and the revival of the radically thetic interpretation of social institutions likewise char-

[2] Above IV.A.I.

[3] Smith, *The Wealth of Nations* 1.2, ed. Cannan, p. 13 (propensity to barter).

[4] *De uno principio*, in *Opere*, 2:32: "Quare adhuc Carneadem de iustitia an sit in rebus humanis, aequis momentis in utramque partem dissertare, adhuc Epicurum, Nicolaum Macchiavellum *De principe*, Thomam Obbesium *De cive*, Benedictum Spinosam in *Theologo politico* et nuper Petrum Baylaeum in magno *Dictionario* gallice conscripto, illa obstrudere vulgo audias: ius utilitate aestimari, temporique locoque servire; imbecillos postulare ius aequum; at 'in summa fortuna,' ut Tacitus ait, 'id aequius quod validius.' Ex quibus colligunt et concludunt metu contineri societatem humanam, et leges esse potentiae consilium, quo imperitae multitudini dominetur." Cf. also *SN* §1109 where Vico sees the belief in Providence guiding the course of human institutions vindicated and Epicurus, Hobbes, and Machiavelli, who believed in "chance," and Zeno and Spinoza, who believed in "fate," refuted. The distinction is important: Hobbes, determinist as he might have been, appeared to Vico as a philosopher of "chance" because of his insistence on the arbitrary character of human institutions. Cf. Hugo Grotius, *De iure belli ac pacis* 1, prolegomena 2, p. 2; 16, p. 10.

[5] Guthrie, *History of Greek Philosophy* 3:55–147.

acteristic of the Atomists.[6] A superficial comparison between Epicurus and Hobbes cannot fail to reveal the congruence between some of their main positions. These similarities should be stated before they are discarded (and justly so) as peripheral. Epicurus's ἀνάγκη and Hobbes's determinism have the same positional value in their respective systems; the elimination of theological considerations. Hobbes, however, could do without the complementary assumption of the slight original declination παρέγκλισις of the atoms to account for chance or free will,[7] much as he could abandon the strictly corpuscular theory itself. Both ascribe an actual and denumerable material substrate for every discrete entity, including the soul (Epicurus) or even God (Hobbes). The epistemology of both consists in the sensualistic interpretation of impression, "images" (*simulacra*), and their mechanical association. A similar account of man's emotional economy leads both systems to an abrogation of man's social nature.[8] To their grim description of the natural state corresponds the rejection of idealizations of primitivity as an *aurea aetas* and the famous, detailed description found in Lucretius of the bestial, anarchic state of primordial man and the rise of religion out of dream images and frights.[9] The strictly contractual origin of laws is likewise a part of Epicurus's doctrine, the stress on the function of political order as a crime preventing agency only, and even the demand for an unconditional submission to authority.[10]

Hobbes had direct access to both Diogenes Laertius and Lucretius, and we can only speculate as to the depth of his doxographic knowledge (which he must have possessed even if we see a grain of

[6] Cf., however, Strauss, *Natural Right and History*, pp. 188ff, who sees the main differences as (i) that between merely ethical and mainly political interests, and (ii) Hobbes's adherence to a new concept of natural right. The first difference is merely of intention; the second is not convincing.

[7] Bailey, *The Greek Atomists and Epicurus;* Sambursky, *Das physikalische Weltbild*, pp. 328–35, esp. p. 334; Zeller, *Dei Philosophie der Griechen*, 3.2:390–429 (esp. pp. 408, 421 and n. 5).

[8] Zeller, ibid., 3.2:455, 471 n. 1; cf. Gassendi, below n.12. On the theory of motivation see also Schwarz in *Charakterköpfe aus der Antike*, ed. Stroux, pp. 149ff. (Ethics); and Kafka and Eibl, *Der Ausklang der antiken Philosophie und das Erwachen einer neuen Zeit*, pp. 58–67 (the assumptions common to Epicurus and the Stoics; these are the reasons that it was easy to mistake Hobbes's doctrine of affections to be of Stoic origin).

[9] Above II.C.n.7.

[10] Diogenes Laertius, 10.150, in H. Usener, *Epicurea* (Leipzig 1887), p. 78.8. On the history of conventionalism in Antiquity see Strauss, *Natural Right and History*, pp. 81–119.

truth in Clarendon's ironic remark that he was a "man of . . . some reading, and somewhat more thinking")[11] and as to the part, if any, this knowledge had in the formation of his thought. Nevertheless it is important to note that a revival of Epicurus's doctrines, even *in politicis*, was conceived by Gassendi and was well underway when Hobbes elaborated his political ideas.[12] Hobbes knew of Gassendi after his third visit to France; we may at least assume some mutual stimulation. Again, a superficial examination will reveal a number of common features to Hobbes's and Gassendi's systems even apart from the Atomistic substratum of both. Both shared an interest and conviction in the new mechanics. Gassendi's formulation of the inertial principle is earlier (at least in publication) than Descartes's and closer to the formulation adopted by Hobbes.[13] Two points in Gassendi's ethics and political theory merit particular stress in light of Hobbes. Defending Epicurus's alleged abrogation of all natural human ties, Gassendi stressed the advantage of seeing man's inclination toward his fellow men, or even toward his family, as an outcome not of blind instinct (*caeco quodam impulsu naturae*) but of education and self-elevation.[14] Social attitudes, Gassendi seems to indicate, are not innate, but a product of man's labor on himself. This is the essence of Hobbes's (and Vico's) opinion later. Indeed, a long section of Gassendi's *De iusticia, iure et legibus* may already

[11] Edward, Earl of Clarendon, *A Brief Survey of the Dangerous and Pernicious Errors to Church and State in Mr. Hobbes' book entitled Leviathan*, p. 2.

[12] Gassendi's literary plans to make Epicurus acceptable date back to 1631: see Rochot, *Les Travaux de Gassendi sur Épicure et sur l'Atomisme*, pp. 31ff. In 1647 his *De vita et moribus Epicuri* appeared, and in 1649 the commentary on the tenth book of Diogenes Laertius. His main positions and plans were well known during this time. Hobbes's first encounter with Gassendi (through Mersenne) occurred during his stay in 1634–1637. The "first tract" may have been written before; but the interest in problems of sensation, his first reading of Galileo's *Dialogo*, occurred during this crucial time. Some share might be attributed to the encounter with modern Epicureanism. For the date of the composition of the "tract," see Watkins, *Hobbes' System of Ideas*, pp. 40–46. On Gassendi's (and Epicurus's) influence in France, see Spink, *French Free-Thought from Gassendi to Voltaire*, pp. 85–168. Cf. also above n. 11.

[13] Cf. Brandt, *Thomas Hobbes' Mechanical Conception of Nature*, pp. 282–85, 327; Dijksterhuis, *The Mechanization of the World Picture*, pp. 429–30; and Lasswitz, *Geschichte der Atomistik vom Mittelalter bis Newton*, pp. 150–54, 172–73.

[14] Pierre Gassendi, *Syntagmatis philosophia* t. 2 par. 3, p. 754. Gassendi's social philosophy is thus much clearer than as interpreted by Borkenau, *Der Übergang vom feudalen zum bürgerlichen Weltbild*, pp. 430–34. Cf. Sarasohn, "The Influence of Epicurean Philosophy on Seventeenth-Century Political Thought: The Moral Philosophy of Pierre Gassendi."

show the influence of the confrontation with Hobbes. A review of the pessimistic descriptions of man's natural state concludes with the remark that the brutal *status naturalis* is not necessarily a historical reality but a necessary fiction.[15] Its purpose is to understand society as it would be now without legal agreements, rather than to describe society as it once was;[16] in other words, it is a limiting case, much as the (mechanical) principle of inertia is not a description of any existing motion, but an imaginary limiting case of motions considered under the ever-diminishing impact of outer forces.

Unlike Gassendi, Hobbes did much more than justify or revive a broken (if not forgotten) tradition. The many affinities to Epicurus's ethics do not bear on the determining concepts and concerns of Hobbes's political thought. Hobbes is first and foremost concerned with political power and collective security, that is, in the state rather than in the happiness or autarky of the individual. But beyond the differences in interest, (i) Hobbes's contraposition of "nature" and "convention" is far more radical and far more methodical than any of the doctrines of his forerunners. And Yet, (ii) precisely this elaborated contraposition enabled Hobbes to anticipate a mediating formula. Finally, (iii) Hobbes's methodological and conceptual borrowings from the new science of mechanics were sufficient to lend his concepts of both "nature" and "society" a new appeal and greater precision.

3. Mediating Nature and Convention

No classical or medieval author ever drew the line between the realm of nature (matter-in-motion, sensation) and the realm of convention, of artificial constructs, as sharply as Hobbes did. And yet

---

[15] Gassendi, *Syn. phil.*, p. 795: "Itaque quicquid sit de illa seu suppositione, seu fictione status, in quo seu Epicurus, seu alii vixisse aliquando discunt primos hominos, tam esse protecto vidatur ipsa societas hominum, quam illorum est origo, antiqua; ac non eo quidem solum et modo, quo bruta generis eiusdem sociabilia inter se sunt; verum illo etiam, quo quatenus sunt et intelligentes, et ratione praediti, agnoscunt non posse ullam inter se societatem esse securam, nisi ea conventionibus, pactique mutuis constabitur." Gassendi maintained, even after his turn from skepticism to Epicureanism, the insistence on the *hypothetical* structure of science, as did Hobbes, albeit not an Atomist. Cf. Gregori, *Scetticismo e Empirismo: Studio su Gassendi*, esp. pp. 179ff. Cf. also Popkin, *Scepticism*, pp. 100–12, 145–54.

[16] Among modern interpreters of Hobbes, Macpherson, *The Political Theory of Possessive Individualism: Hobbes to Locke*, pp. 17–29, stresses this aspect of the natural state most.

he sought to mediate between them—in his theory of science as well as in his political thought. With great vigor Hobbes set out to prove that even though all human institutions—language, religion, law— have their origin *in* nature, they should nonetheless be understood as artificial constructs through and through. They belong to nature inasmuch as "foresight" can be reduced to sensations and sensations to matter-in-motion; yet they derive, I shall argue, their *validity* not from nature, but from human imposition. Only a deliberate, volitional act, and not any aetiological rationale, will lend to human institutions—beginning with language itself—what they need in order to be absolutely valid and endure: absolute univocation.[17]

The first instance in which Hobbes can be shown to have sought a mediation between nature and imposition is his theory of *conatus* (endeavor). Endeavor is "the beginning of motion," a measure of motion in an instant before it translates into distance.[18] It is not motion or action, but the tendency to act or react—a force. In simple bodies under impact, *conatus* translates into a preservation of motion with due changes in direction, just as anticipated by Descartes's laws of motion. In elastic, complex bodies it translates in part into a complex pattern of inner motions, which result in the acquisition of force. With this somewhat vague but fruitful notion, Hobbes thought he had solved Descartes's inability to account for either elastic bodies or other forms of delayed reaction. Leibniz admitted his indebtedness to Hobbes. The particles of motion that hit or pen-

[17] Hobbes's pronouncements to the effect induced much-discussed revisions in the interpretation of his ethical and political doctrines. Taylor, "The Ethical Doctrine of Hobbes," pp. 406–24, interpreted Hobbes's theory of obligations, and correspondingly, his insistence on *mala per se*, as almost foreshadowing Kant's categorical imperative, while Warrender, *The Political Philosophy of Hobbes: His Theory of Obligation*, chose to see in Hobbes's natural laws divine precepts. For a discussion, see S. M. Brown, "The Taylor Thesis," in *Hobbes Studies*, ed. K. C. Brown, pp. 31–34, 57–71; and Watkins, *Hobbes' System of Ideas*, pp. 85–89. My suggestion is, again, that both the absolute and the relative character of obligations are but *aspects* of the notion—the natural genesis of obligations both demands their imposition as absolutely binding and relativizes their validity *in praxi*. This view of Hobbes's mediation between teleological and mechanistic models is close to that of Polin, *Politique et philosophie chez Thomas Hobbes*, pp. 7ff., 12–23, 51ff., 176ff.

[18] Thomas Hobbes, *De corpore* 3.15–16, in *Opera Philosophica quae Latine scripsit omnia*, ed. Molesworth, vol. 1. On the formation of the notion in Hobbes's writings before *De corpore* cf. Brandt, *Thomas Hobbes' Mechanical Conception of Nature*, pp. 294–303. Cf. also Watkins, *Hobbes' System of Ideas*, pp. 120–37 (causality and volition). The link to Leibniz was already seen by Hönigswald, *Hobbes und die Staatsphilosophie*, pp. 81–83.

etrate our organs of sensation likewise cause "inner motions." These translate into phantasms, a complex local pattern of inner motion, and thought. They cause another mode of inner or initial motion—force—that we know as volition. Volition is thus both dependent and relatively independent of the outer world. It receives its formative challenge from matter-in-motion and translates back into the same, but it does so through the mediation of a complex pattern of inner motion—harder to predict the more complex it is. The dominating role of motions and *conatus* is the preservation of the state which a body is in—*conatus suum perseverare motum*—which pertains not only to motions, but also to the balance of motions within complex bodies.[19] In short, will and mechanical causality are not at odds.

Hobbes's commitment to the basic matter-in-motion character of our "phantasms" forbade him to view them as a "picture" of reality in our imagistic sense. They rather represent reality inasmuch as causal regularities in the *sensatum* cause regularities in the sensation. He never assumed a one-to-one correspondence between phantasms and things, nor did he doubt the overall validity of our sensations and memories of sensations. Man and beast alike have them, and even higher animals can associate an impression to another if they succeed each other regularly. One then turns into a "natural sign" of the other. The association of impressions is the basis of all intellectual activity. But even the animals with the highest intelligence are incapable of converting natural or "inner" signs into arbitrary, conventional, and hence communicable signs. Rain may always follow clouds, and an animal may run for shelter when it sees clouds. But clouds do not predict rain. Only an arbitrary act can make clouds into a sign for rain; which would then be unambiguous only if "clouds" would signify *only* rain, and not clouds.[20] Better

[19] He did not, however, distinguish circular from rectilinear inertia; Brandt, *Hobbes*, pp. 303ff. Neither did Galileo.

[20] "Natural signs," we learn in *De corpore* 1.2.2, are such things that can be used to signify other things that follow them regularly. But Hobbes warns against any inference from signs as to such regularities (ibid. 1.5.1): clouds may be a sign for rain, but they do not predict it. Only an arbitrary act makes such (ambiguous) signs into univocal "names" (ibid. 1.2.4), and yet the objective causal nexus guarantees a rough approximation of our system of names to the order of things. The very same structure permeates Hobbes's epistemology and theory of science. We do not perceive things in themselves, only their phantasms; yet Hobbes's belief in the strict chain of material causes guarantees

yet is an arbitrary sign which by nature is tied neither to clouds nor to rain. The unequivocation of arbitrary signs is the basis for man's ability to calculate with them, to anticipate the future. "Foresight" is grounded on the human capability to construct an artificial system of signs. An unequivocal language is the best guarantee for society, for the body politic, for science.

The most interesting feature in Hobbes's theory of language—which assimilates reminiscences of the medieval theory of suppositions with Nominalistic traditions—is, I believe, its anti-substantial character, diametrically opposed to current meta-linguistic theories, be they rationalistic or empiricistic. "SeP" merely means that S is the name of P whether S stands for an element in the phantasm or for another name. "S is" or "S exists" is a meaningful proposition only if understood elliptically, as a shorthand for "S is something," that is, "S is a body" (body is the most universal category, or name of names, of all enduring elements in sensation).[21] There is no bigger danger in science than the danger of hypostatization, of endowing names of names with properties of names or the converse. Our world, that of the common man as that of the scientist, is a construct. This is our confinement as well as our opportunity, because truth lies only in propositions, and to be true is the same as to be constructable. Truth is fact in the original sense of the latter—that which can be done. Earlier I tried to show the important critical functions that the "method of annihilation" had in late medieval Scholasticism. For Hobbes, the imagination of the self *toto mundo destructo* became a prerequisite for the reconstruction of a termino-

___

our knowledge of the outer world. Truth is sometimes seen as a property of sentences only in virtue of the consistent usage of names; at other times, it rests on basic intuition. His theory of science emphasized at times the thetic-arbitrary beginnings of science (from definitions), at times the hypothetical-experimental beginnings. Cf. Pacchi, *Convenzione e ipotesi nella formazione della filosofia naturale di Thomas Hobbes*, esp. pp. 194–215. Rather than seeing such assertions as expressions of very distinct phases of development, we can regard them, without denying changes of emphasis, as aspects of the following state of things: although there is a general congruence between phantasms and things, it is not a one-to-one relation; the congruence is guaranteed by the strict material causation in the universe; the task of systematic knowledge is at times to remove mistaken or abundant "names" and connections between names, at times to anticipate (infer) new connections. Language and science have their origin *in* nature, although they are arbitrary throughout. Only when made arbitrarily univocal can language help to "understand" nature.

[21] *De corpore* 1.3.2–4 (proposition as relation between two natures—whereby the *copula* is redundant).

logically consistent, scientific world view out of the remaining *phantasmata.*[22]

4. The Source of Absolute Obligations

On the one hand, then, language has its own origin in "natural signs"; on the other, the natural connections among regular sequences of events are always ambiguous, and no "sign" is actually natural. Inasmuch as they are part of our language, they are always arbitrary, disregarding their origin. Whatever is, is by "nature" ("nature," however, is again a mental construct); but the definite meaning of human institutions cannot be derived from an appeal to the laws of nature, nor do these suffice as a condition of their validity. Natural necessities (the weakness and equal vulnerability of all men in the state of nature, and hence the fear of death) may have generated our submission to contractual obligations, conferring our "rights" to the sovereign; but necessities of nature do not and cannot explain why such obligations must be severed altogether from considerations of expediency; in other words, why they must be *absolutely* binding.[23] The absolute allegiance to the sovereign knows no exception, allows no rebellion. And yet, should a rebellion succeed, and Hobbes sees most states as dominions by acquisition, the very same allegiance is due to the new, illegitimate sovereign.

This peculiar, almost dialectical relationship between nature and convention, as a clue to the understanding of the structure of all human artifacts, is the core of Hobbes's argument. The argument draws its strength precisely from the fact that it is not utilized in the analysis of the state only, but to explain all manifestations of sociality, actual and potential. The most obvious instance in which Hobbes can be shown to seek for a mediation between "nature" (necessity) and "imposition" (volition) is his theory of *conatus*. Only a few among Hobbes's contemporaries, and fewer still among his later interpreters, have grasped this consistent dualism and its me-

---

[22] *De corpore* 2.7,1. On the principle of annihilation in the Middle Ages above III.B.3. The importance of annihilation both in Hobbes's epistemology and social theory—as a precondition for reconstruction—has been stressed by Goldsmith, *Hobbes's Science of Politics*, pp. 16–17, 84–85. Descartes: above, III.D.3. To Boyle, conceiving "that all the rest of the universe were annihilated" rendered a criterion to distinguish primary from secondary qualities: *Origins of Forms and Qualities According to the Corpuscular Philosophy*, in *Works*, 3:22–23.

[23] Above n. 17.

diation; most of them tend to stress either the materialistic-egotistic or the "Nominalistic"-voluntaristic aspect of his thought. Hobbes's concept of nature, natural laws, and language allowed him to deny a one-to-one correspondence of words and things, of imposed constructs and the fully determined chain of physical motions, and yet to insist on an overall congruence between both. So also the state. It is, we believe, neither a simple mechanical compound nor an organism; it has no discrete body. Its "matter" and "artificer" are the diverse human beings who constitute it.[24] Bodies, simple or complex, are distinguished by their force, or "natural resistance," which allows them to endure and maintain their integrity. The state has no such natural durability: its endurance cannot be calculated with the help of any physical laws. The contract on which it is founded is an *ens rationis*. The state is a *creatio continua* and always in danger of dissolution. It is a condition of man fighting *against* nature (including his own) with the aid of devices and designs taken *from* nature and driven by natural necessities.

It might be objected that the classical question of whether social entities exist φύσει or θέσει was in Antiquity raised not only *in politicis* but also in all spheres of human activity, including language. Yet, never before was the disjunction elaborated or its dialectical structure stressed so consistently as by Hobbes. One important difference between the cultural philosophy of Epicurus and that of Hobbes is that the former was interested in the natural substrate of language;[25] Hobbes, while not denying it a natural origin, is rather interested in the *validity* of language. In the concept of "natural signs" as a product of both the natural sequence of images and an artificial function, even though this function is itself a product of causal necessity, Hobbes finds another model for analyzing other human constructs, including "natural laws" and obligations.

---

[24] Thomas Hobbes, *Leviathan*, ed. Oakeshott, p. 5. The analogy of the human body, with which Hobbes commences, is not to indicate that the state is a natural body, but that it is an imitation thereof, much as its constitutive act, the covenant, "resembles that *fiat*, or the *let us make man*, pronounced by God in the creation" (ibid.).

[25] Whereby "nature" differed from society according to its different impressions. Steinthal, *Geschichte der Sprachwissenschaft bei den Griechen und Romern*, pp. 325–29; for a detailed description of the genesis and background of the controversy the book is still most useful, in spite of its strongly Hegelian vocabulary. See also Cassirer, *The Philosophy of Symbolic Forms*, trans. Mannheim, 1:148.

Hobbes's state is likewise both: a product of natural (determined) causality *and* an arbitrary construct.

As against the traditional way of posing the question, Hobbes is interested both in the physical mechanism by which the social order is constructed and in the unique laws that govern these constructs qua constructs.[26] In the main tradition of political thought, those who did not derive social order from necessity or nature or natural law did not investigate much further; having once postulated the thetic character of legal institutions, they went no further. They knew that human laws were not merely a reflection of nature, but they did not inquire into the question of what devices could be applied for their validation. Hobbes's inquiry begins where the classical positivistic tradition ended.

The contraposition, albeit for the sake of an ultimate mediation, of nature and society could not have been defended without the new laws of motion and the new method of eliciting them represented by (if not originating in) Galileo's mechanics. The new laws of motion had, since the seventeenth century, two paradigmatic functions. Hobbes's theory of science as a systematization of mental constructs corresponds immediately to his theory of society as an autonomous (though necessitated) human construct, and his concept of the state of nature is nothing but a limiting case analogous to the inertial principle. Both are derived by severing a phenomenon (body, society) from its actual context and seeing it "in itself."[27]

At the same time, the laws of motion could be taken as the ultimate laws to which social phenomena could be reduced if, indeed, there is nothing but matter-in-motion; or at least they could be taken as a material metaphor or paradigm of the laws governing society. The conception of the relationship between physical and political bodies will, throughout the seventeenth century, oscillate between the merely *methodological* and the *material* analogy. Hobbes, in our interpretation, tended toward the former, which does not exclude the *metaphorical* use of mechanical terms. Whether we read

[26] *De cive* preface, *English Works*, 2:xiv: "For everything is understood by its constitutive causes. For as in a watch, or some such small engine, the matter, figure and motion of the wheels cannot be known, except it be taken assunder and viewed in parts; so also to make a more curious search into the rights of states and duties of subjects, it is necessary . . . that they be so considered as if they were dissolved."

[27] Above III.c.n.10 (Clauberg).

him correctly or not, Hobbes himself based his claim to have founded "civil philosophy" on his methodological ingenuity, not on the novelty of his opinions.

5. The State as a Physical Body: Spinoza

Hobbes's provisional argument for the separation between natural and social phenomena could be reversed on his own premises. Spinoza agreed with Hobbes on many physical and anthropological principles. He reversed the emphasis, however. A more subtle account of man's emotional balance (enabling him to dethrone *fear* from the almost absolute monopoly it held in Hobbes's system) and a more realistic outlook of the state (as only one of the many social formations) enabled him to place society once more within nature, to deal with the state as a complex physical body (or balance of power), and to ascribe also to the state the *conatus suum conservandi motum* which Hobbes seems to have denied that it possesses.

Although, properly speaking, extension (or substance as extended) is but one in number, Spinoza insists on the existence of definite physical bodies—on the right to see not only in simple but also in compounds or compound of compounds discrete physical entities.[28] Physical bodies are defined by the *conatus suum conservandi motum*, which depends in part on their individual *quantitas motus*, that is, balance of motions. Descartes's general formula of conservation of momentum in the universe ($m \cdot v$), is used, we recall, as a definition of *single* bodies, as a physical *principium individuationis*. As long as the body retains its basic "proportion," its specific balance of motions, it remains *one* body (an *individuum*) even if parts of it are replaced by others.[29] Organic bodies are mechanical complexes capable of assimilating other bodies or regenerating substitutes (not necessarily copies) of a lost element. Spinoza did not abandon altogether the Cartesian relativization of single bodies or movements; he gave it a positive meaning, much as he did in the case of corresponding ideas. Every piece of extension in motion may be seen as one, or many, bodies, depending on our point of view. A worm in our blood will recognize the blood, not as one body, but as a universe full of many (microscopic) entities; neither do *we* recognize

---

[28] Above II.F.I; Gueroult, *Spinoza*, 1:529–56.
[29] Spinoza, *Ethics* 2 prop. 13 lemma 4; and above II.F.I.n.12.

the *facies totius universi* as what it is—one huge organism. The *unity* of an *individuum* is its form of organization and consciousness thereof.[30]

These reflections have grave implications—psychological, anthropological, political, and ethical. Some of these Spinoza shared with Hobbes; in many cases he is more consistent. Against Hobbes, he can view society literally as an organism of sorts, as a body.[31]

(i) He agrees with Hobbes that man's emotions and actions are a function of his *conatus esse conservandi* only, but he draws a different picture of the emotional economy and its political relevance. Spinoza emphasizes the inner conflict of affections due to their origin in inadequate self-consciousness. Hobbes regarded man as one soul, one person, one body; Spinoza's definition of a body and its image allows him to see in man many bodies, many sometimes conflicting ideas, and souls.[32] For Hobbes, fear of death was the only emotion powerful enough to force man into the social contract and keep him within the state; Spinoza allows for a whole gamut of sometimes contradicting manifestations of self-interest—ambition, gain, fear, care of others—to participate in the creation and maintenance of the *res publica*.[33] Society cannot be maintained by fear alone, especially abstract fear as the realization of what would happen without laws. Society is a balance of more or less enlightened self-interests, and only if they converge is the body politic durable. The prudent sovereign is he who does not attempt to eliminate such private interest and opinions (which cannot even be achieved through terror) but

---

[30] Above II.F.n.13.

[31] Spinoza, however, does not explicitly declare societies to be physical bodies; the nearest expression to this effect is in the *Tractatus Theologico-Politicus* 3, Van Vloten-Land, 2:124: "Ad quod nullum certium medium ratio et experientia docuit, quam societatem certis legibus formare certamque mundi plagam occupare, et omnium veres ad unum quasi corpus, nempe societatis, redigere." Yet, our analysis has shown that every single corpus short of the whole universe is only a body in a relative sense.

[32] *Epistula* 24, Van Vloten-Land, *Opera* 3:107: "Unde sequi videtur, sicut corpus humanum ex millenis compositum est corporibus, ita etiam humanam mentem ex millenis constare cogitationibus; et, quemadmodum humanum corpus in millena resolvitur, unde componebatur, corpora, sic etiam mentem nostram, ubi corpus deserit, in tam multiplices, ex quibus constabat, cogitationes resolvi."

[33] On Spinoza's theory of passions and its consequences on his political theory, see Wernham, *Spinoza: The Political Works*, pp. 6–11. Lack of any reference to Spinoza's theory of bodies is the only omission in this excellent analysis. See also Wartofsky, "Action and Passion: Spinoza's Construction of a Scientific Psychology," in *Spinoza: A Collection of Critical Essays*, ed. Grene, pp. 329–53.

utilizes these private interests, without his subjects' awareness, to his own purpose. In this part of his doctrine Spinoza perhaps found use for Maimonides' "cunning of God," in a manner reminiscent of Mandeville's "private vices, public virtues," or Vico's "providence."[34] This is true for monarchies, aristocracies, and republics alike: the quality (or power) of a constitution depends not on the form of government so much as on the coordination of private interest it achieves, its delegation of power, and systems of mutual control of instances.

(ii) Society, both Hobbes and Spinoza assume, is antagonistic. Yet Hobbes's sovereign is exempt from this antagonism: either the state has the power by delegation of natural rights or it is no state at all. Such delegation by social contract, Spinoza seems to maintain, has no absolute meaning. The state has, identically with its subjects, as many rights as it has the power to enforce them. The sovereign merely participates in the antagonistic, permanent *libido dominandi*.[35] Of course, he can enforce his will by sheer terror; such permanent slavery Spinoza sees as the basis of the endurance of the Turkish Empire.[36] Yet, when such a state collapses, it collapses totally; no change of design, no accommodation is possible. If not a reign of terror, a constitution has to, in order to endure, permanently reconcile all self-interests of all individuals, groups, and the sovereign himself. At any rate, Spinoza's state has no surplus dignity over its constituents; nor is its unity of a higher ontological level than that of individuals, groups, or estates. A complex physical body in general may be stronger or weaker than each of its parts, more primitive or better developed.

[34] Above IV.A.1.

[35] Wernham, *Spinoza*, pp. 28–35. In his *Political Theory of Possessive Individualism*, Macpherson presents Hobbes's analysis of the state as close to the Marxist analysis of the antagonistic structure of the bourgeois society. Certainly, Spinoza offers even a better analogy. The Marxist analysis of the modern state insists on the inclusion of the state within the antagonistic members of society. The bourgeois state only *appears* as an embodiment of collective aspirations and as being above the antagonism of society. In fact, it is the guardian of the right to antagonize, that is, exploit, radically as never before. The separation between state and society is a bourgeois myth. Marx, *Zur Judenfrage*, in *Die Frühschriften*, ed. Landshut, pp. 171–207 (with a strong leaning on Hegel's dialectics of being and appearance—*Sein und Schein*).

[36] Spinoza, *Tractatus Theologico-Politicus* preface and ch. 7.

(iii) The more enlightened a society is, the clearer its self-image and the consciousness of its deficiencies will be.[37] An enlightened society will always seek to achieve greater consistency by complying with its basic "pattern" or by changing its pattern if necessary. Spinoza's *Tractatus Politicus* is a reflection on how to define these patterns for existing constitutions. Aristocracies, for example, consist of a given proportion between patricians and subjects, which must not be smaller than one to fifty;[38] Spinoza implies that De Witt's administration failed because of a lack of this proportion. In order to achieve the greatest consent, an aristocracy must keep the patriciate mobile—a mercantilistic oligarchy suits this idea best—and highly decentralized. Spinoza's ideal constitutions were, in Vico's language, polities of merchants and shopkeepers.[39]

(iv) Rigid, completely unified political organisms are not necessarily more durable. The simpler a body, the more dependent on outside circumstances; a balanced, closed system becomes all the more durable the richer its inner movements, since its power of regeneration, or "replacement," is increased. Tyrannies, or rigid constitutions, could last longer than complicated systems of balanced particular interests, but for outer pressures and circumstances; in the face of those, a rich, flexible system has the better chance. A rigid "unification" is as useless as total anarchy. A stone might last longer than an organism, yet if it does collide and break, it will have lost its identity totally. If a human organism suffers the loss of a hand, it can replace its function through other organs. A tyranny that loses its tyrant is gone; a republic, a monarchy, or a polity with enough organized delegation of powers and active groups has a better capacity to withstand outer crisis. This is, in the long run, why the Israelite theocracy failed: it was too rigid; it collapsed when external pressures loosened. In short, Spinoza's political theory is a commentary on his fundamental axiom *ordo et connexio idearum idem est ac ordo et connexio rerum*.[40]

---

[37] Spinoza, however, assumes (*Ethics* 3 prop. 58; 4 props. 35–37) that the wiser a man is, the more he recognizes the similarities and similarities of interest between himself and his fellow men.

[38] *Tractatus Politicus* ch. 8, ed. Wernham, *Spinoza*, p. 379.

[39] SN §335, quoted in Wernham, *Spinoza*, p. 343.

[40] Above II.A.n.6.

6. *"Mechanical Philosophy" and Revolutions*

The history of social theories since Antiquity is not devoid of fresh starts and novel approaches. Nor were the best of them short of systematic arguments. The constant search for a new and specific method to ground every new interpretive effort is nevertheless an effort peculiar to modern times. The methodological consciousness of political theories of the seventeenth century and long after took the natural sciences as a model. Not only was the *metodo risolutivo e compositivo* or its equivalents copied; the vocabulary of post-Galilean mechanics was applied to "political bodies." Instead of the organological metaphors describing society and its institutions in the early tradition of political thought, or the corporational terminology that replaced them in the later Middle Ages, mechanical metaphors start their eminent career. In the seventeenth century Hobbes and Spinoza stand for the most consistent attempts to integrate natural sciences and social theory. Their argument is interesting in itself. It also teaches us something about the possibilities and limitations of scientific analogies. For historians of the age, they are instructive in yet another way. They reflect a persistent and influential mood of the times.

The mechanical analysis (methodological or material) of "social bodies" assumed more than it could prove. Mechanical terms were, in their use for political theory, in no way better grounded than the organological-biological or corporational terms they replaced. But they served as a most adequate expression for a new conviction. Society and its institutions ceased to be an *immediate* product or reflection of nature. They became artificial bodies, a product of man's deliberation and labor, not of his alleged "social instincts." As such, they were believed to be capable of a thoroughly rational design.

In a certain way, the motif was inherited from the political thought of the previous centuries. *Der Staat als Kunstwerk* may or may not have been, as Burckhardt believed, the reality of Renaissance Italy; it certainly was a typical Renaissance dream.[41] The last in a certain kind of utopias, and the most interesting, was Campanella's *civitas solis*, a design for the perfect *polis*. The utopian city is artificial in all respects. It represents the perfect environment, ca-

---

[41] Manuel and Manuel, *Utopian Thought in the Western World*, pp. 150–80; Feldon-Eliav, *Realistic Utopias.*

pable of transforming man's nature by transforming his circumstances. The reformation or transformation of man is its goal. In summing up this tradition, Campanella's utopianism has become at the same time a prototype of future utopian ideologies. Ideologies can often be defined not merely by their positive contents, but also by a systematization and forced unification of the adversary's point of view: "dialectical materialism" in the vulgar version of Lenin most clearly reveals its logic through the alleged unity and continuity of "idealism" that it postulates. Campanella found a continuity of diabolic logic in the "Aristotelian" point of view: *exiit Machiavellismus ex Peripatesimo*.[42] Aristotle, believing in the eternity of the world and hence in the periodical repetition of all its constellations, could not but assume the unchangeable nature of man also. To accept mankind "as it is" means to confine political theory to a "realistic" and deplorable logic of *raison d'état*.[43] But science can prove that the world progresses to its goal, and mankind must be perceived as capable of perfection.[44] The polity that can and will breed a new type of man, not least by planned eugenics, is a true outcome of *poiesis*, an act of designed creation after the pattern of eternal ideas. Whether Campanella drew these conclusions or not, his *civitas solis* is a true picture of the cosmos, shaping matter by imitating the ideas.

Hobbes and Spinoza, to name only them, when negating utopian thinking, did not simply return to the Aristotelian and medieval tradition of political realism. True, they insisted again on the assessment of "man as he is" as the only basis of any possible political theory; nor did they believe in a possible transformation of human nature or of the anthropological conditions of society. But they retained, against the Aristotelian or medieval tradition, the belief in the power of man to affirm the state or to negate it, to design all its

---

[42] Campanella, *Atheismus triumphatus*, p. 20; cf. the introd. to *Metaphysica*, ed. Di Napoli, 1.22 and also 3.114: "Qui vero negant religionem, sunt indocti, sophistae, ac scelesti, exitium mundi, ut Aristotelici, Sadducei, Averroistae, Epicurus et Machiavellus." Cf. Vico's list of the adherents of *raison d'état* (above IV.A.n.3). It should be added that Campanella's *Metaphysica* had some influence on Vico; cf. K. Werner, *Giambattista Vico als Philosoph und gelehrter Forscher*, p. 145 (the doctrine of *primalitates: posse, nosse, velle*).

[43] Meinecke, *Die Idee der Staatsraison in der neueren Geschichte*, pp. 115–29.

[44] On Campanella's "scientific" eschatology see also Doren, "Campanella als Chiliast und Utopist," in *Kultur und Universalgeschichte*, pp. 242–59; and Bock, *Thomas Campanella, politisches Interesse und philosophische Spekulation*, pp. 229–98, esp. pp. 265ff.

institutions or to change them deliberately. Society is an artifact and only inasmuch as man's mind is a part of nature and obeys natural laws can society be called natural. Some implications of this new sense are obvious; some were actually drawn, some were not. If the state originates in a necessary fiction and thus has neither a body nor an innate *conatus suum esse conservandi*, then it takes more than a virtuous monarch nourished by a wise instruction (*speculum regale*) to sustain it. Maintaining society must be, for Hobbes, a continuous effort of all society; the making of the state is a *creatio continua*. Or, if we regard a polity, with Spinoza, as a complex body or a self-correcting balance of adverse forces, then the sovereign is hardly a demiurge, but he should certainly be a decent mechanic. The determining force of society is not nature (as instinct or climate) but a man-made constitution. Certainly constitutions, since only of functional value, might be changed when appropriate. A possible implication of the mechanization of politics was the preparation of the rational idioms of revolution. Political theory began to conceive revolutions as neither illnesses of the organism nor as quasi-natural catastrophes (*revolutiones*),[45] but as deliberate changes of design.

7. Teleology and Mechanism

The description of states as "balance of power" is not far from "invisible hand" explanations discussed in the last chapter. The latter were rather evolutionary, the former static, and both were instances in which a genuine mediation between causal ("blind") mechanism and goal-oriented structures was sought during the seventeenth and eighteenth centuries—and earlier. The need for such a mediation was, of course, felt urgently in many other domains—certainly by those who subscribed to Descartes's program to resolve all organic phenomena into "matter in motion." That an automaton must be a self-regulating, self-correcting, and self-reduplicating mechanism was vaguely understood by Spinoza (who tried to specify the conditions under which elements of a "complex body" turn into functional elements that can be replaced by other members of the body) and Leibniz (who assumed a dual, teleological and mechanical, structure to every representation of forces). Yet, none of these attempts amounted to a viable cybernetic model, with identifiable

---

[45] Above I.C.n.I.

and constructable material elements and precise recursive rules of transformation. Such models have become today a powerful heuristic tool in understanding language, life, and society; they may yet enable us to understand organisms by constructing simple ones out of inorganic compounds. They are a precise mediation between mechanical causality and functional, quasi-teleological structuredness. The "mechanical philosophers" since the seventeenth century lacked such models; at best, they hoped for them. Some, like Vico, restricted their search to the social domain only. Others, like Kant, set the limit of mechanical explanation to lifeless matter: we may be able to account precisely for the past and present state of the machine of the heavens; we cannot describe the mechanisms of even the smallest leaf. And Kant has also defined the elusive structure: we assume, but cannot construct, a *Zweckmässigkeit ohne Zweck*, a goal-orientedness without a goal.[46] A *teleological* explanation that would be *mechanical* at one and the same time is denied to our discursive intellect (*intellectus ectypus*); it is the mark of such an intellect that it assumes a given multiplicity to a rule. The multiplicity is, then, contingent. It could be otherwise, and the difference between the real and the possible is the driving force of our scientific investigation. A superior intellect (*intellectus archetypus*) may be free of such impediments. It may conceive of the parts as constituting the whole and the whole nonetheless as determining the parts: no gap exists for it between the possible and the real, because it *creates* that which it *thinks of* in the immediate sense of the term.[47]

The only domain, then, in which a serious effort was made to describe such quasi-cybernetic systems was the consideration of social—later of economic—processes. The emerging idea of such systems distinguishes, we can now say, between the medieval version of "the cunning of God" and their modern counterparts—the "hidden plan of nature," "the cunning of reason." The task was thus defined: to discover mechanisms of "accommodation," of adjustment and survival. *Tantae molis erat Romanam condere gentem.*

[46] Immanuel Kant, *Kritik der Urteilskraft, Werke*, ed. Weichschedel, 10:299, 319 (beauty), 325, 473, 544.

[47] Ibid. §77, 10:526.

VI

CONCLUSION:
FROM SECULAR THEOLOGY TO
THE ENLIGHTENMENT

A. KANT AND THE DE-THEOLOGIZATION
OF SCIENCE

1. Justifying a Digression

Whether or not one favors the assumption of a secular theology in the seventeenth century, I believe that I have at least proven the existence, in that century, of a peculiar idiom, or discourse, in which theological concerns were expressed in terms of secular knowledge, and scientific concerns were expressed in theological terms. Theology and other sciences became almost one. I have tried to shed some light on the unique texture of this common idiom by comparing it to past theological interests or modes of reasoning.

Eventually, the secular theologians of the seventeenth century (if, indeed, they merit this name) gave way to a new generation of *savants* whose posture was often anti-theological, sometimes also anti-religious, occasionally even atheistic.[1] It seems as though the secular theology of the seventeenth century was bound to dig its own grave, because it often stressed, however ambiguously, the self-sufficiency of the world and the autonomy of mankind. Immanuel Kant was neither the most radical nor the most representative among the philosophers of the "Enlightenment." Yet his relentless endeavor to emancipate metaphysics and science from its theological baggage, and to develop an ethical theory in which human beings are their own supreme law-givers, was the most sys-

---

[1] During the seventeenth century, "atheism" was still a label imposed on others. "Car c'est la mauvaise coustume des ignorans d'appeler Atheés tous ceux qui ne se rendent pas à tous prejugés, et quand on aime la veritable liberté, on n'est pas republiquain pour cela . . ." (Leibniz to Burnett, 1701, on Toland, GP, 3:279). Self-proclaimed atheists appear only later in the eighteenth century. Even then the enlightened Mendelssohn, though eager to secure freedom of religion, excludes them from the absolute tolerance he demanded toward religions.

tematic and complex. Although it falls outside the chronological scope of our study, the temptation to add a few remarks about his critical enterprise can be justified in that he not only removed God from his methodological offices, but he was also the first to have fully grapsed and articulated them. I do not read the history of seventeenth- and eighteenth-century thought as a prelude to Kant. Sometimes Kant failed to follow some of its most promising avenues. Yet, in pursuit of his program of the de-theologization of science, Kant articulated many assumptions and aporias of the secular theologians better than they did.

2. Thoroughgoing Determination

Kant's refutation of all proofs for God's existence is based on the demolition of the methodological functions of the concept of God in any future understanding of nature. Like some of his rationalistic predecessors, Kant recognized that the idea of the *absolute* rationality (or intelligibility) of the "totality of things" demands, entails the so-called principle of "complete determination" (*durchgängige Bestimmung*) of everything.[2] Kant's distinction between the logical "determinability" of a *concept* and the consistent, transcendental "determination" of a *thing* may be illustrated as follows: a rational number is always completely constructed by two integers. An irrational number, though it can never be completely constructed, is always constructable to any desired precision. To the question "Is the nth number after the digit the number four?" the answer is always yes *or* no; I can construct the irrational number up to (n) and determine its value. But the irrational as a whole is never completely determined (at least from an intuitionist point of view), that is, completely constructable. Or consider another illustration. To the question whether Napoleon had a Muslim ancestor the answer is already determined as yes or no, even if in practice I may never be able to ascertain it. But the same question asked about Stendhal's Julien Sorel is neither yes nor no until I asked and answered it arbitrarily, be-

---

[2] Kant, *KdRV* B599–611, *Werke*, 4:515–23. In a different terminology: Kant denies the existence of "incomplete objects." These are not objects, but "concepts" only. Cf. T. Parsons, *Nonexistent Objects* pp. 20–21. Kant, however, attributes such non-existents with various degrees of "reality"—which Parsons cannot do.

cause Stendhal was mute on that point. Julien Sorel is *only* a concept because he is not "completely determined."

Completely determined means determined against all possible simple predicates, be their number finite or infinite. Simple predicates are those that neither imply nor exclude each other. Leibniz's monads were construed from such simple predicates. Kant called them, *pace* Baumgarten, "realities" rather than "perfections." Both terms refer to a long tradition according to which the positive predicates (attributes) of a subject add to its reality: the more positive attributes it has, the more real it is.[3] As just stated, all of them are compatible by definition; wherefore, Kant says, a hypothetical subject is indeed conceivable in which all simple predicates inhere, and it would then be "the most real thing," an *ens realissimum* that embodies the idea of the unity of all realities (perfections)—*der Inbegriff aller Realitäten.* But if everything is throughout determined, it does not follow that whatever is thoroughly determined is a thing. We tend to form an hypostatized version of the assumption of a complete determination of every "thing." Claiming that such a thing is conceivable is far from saying that this *ens realissimum*—which our reason "hypostatized, thereafter personified"—must be conceived as existing, that it is a necessary being. Existence is not a predicate, hence not one of the realities attributable of necessity to the "most real being." Kant admits that another concept, the concept of a necessary being (*ens necessarium*), does entail existence; but it is a vacuous concept, a concept without further content. The fallacy of the ontological proof of God's existence, even in its most sophisticated elaboration (such as Leibniz's), is not that it understood existence as an attribute (this is how Kant's refutation is usually rendered). It rather lies in the arbitrary and hence mistaken identification of two concepts of reason—the concept of the most real thing being the concept of the necessary being. The ontological proof is a case of mistaken identification of two ideas of pure reason.

More important to our concerns is the circumstance that Kant set out to prove that the principle of complete determination, and with it the methodological concept of God, is at best a regulative ideal of reason and has no bearing whatsoever on our actual interpretation of nature by our understanding. In a more recent idiom we might

[3] Above II.A.n.18; II.H. 3–4.

say that they are not theoretical, but at best metatheoretical assumptions. They are the means by which pure reason (*Vernunft*) conceives of totalities in themselves and in their ultimate, discrete, but altogether abstract components: things in themselves are conceived as completely determined. But neither the principle of complete determination nor the concept of a "sum total of all realities" is a part of our experience of nature (*Erfahrung*). Nor are they necessary to understand experience, to structure experience, as are the categories of understanding. The interpretation of experience does not demand the construction of absolutely simple predicates and of things-in-themselves in which such simple predicates could inhere. Even Leibniz admitted that much. Simple predicates and their sum total are pure abstractions. The rationality and coherence of our experience—that is, of nature—and of the spontaneous categories with which we grasp and pattern it are grounded not on the principle of complete determination, but rather on the "synthetic unity of our consciousness." In other words, the coherence and consistency of our experience do not demand or entail the assumption of the ultimate coherence of the "totality of all things." It suffices to assume (as we must) that, *if* an entity existed which has no orderly link to any other members of experience, such an incoherent something could not be perceived; just as a mathematician cannot deny the existence of totally random sequences of numbers, even though he must insist, by definition, that there can be no *formula* to construct such sequences. If there were such a formula, the sequence would *eo ipso* not be a random one. Our experience is an a priori patterned, coherent experience. This is a much more modest claim than that for the unity and coherence of the "totality of everything," which is unprovable and perhaps even self-contradictory.

All this is not to say that the methodological concept of God—and principle of thoroughgoing determination—are not *linked* in some ways to our understanding (*Verstand*). The principle is indeed a projection of, or extrapolation from, a very basic figure of logic from which one important category of understanding is also derived, namely that category that permits us to quantify qualities and to discriminate the real from the unreal. It was also the very starting point of the neo-Kantian theory of science. A more elaborate explication is needed for Kant's distinction between "negative" and "infinite" judgments and the category of "limitation" which corresponds to that distinction.

3. Negation and Privation

Lies may lack a leg to stand on, but they have many faces. Among the many complicated problems of negation is also this one: Can a logical system, as formalized as can be, do with only *one* form of negation? The answer appears to be negative; we must, so it seems at first sight, distinguish various modes of negation—so as to distinguish between negation within a statement and the negation of a statement, between meaningless and false propositions, between well-formulated formulas and nondemonstrable formulas, between factual and category-mistakes, and so on. Or should we argue, *pace* Prior and others, that negation is always one and the same, while its causes may be various? A statement may be negated because it is counterfactual, or because it involves a category-mistake, or because it is meaningless; yet in any one of these cases the negation functions in the same manner. The statement "Napoleon won the battle of Waterloo" is false to the same measure and in the same sense as the statement "Admiration is triangular" or even "Within not dances even," though for different reasons. Or again: each of them denies another attribute of a statement (or of its parts)—say, factuality, possibility, or meaning; but the negation has always the same meaning, even if incapable of further meaning.

I very much doubt the validity of arguments like this. Yet I refrain from elaborating on the point for fear of entering a semantical maze from which there is no return. This much, however, can be proved: some distinction, either between modes of negation, or causes for negation, or any other binary disjunction of negative propositions, is logically necessary. Its necessity does not emanate from usages of language which may result from an erroneous logical intuition; we are led to it by purely formal considerations. Even in a well-formalized system of propositions, if it is to be consistent and rich enough to express at least numerical relations in our world, we cannot reduce negation to only one form, or mode, or cause, or interpretation. As is well known, it has been mathematically proven that every formal system which suffices to derive the propositions of arithmetics from a finite subgroup of well-formulated formulas (axioms) with the aid of syntactic rules (mechanical rules of substitution) must admit propositions that are properly formulated and, in a sense, even true, yet unprovable. Had Gödel not proven his incompleteness theorem, we might have been able to claim that even in a large formal system one need not distinguish in principle between formulas that are not well-formulated and formulas that are not demonstrable, since both lead to a contradiction of a proven formula. We might have been able to argue further that one need not know in advance that a formula is not well formulated or need not establish, in addition to the rules of inference, special syntactic criteria for the propriety of formulas, since the rules of inference suffice to show, though not at first glance, that an improper formula is self-contradictory. Gödel's theorem of incompleteness made any such argument impossible, since it proved the existence of well-formulated formulas that are not demonstrable, and hence also the necessity to distinguish among various modes, levels, or causes of negation.

Yet precisely because such a distinction is necessary, it does not operate on one level of discourse. Not only is the one class of negative propositions not reducible to the other, they cannot be combined with each other on one level of discourse without abandoning the principle of the excluded middle. This, I argue, is the true source of the insecurity of classical logic since Aristotle, whenever it sought to distinguish among various modes of negation.

4. Infinite Judgments, Understanding, and Reason

Kant had good reasons for abandoning the Aristotelian distinction between simple negations and privations in favor of another Aristotelian distinction, that between determinate and indeterminate negations.[4] "Privation" ($\sigma\tau\acute{\epsilon}\rho\eta\sigma\iota\varsigma$) is a state in which a subject lacks a predicate (attitude, form) which it could possess "by nature." In a strict sense, the term stands only for one of a pair of contrary qualities, as when I say of Homer that he is blind. Aristotle felt uncomfortable with the negations of some privations, and rightly so, because a privative judgment both denies (that Homer can see) and affirms (that to say of Homer that he sees or that he is blind does not constitute a category-mistake). What, then, does a negation of such a privative judgment state? It either negates the affirmative aspect of

[4] Kant, *KdRV* B95–98, *Werke*, 3:112–13, 122.

a privation or its negative aspect, but not both. It is, of necessity, ambiguous—because it combines negations on two levels of discourse.

Moreover, even if, in the case of genuine contrarieties, the negation of a privation amounts to an affirmation (as when I say that Homer is not blind), yet if a range is involved within a quality—the Middle Ages spoke of a *latitudo formarum*—the negation of one extreme does not imply the other; not-cold does not imply hot. Such a negation constitutes "indeterminate notions" (ὄνομα ἀόριστον)— much as any negation of a quality that is not one of a pair of contrarieties. Genuine privations had, for Aristotle, a distinct ontological status. Form and privation, as contrarieties, are "causes," that is, constitutive principles of any being (οὐσία), and they presuppose a third cause underlying both—namely matter—which permits a being to assume or not to assume a given form "natural" to it. It is not of the nature of paper either to see or to be blind.[5] In many ways, the abandonment of matter as a principle of individuation—be it with the introduction of individual forms (Duns Scotus) or with the abandonment of the need for such a principle altogether (Ockham)—already undermined the ontic status of privations and enhanced the interest in indeterminate negations. At any rate, from the seventeenth century the ontological meaning of privation became altogether untenable for natural philosophers, for whom nature was uniform and homogeneous; they exchanged Aristotle's hierarchy of "qualities" and "natures" for quantitative, universal laws of one "nature" which are applicable to all beings. To them, the "nature" of a physical and even a metaphysical subject is nothing but the sum total of its predicates. On the other hand, the quantification of qualities became, in the seventeenth century even more than in the thirteenth, a problem of prime physical importance, for example, in the estimation of forces and the dispute over the *vis viva*. With it also grew the interest in indeterminate (or "infinite") judgments.

Kant removed infinite judgments from the domain of formal logic. "Logic," for Kant as for Aristotle, was logic of terms; "for-

---

[5] Aristotle, *Metaphysics* Δ22.1022b22–1023a7; Wolfson, "Infinite and Privative Judgments in Aristotle, Averroes, and Kant," pp. 173-87. On the further employment of "infinite judgment" as a universal methodical principle in the neo-Kantian interpretation of science see my article, "The Persecution of Absolutes."

mal" logic disregards any content of the terms in a proposition—we would say: it treats all categorematic terms as variables. The infinite judgment, he claimed, belongs rather to the domain of transcendental logic, a logic that does not abstract from all content but is concerned with all the possible contents of terms: not with any concrete content, but with the preconditions for having content. It constitutes, one might say, a first-level "interpretation" of formal logic. As a term, the expression "non-P" can be handled as any positive predicate: there is no need to single out, from a formal point of view, judgments containing such terms. But in view of possible content, "S is non-P" instructs us (i) S is *not* P and (ii) that S is a proper subject which belongs to the (possibly infinite) set of all subjects of which P cannot be predicated. The predicate non-P is complex: positive in its form and *limitative* in its meaning. It denies of S at least one predicate but leaves as a possibility all possible predicates, so that $[(\sim p_i) \wedge (p_1 \vee p_2 \vee \cdots \vee p_{i-1} \vee p_{i+1} \vee \cdots p_n)] \equiv$ non-P. Again, as in the case of Aristotle's privation, a simple argument suffices to show that, against his explicit wish, Kant in fact abandons the principle of the excluded middle. The negation of an infinite negation (S is not non-P) negates either that S is not P *or* that S is a proper subject at all, but not both: wherefore it is not tautologically true that not-non-$P_i(x) \equiv P_i(x)$. But Kant was unaware of this, in spite of his observation that the infinite negation unites affirmation and negation. (This unity of two contrary forms of judgment within a third repeats itself in all four classes of judgment and their corresponding categories. The singular, infinite, disjunctive, and apodictic judgments are only necessary for transcendental logic, unite the preceding disjunction, and thus may be seen as an anticipation of Hegel's dialectical method in a precise sense.)

The infinite judgment is the paradigm and the source for a most important structuring pattern of our experimental knowledge, namely the category of limitation. This category will allow us eventually to quantify qualities, inasmuch as we regard the full presence of a quality in a subjct as "reality," its total absence as (simple) negation—zero value—and any partial presence of it, or degree, as "limitation." "Quality" thus defines a continuous range and becomes quantified. The category of limitation allows us to synthesize phenomena—say, the force of attraction between bodies—as *intensive magnitudes*. It is obvious that Kant here systematized the dispute over the nature of intensive magnitudes, a dispute with which he once started his academic career. Since Leibniz, the dispute focused on the notion of force as *vis viva* but, as already mentioned, had its origin in the inclination of Scholastics to find a proper mathematical description for the quantification of qualities (*latitudo formarum*).[6] Herein lies also the origin of quantifying qualities with the infinite judgment. Here, as in other instances, Kant's position is a creative systematization of an ongoing dispute with a long history.

Kant moves on to prove that the categories of quality, like all others, are not only capable of structuring our experience, but that they actually do so. The category of reality can be "mapped into" our (inner) experience of time. The act of perception in time implies a scheme of the desired category, for example, the more intensive a perception, the more it is experienced as real; and perceptions cannot but change gradually. The "scheme" of a category within the sense of time links elements that seemingly have no mediation: judgment and sensuality. Yet, without mediation there would be no cognitive ordering of experience.

The infinite judgment is thus a necessary principle for every intelligible experience from mere sensation to the formulation of laws of nature. After many transformations from a form of judgment, through categories, and then a schema, into a principle of interpretation of nature, it permits us to formulate "synthetic a priori" propositions concerning acceleration and force in physics, such as Newton's first three principles. So much, then, for the role of the infinite judgment in understanding, that is, science. Yet the same infinite judgment governs not only our understanding of experi-

---

[6] A. Maier, *Kants Qualitätskategorien*, pp. 8–23, and above v.B.3. I hope to develop further, on another occasion, the suggestion to interpret the schematism as a "mapping" of the categories *into* time rather than as a rule for generating concepts only, as understood by Bennett, *Kant's Analytic*, pp. 141–52. Kant's problem of homogeneity is (against Bennett *pace* Warnock, "Concepts and Schematism," pp. 77–82) a genuine problem because it does not pertain to the application of concepts to concepts, but to concepts on (undefinable because unisolatable) sense data. The "modest point" of which Bennett says (p. 151) that it "might have come to something" is, in fact, the heart of the doctrine if my interpretation is correct. It does, though, elevate intensive magnitude ("reality") to a central position. *That* we are able and justified to apply concepts on an (undefinable) nonconceptual substrate, says Kant, is because they are foreshadowed qua schemes in the sense of time. By "mapping" I mean a procedure akin to the mapping of metamathematics into mathematics with the aid, e.g., of Gödel-numbers—or of a three-dimensional figure on a two-dimensional surface. Cf. also above II.H.5.

ence, but also our understanding of understanding itself—the reflexive effort of reason (*Vernunft*). This effort can be critical (as when we order and legitimize our categories of understanding) or noncritical and speculative, thus leading to hypostatizations.

Kant construes a direct link from the infinite judgment and the category of limitation to the principle of complete determination. The former is transformed almost of itself into the latter. Complete determination assumes a set of all simple perfections; Kant, we recall, agreed to call them "realities," since every quality represents a reality sui generis and in full presence corresponds to a sense of reality. Analogously, "reason" is led almost naturally to regard the whole range of simple predicates as the sum total, the maximum of reality, and each predicate as one of its degrees—in the very same way in which "understanding" detects degrees *within* each quality (reality). "Reality" is transformed, in the speculative extrapolation of reason, from a category applicable to qualities into a quality in itself. Rather than being, as it was in the effort to understand nature, a common denominator of qualities, "reality" has itself become a quality, of which the various qualities—the simple predicates, the "realities"—are degrees. Once the sum total of all simple predicates is thought of as a hypostatized entity possessing the maximum of reality, all other realities can be compared to it and measured against it as various partial degrees of the same reality in comparison with "the most real being." Here we have the best example of how the procedures of understanding are "objectivized, then hypostatized, finally personified."[7]

Here, as elsewhere, one is struck by the architectonic precision of Kant's system. The same figures of interpretation reappear at all levels of discourse. The "infinite judgment" appears first as a figure without interpretation (and therefore without a function) in formal logic. Transcendental logic endows it with an interpretation. In the tables of categories the infinite judgment is interpreted as a limitation, the basis for quantification of qualities. In the act of sensation, as manifested in the pure form of intuition (time), it is "pictured" or "schematized" in the intensity of a sensation. This leads to the objective rule of the infinite judgment in the interpretation of nature by way of the principle of intensive magnitudes. ("The real element

[7] Kant, *KdRV* B611, *Werke* 4:523 (note).

which is the object of a sensation always has an intensive magnitude, i.e., a degree.") Within the domain of pure reason, the infinite judgment permits the transformation of the concept of a "sum total of all possibilities" into that of a "most perfect being" or most real being. The methodological concept of God was banned from the interpretation of nature, but it still retained a certain role as a regulative ideal of reason (as the principle of complete determination). In our language we might say: God remained, even in the *Critique of Pure Reason*, a metatheoretical assumption, an assumption that, albeit redundant in the explanation of nature, is nonetheless almost "natural" to our reason. Kant expelled the methodological concept of God from the theory of science and grounded the universality of natural law and uniformity of nature without it; but its shadow persisted. The concept of God, he argued, is a natural shadow or projection of principles we use to structure nature. The shadow, Kant seems to have claimed, is virtually inescapable. But it is only a shadow.

To the ideal of a most perfect being, espoused "naturally" by pure reason, corresponds—in the realm of morality—the ideal of a "highest good," *summum bonum*, to guarantee the realizability of the commands of morality, not to legitimize them. Reason alone is their legitimation. And, just as teleological explanations are needed because beyond the minimal demands of noncontradiction of our experience the natural scientist needs a good many more "specifications of nature" to render experience intelligible as a whole, the whole of nature may have the perfection and happiness of man as a goal toward which even antagonism and adversity contribute as driving forces.

All of the four themes of our book are central to Kant's critical endeavors: knowledge by construction, the construction of nature, and, preceeding it, the construction and legitimation of our constructive ("synthetic") conceptual tools; the less-than-logical necessities (synthetic a priori judgments); the methodical ideals of reason; and the intrinsic, goal-oriented mechanism of nature and society. Kant explicated all of these themes without the theological baggage previously attached to them. The de-theologization of the foundations of knowledge was doubtless, in his eyes, his contribution to the "Enlightenment," that is, to the emancipation of humanity from its "self-inflicted bondage."

B. ENLIGHTENMENT AND EDUCATION

Many philosophers of the Enlightenment shared with Kant his distaste for theology. The posture of some was (unlike Kant's) also militantly anti-religious, if not atheistic. It seems as though the Enlightenment, notably in England and France, sometimes broke with the Christian past deliberately and thoroughly. A recent interpretation discovered at the core of the Enlightenment "the revival of paganism."[1] Paganism is indeed a generic term invented and hypostatized by Jews, Christians, and Moslems. Whether or not one can call most Enlighteners "pagans" depends not on the number of gods they recognized (in this sense, William James was the only modern pagan I know),[2] but on the texture of their ethical-social doctrines. In these I detect a basic concern that places them much closer to the mainline history of Christianity than even the secular theologians ever were: it was their missionary and educational zeal. In many countries, the *illuminati, Aufklärer, philosophes* set out to reform humanity and society through knowledge and reasoning.

Peter Gay rebelled against the overemphasis given to alleged eschatological-utopian elements in the Enlightenment since Carl Becker. In this respect his point of view is helpful, except that he threw out the baby with the bathwater. The Enlightenment inherited from Christianity not its apocalypticism, but rather its social and pedagogical drive. The ideals of the Enlightenment were secularized, inverted Christian ideals through and through. From Christianity the Enlightenment inherited its missionary zeal—not from any pagan religion of classical Antiquity, for none of them possessed it. The Christian tenet that "there is no salvation outside the Church" was matched by the new belief that there is no salvation except through the use of reason. Superstition and ignorance became the original sin of mankind. Withholding knowledge became, as *superbia* to the Catholic, a cardinal vice. Freemasonry was, in its way, a counter-church with its counter-symbols and counter-sacraments. It preached the brotherhood of mankind. This dialectical relation of the Enlightenment to Christianity overshadows all

---

[1] P. Gay, *The Enlightenment: An Interpretation*, 1: *The Rise of Modern Paganism*, esp. pp. 8–10, 308ff., 368ff.

[2] James, "The One and the Many," in *Pragmatism*, pp. 89–108; *id., The Varieties of Religious Experience*, Postscript.

vestiges one might find in it of the spirit of classical Antiquity, Stoic or Epicurean included.

Salvation through knowledge only was not an altogether alien theme to Christianity. The dangers of extreme intellectualization of its doctrines accompanied the Church from the time of the ancient Gnostics. The danger arose whenever intellectuals nurtured an exaggerated perception of their own importance and value. The Amalricans whom we mentioned earlier[3] taught that only philosophical knowledge saves: and it saves Jews, Moslems, and Christians alike. Their worship of knowledge was as heretical as it was exceptional; but it testified to the elitist character of medieval science. It may have been stronger on the Moslem or Jewish horizon, but was not absent in Christian Europe. True knowledge was esoteric, objectively as well as subjectively. Of the few who were literate, only a few dedicated their careers to theoretical pursuits; those who did tended to view the *vulgus* as incapable of understanding and always doomed to ignorance.

In this sense, medieval knowledge was *closed* knowledge. In another sense, it was not. The dialectics of open and closed knowledge characterizes the history of science from its outset. Knowledge relevant to a society was more often than not in most ancient societies secret knowledge, handed down selectively from one generation to the other in closed circles, without clearly articulated criteria to distinguish false from true knowledge, except by the very *act* of transmission.[4] Greek philosophy was one of the few cultures I know of to espouse the ideal of open knowledge, knowledge accessible by all and open to criticism by all. This is the social basis of the notion of a *proof*. In another sense, of course, the *theoria* was confined to the leisure class. Medieval schoolmen, at least since the thirteenth century, were intoxicated by the ideal of strict proofs, and their knowledge was an open one within the confines of those permitted to participate in its administration: it was, so to say, open horizontally, but closed vertically. The various barriers against access to all knowledge were removed, at least ideally, by many of the seven-

---

[3] Above II.c.1.

[4] On the relation of the emerging notion of proof to society in the *polis*, see Lloyd, *Magic, Reason and Experience: Studies in the Origins and Development of Greek Science*, esp. pp. 246–64. Together with A. Steinsaltz, I have completed a short study, *Sociology of Ignorance*, that will soon appear in Hebrew (an English translation is in preparation). It deals in more detail with the dialectics of open and closed knowledge.

teenth-century thinkers discussed here; wherefore many of them wrote deliberately in the vernacular. Only during the Enlightenment did open knowledge become a militant, missionary ideal.

Belief in the open character of systematic knowledge was already part of the intellectual profile of many seventeenth-century thinkers; the Enlightenment added to it, in a manner of speech, the demand for social action, for the deliberate preaching of knowledge as the only means for the amelioration of the human condition. In contrast to their medieval counterparts, the rationalists of the seventeenth and eighteenth centuries ascribed to each and every individual, irrespective of his or her formal education, a "common sense," "bons sens," "gemeiner Menschenverstand" which suffices to make us all capable of being educated, that is, of being raised to the level of philosophers. Even the shift in the connotation of the term "common sense" tells a good part of this story. In the technical terminology of the schools it stood for that additional, underlying capacity to coordinate the data that flow to the mind from the five sense organs; without it, we could not idenitfy a common source to given perceptions. But since the seventeenth century—perhaps under the influence of Stoic usage—the term came to mean the innate capacity of every person to reason and judge correctly.

It may appear ironic that the medieval, elitist image of knowledge was coupled with Aristotelian philosophy—basically a common-sense philosophy that aims to explicate "what everyone knows, only better";[5] while the new, egalitarian image of an open, systematic knowledge was coupled with sciences that in part were now derived from counter-intuitive premises and soon proliferated and became so technical that they could hardly be mastered by the educated layperson. The tension was not as pronounced in the seventeenth century as it became in the eighteenth; and it found temporary relief in the slowly emerging image of a common "culture" or "education." A new entity, "culture," connoted more than "mores" and less than "learning." It became the middle ground between specialized knowledge and ignorance. Most *philosophes* saw their task not so much in the generation of new science as in the translation of science into an accessible idiom, opened to all cultured persons. Not all protagonists of the Enlightenment believed in a

---

[5] Aristotle, *De caelo* Δ1.308a24: ὅπερ καὶ οἱ πολλοὶ λέγουσι, πλὴν οὐχ ἱκανῶς. Cf. *Ethica Nic.* Θ1.1145b2–6.

steady, progressive *Erziehung des Menschengeschlechts.*[6] But all of them believed in the social function of science. They set out to create the largest possible common denominator of necessary and accessible knowledge by "education"—the true "formation" (*Bildung*) of humankind.

C. THEOLOGY AND SCIENCE

These sketchy remarks concerning the eighteenth century would be downright misleading if they left the impression that the enlightened world of letters avoided or condemned theology always and everywhere. A new tradition of enlightened theology was also established, a historical-critical mode of theologizing, particularly in the Protestant world. This and subsequent movements in theology evidently lost the genuine interest in the natural sciences that was the mark of the secular theologians in the seventeenth century. Physics ceased to be an integral part of theology proper, and I doubt whether ever again we shall witness that God spoken of "the discourse of whom, from the appearance of things, does certainly belong to Natural Philosophy."[1] History rather than nature became the discourse of theologians. Perhaps one might say that the theocentric theologies of the Middle Ages gave way to cosmocentric theologies in the seventeenth century, which again were superseded by a variety of anthropocentric theologies down to our century. Perhaps one could detect in this progress the road leading to an "atheistic theology."[2] But all of this lies outside the scope of this book.

[6] Moses Mendelssohn, *Jerusalem oder über religiöse Macht und Judentum*, ed. Mendelssohn 3:317–18: "For my part, I cannot conceive of an education of the human race as my late friend Lessing imagined it under the influence of I don't-know-which historian of mankind." Indeed, the progress of individuals demands occasional retrogression of the collective. For many further references, and for a thorough account of his philosophy of history, see Altmann, *Moses Mendelssohn: A Biographical Study*, pp. 539–43. Lessing's "Erziehung des Menschengeschlechts" epitomized the enlightened version of the principle of accommodation as a theory of progress.

[1] Newton, *Principia*, "General Scholium," trans. Cajori 2:546. In the earlier edition, Newton even spoke of "experimental philosophy": *Principia* ed. Cohen and Koyré 2:529; Cohen, *Introduction* p. 244.

[2] The title of a penetrating critique of Christian and Jewish religious thought at the turn of the century written by Rosenzweig, *Kleinere Schriften*, pp. 278–90. It was rejected for publication. On its place in the development of his own theology, see my article, "The Genesis of Rosenzweig's 'Stern der Erlösung': 'Urformell' and 'Urzelle,' " pp. 17–29.

Aside from the state of theology, sad or happy, the very loss of religious commitments has caused some to fear for the whole fabric of society and culture. A medieval anecdote tells us of a kingdom where power is justice, day is night, warriors flee from battle, one is two, the friend is an enemy, evil is good, reason and licentiousness are one, thieves rule, doves become eagles, will is a counselor, money judges, and "God is dead." It is entitled, "On the Actual State of the World," but it does no more than to visualize an ancient *topos*—"the inverted world."[3] Its last phrase meanwhile turned, on occasion, into a solemn declaration: we may be already living in that kingdom. Whether the loss of religiosity is beneficial or not is not for me to judge. I do, though, feel obliged to distance myself from extravagant claims in respect to the ties between religion and the fortunes of rationality. Stanley L. Jaki argued time and again that modern science *could not* have emerged nor can it be sustained without "rational theism." "Happy the one who is able to recognize the causes of things": Jaki *knows* why eastern cultures never established a technological society or a coherent body of science; why the Greeks never developed a "viable science" or why, if they did, it "failed"; and why Bohr's methodical and epistemological foundations of quantum physics are bad:[4] science and "rational theism" draw from the same source and are propelled by the same force, namely "the search for the ultimate." Pantheism and eastern acosmism can not engage in this search; their resignation spells death for science. He admires Duhem's uncritical idealization of medieval precursors to modern physics and resents the more balanced picture gained since Anneliese Maier and Alexander Koyré, a picture in which not only medieval achievements are stressed, but also their limitations.[5] Should a modern scientist profess (like Planck and Ein-

---

[3] *Gesta Romanorum*, ed. Osterley, c. 144; I have contracted the four sets of answers and used the translation of Swan and Hooper, *Gesta Romanorum or Entertaining Moral Stories*, p. 251. The story is not found in some of the older English MSS: Dick, *Die Gesta Romanorum nach der Innsbrucker Handschrift vom Jahre 1342*. On the *topos* "inverted world," see Curtius, *Europäische Literatur*, pp. 104–108. It may have been the source of Jean Paul's exclamation that God is dead—if a source is needed at all. Nietzsche needed no source either. The *gesta* are not mentioned in Von der Luft, "Sources of Nietzsche's 'God is Dead!' and Its Meaning for Heidegger," pp. 263–76.

[4] Jaki, *The Road of Science and the Ways of God*, pp. 14ff. (eastern cultures), 19ff. (Greek science), 197ff. (Bohr, Heisenberg, complementarity).

[5] Ibid., pp. 34–49, 230–34 (Koyré). That Ockham wanted to eradicate all generalizations (pp. 41–42) is downright false; he merely forbade hypostatizing them. The ultimate source for many of Jaki's interpretations is, perhaps, the belief that the ways of God in

stein did) pantheistic inclinations he can be shown to be a good theist *malgré lui*. Those who are not build upon sand.

Confusing "after" with "because" and hypostatizing cultural features are some of the springs of these and similar biases. That one can draw many meaningful connections between medieval theology and early modern science is certain. That without the former, the latter would never have emerged or advanced in any guise is neither demonstrable nor plausible. Skepticism, pantheism, and atheism seem to me as strong companions of the advance of science as the search for "the ways of God." I do not know that a nontheistic society could not generate a rational-technological culture similar to ours; that the Greeks could not have invented the calculus of the science of dynamics or bridged the gap between theory and praxis due to their mentality (Spengler) or slave economy (Farrington) or religious propensities (Jaki); I only know for certain that they did not.

"Die Entzauberung der Welt"[6] has also been a constant background concern of this book. Weber saw the origins of this long and arduous process in Greece and ancient Israel. Reasons can always be found along its way why it slowed down or why it turned in one direction rather than another. They will never be sufficient and very seldom necessary. New beginnings are recognizable as such only in retrospect. We do not know how often the wheel was invented, how many inventions or ideas were never followed up; we know of several that were followed up only much later:

> In the case of all discoveries, the results of previous labours that have been handed down from others have been advanced bit by bit by those who have taken them on, whereas the original discoveries generally make an advance that is small at first though much more useful than the development which later springs out of them. For it may be that in everything, as the saying is, "the first start is the main part," and for this reason also it is the most difficult; for in proportion as it is most potent in its influence, so it is smallest in its compass and most difficult to see: whereas when this is once discovered, it is easier to add and develop the remainder in connexion with it.[7]

---

the Scriptures are "simple"—read rational. Alas, "the ways of God" according to the Scriptures are "mysterious" and complicated; and as for knowledge or wisdom, "it cannot be found in the land of the living" (Job 28:13).

[6] Weber, *Gesammelte Aufsätze zur Religionssoziologie*, 1:513.

[7] Aristotle, *De sophisticis elenchis* 34.183b17–184b9, trans. W. A. Pickard-Cumbridge (Oxford, 1912); cf. Kapp, *Greek Foundations of Traditional Logic*, pp. 5–7.

To detect and explain the "small beginnings" of which Aristotle spoke is difficult: they are clothed in the guise of that which surrounds them and yet they are new. The new, though inevitably expressed in an inherited idiom, has also the quality of creation out of nothingness, inexplicable and unexpected. No matter how well-prepared and well-suited to its culture a decisive step that founded a new discipline, a new theory, may seem in retrospect—it could have come much later or not at all. The essence of every creative achievement is its freedom: "Es geht in der Wissenschaft so stark und unbekümmert und herrlich zu wie in einem Märchen."[8] Aristotle, always ready to acknowledge forerunners, concluded his "Topics"— one of his earliest courses of lectures as an independent teacher— with a claim for absolute novelty; it is the continuation of the passage just quoted:

> This is in fact what has happened in regard to rhetorical speeches and to practically all the other arts: for those who discovered the beginnings of them advanced them in all only a little, whereas the celebrities of today are the heirs, so to speak, of a long succession of men who have advanced them little by little. . . . Of this inquiry, on the other hand, it was not the case that part of the work had been thoroughly done before, while part had not. Nothing existed at all.

The problems we raised in this book were not all new, the answers sometimes only in matters of nuances. If, however, "It seems to you after inspection that . . . our investigation is in a satisfactory condition compared with other inquiries that have been developed by tradition, there must remain for you all . . . the task of extending us your pardon for the shortcoming of the inquiry and for the discoveries thereof your warm thanks."[9]

[8] R. Musil, *Der Mann ohne Eigenschaften* (Hamburg, 1952) c. II, p. 41.
[9] Aristotle, *De soph. elen.*, ibid.

BIBLIOGRAPHY

Abaelard, Petrus. *Dialogus inter Philosophum, Judaeum et Christianum*. Migne, *PL* 178: 1609ff.

———. *Expositio in Hexaemeron*. Migne, *PL* 178: 732ff.

Abarbanel, Yitschak. *Perush hatora*. Warsaw, 1862.

Abelson, J. *The Immanence of God in Rabbinical Literature*. London, 1912.

Abraham bar Hiyya. *Sefer megillat ha'megalle*. Edited by A. Poznanski. Berlin, 1924.

Abraham ibn Ezra. *Perush hatora*. Edited by A. Weiser. 3 vols. Jerusalem, 1976.

Adams, Marylin M. "Intuitive Cognition, Certainty, and Scepticism in William of Ockham." *Traditio* 26 (1970): 389–98.

———. "Universals in the Early Fourteenth Century." *CHM*: 411–39.

Adams, Robert M. "Leibniz's Theories of Contingency." *Rice University Studies* 63 (1977): 1–41.

Aiton, E. J. *The Vortex Theory of Planetary Motions*. London and New York, 1972.

Alanus ab Insulis. *De fide catholica contra haereticos libri quattuor*. Migne, *PL* 210: 305ff.

Alexander of Hales. *Summa theologiae*. Edited by Quaracchi. 4 vols. Florence, 1924.

Alfunsi, Petrus. *Dialogi*. Migne, *PL* 157: 535ff.

Alt, Albrecht. *Die Ursprünge des israelitischen Rechts, kleine Schriften*. Munich, 1953.

Altmann, Alexander. *Moses Mendelssohn: A Biographical Study*. University, Ala., 1973.

Ambrosius of Milan. *De Iacobo*. CSEL 32.2.

———. *De Paradiso*. Edited by C. Schenkel. CSEL 32.1.

———. *Hexaemeron*. Edited by C. Schenkel. CSEL 32.1.

Anastos, Milton V. "Porphyry's Attack on the Bible." In *The Classical Tradition: Literary and Historical Studies in Honor of Harry Kaplan*, edited by L. Wallach, pp. 421–50. Ithaca, N.Y., 1966.

Anderson, Carol Susan. "Divine Governance, Miracles, and Laws of Nature in the Early Middle Ages: The De mirabilibus Sacrae Scripturae." Ph.D. dissertation, UCLA, 1982.

Andresen, Carl, *Logos und Nomos: Die Polemik des Kelsos wider das Christentum*. Arbeiten zur Kirchengeschichte 30. Edited by K. Aland et al. Berlin, 1955.

Anscombe, G.E.M. "Aristotle and the Sea Battle." *Mind* 65 (1956): 1–15.

———. "The Principle of Individuation." In *Articles on Aristotle, 3: Metaphysics*, edited by J. Barnes et al., pp. 88–95. New York, 1979.

Anselm of Canterbury. *Cur deus homo*. In *Opera omnia*, vol. 2, edited by F. S. Schmitt. Edinburgh, 1946.

———. *Proslogion*. In *Opera omnia*, vol. 1, ed. Schmitt.

Anselm of Havelberg. *Dialogi* . Migne, *PL* 188: 1139ff.

Apostle, Hippocrates G. *Aristotle's Physics*. Bloomington, Ind., 1969.

Archimedes. *The Works of Archimedes*. Translated and edited by Thomas L. Heath. New York, 1953.

Ardens, Radulphus. *Homiliae in Epistolas et Evangelia Dominicalia*. Migne, PL 155: 1067ff.

Arendt, Hannah. *Between Past and Future: Six Exercises in Political Thought*. Cleveland and New York, 1963.

——. *The Human Condition*. Chicago, 1958.

Aristotle. *Opera*. Edited by Immanuel Bekker. 2 vols. 1831; reprint Darmstadt, 1960.

——. *De caelo*. Edited by D. J. Allan. Oxford, 1936.

——. *Metaphysica*. Edited by Werner Jaeger. Oxford, 1957.

——. *Physica*. Edited by W. D. Ross. Oxford, 1950.

Arnauld, Antoine. *The Art of Thinking: Port Royal Logic*. Translated by James Dickoff and Patricia James. New York, 1964.

Arnobius. *Adversus Nationes*. Edited by Concetto Marchesi. Milan, 1953.

Arnold, Gottfried. *Unparteyische Kirchen und Ketzerhistorie*. Schaffhausen, 1740.

Auerbach, Erich. *"Figura," Scenes from the Drama of European Literature*. New York, 1959.

——. *Mimesis: Dargestellte Wirklichkeit in der abendländischen Literatur*. 2d ed. Bern, 1959.

Augustine of Hippo. *Adversus Judaeos*. Migne, PL 42: 51ff.

——. *Contra epistulam Manichaei*. CSEL 25.

——. *Contra Faustum Manichaeum*. Migne, PL 42: 207ff.

——. *Contra Seceundinum Manichaeum*. Migne, PL 42: 577ff.

——. *De civitate Dei*. Edited by B. Dombart and A. Kalb. CCSL 47, 48.

——. *De diversis quaestionibus*. Migne, PL 40: 11ff.

——. *De Genesi contra Manichaeos*. Migne, PL 34: 173ff.

——. *De libero arbitrio*. Migne, PL 32: 1221ff.

——. *De vera religione*. CCSL 32: 187ff.

——. *Enchiridion*. Edited by O. Scheel. Sammlung ausgewählter kirchen- und dogmengeschichtlicher Quellenschriften, ser. 2, vol. 4. Tübingen, 1930.

——. *Epistulae*. Edited by A. Goldbacher. CSEL 44.

——. *Retractiones*. CSEL 36.

Aureoli, Petrus. *Scriptum in libros sententiarum*. In *Scriptum super primum sententiarum* (proemium-dist. 8), edited by E. Buytaert. 2 vols. St. Bonaventure, N.Y., 1953–1956.

Averroës (Ibn Rushd). *Aristotelis opera cum Averrois commentariis*. 9 vols. Venice, 1562–1574.

Bacon, Sir Francis. *The Works of Francis Bacon*. Edited by James Spedding, Robert L. Ellis, and Douglas D. Heath. 15 vols. London, 1887–1892.

Bacon, Roger. *Questiones supra libros quatuor physicorum*. In *Opera hactenus inedita Rogeri Baconi*, vol. 8, edited by F. U. Delmore. Oxford, 1927.

Baer, Yitshak F. *Yisrael ba'amin* [Israel among the Nations]. Jerusalem, 1955.

Baeumler, Alfred. *Das Irrationalitätsproblem in der Aesthetik und Logik des 18. Jahrhundert.* 1923; reprint Darmstadt, 1967.

Bailey, Cyrill. *The Greek Atomists and Epicurus.* Oxford, 1928.

Bannach, D. *Die Lehre von der doppellten Macht Gottes bei Wilhelm von Ockham.* Veröffentlichungen des Instituts für Europäische Geschichte Mainz, 75. Wiesbaden, 1975.

Baron, Salo W. "The Historical Outlook of Maimonides." In *History and Jewish Historians,* pp. 109–63. Philadelphia, 1964.

Baudry, Léon. *Guillaume d'Occam: Sa vie, ses oeuvres, ses idées sociales et politiques.* Paris, 1950.

Bayle, Pierre. *Dictionnaire historique et critique.* Amsterdam, 1740.

——. *Dictionnaire historique et critique.* Edited by A.J.Q. Beuchot. 5 vols. Paris, 1820–1824.

——. *Pierre Bayle Historical and Critical Dictionary, Selections.* Edited by R. H. Popkin. Indianapolis and New York, 1965.

Becker, Oskar. *Grundlagen der Mathematik in geschichtlicher Entwicklung.* 2d ed. Suhrkamp, 1975.

——. *Mathematische Existenz: Untersuchungen zur Logik und Ontologie mathematischer Phänomene.* Halle, 1927.

Beda Venerabilis. *Super acta apostolorum.* Migne, *PL* 92: 937ff.

Belaval, Yvon. *Leibniz critique de Descartes.* Paris, 1960.

Bengsch, A. *Heilsgeschichte und Heilswissen: Eine Untersuchung zur Struktur und Entfaltung des Hl. Irenaeus von Lyons.* Leipzig, 1957.

Benin, Stephen D. "The 'Cuning of God' and Divine Accommodation." *Journal of the History of Ideas* 45 (1984): 179–91.

——. "Thou Shalt Have No Other God before Me: Sacrifice in Jewish and Christian Thought." Ph.D. dissertation, University of California at Berkeley, 1980.

Bennett, Jonathan. "Analytic-Synthetic." *Proceedings of the Aristotelian Society* 59 (1958–1959): 163–88.

——. *Kant's Analytic.* Cambridge, 1966.

Benoit, A. *Saint Irénée: Introduction à l'étude de la théologie.* Paris, 1960.

Ben-Sasson, Hayyim-Hillel. "Yihud 'am yisrael le'daat bne hame'a hastem esre." *Perakim leheker toldot yisrael* 2 (1971): 145–218.

Bentley, Jerry H. *Humanists and the Holy Writ: New Testament Scholarship in the Renaissance.* Princeton, 1983.

Benz, Ernst. *Ecclesia spiritualis: Kirchenidee und Geschichtstheologie der franziskanischen Reformation.* Stuttgart, 1934.

Benzen, Aage. *Daniel.* Handbuch des Alten Testaments, 1st ser., 19. Edited by O. Eissfeldt. 2d ed. Tübingen, 1952.

Berger, Peter L. and Thomas Luckmann. *The Social Construction of Reality.* Garden City, N.Y., 1967.

Berges, Wilhelm. "Anselm von Havelberg in der Geistesgeschichte des 12. Jahrhunderts." *Jahrbuch für die Geschichte Mittel- und Ostdeutschlands* 5 (1956): 38ff.

368 BIBLIOGRAPHY

Bergson, Henri. "L'Intuition philosophique." *Revue de Métaphysique et de Morale* 19 (1911): 809–827.

Berkoff, Hans. *Kirche und Kaiser: Eine Untersuchung zur Entstehung der byzantinischen und theokratischen Staatsauffassung im 4. Jahrhundert.* Zollikon and Zurich, 1947.

———. *Die Theologie des Eusebius von Caesarea.* Amsterdam, 1939.

Bernard of Clairvaux. *Ad Hugonem de Sancto Victore Epistola.* Migne, *PL* 182: 1031ff.

Beumer, J., s.j. "Der theoretische Beitrag der Frühscholastik zu dem Problem des Dogmenfortschritts." *Zeitschrift für Katholische Theologie* 74 (1952): 209ff.

Biale, David. *Gerschom Scholem: Kabbala and Counter-History.* Cambridge, Mass., 1979.

Bizer, Ernst. *Studien zur Geschichte des Abendmahlstreits im 16ten Jh.* Darmstadt, 1962.

———. "Ubiquität." In *Evangelisches Kirchenlexicon* 3 (Göttingen, 1959): 1530–32.

Blanchet, Louis. *Les Antécédents historiques de "Je pense donc je suis."* Paris, 1920.

Bloos, L. *Probleme der stoischen Physik.* Hamburg, 1973.

Blumenberg, Hans. *Die Genesis der kopernikanischen Wende.* Frankfurt-am-Main, 1975.

———. *Die kopernikanische Wende.* Frankfurt-am-Main, 1965.

———. *Die Legitimität der Neuzeit.* Frankfurt-am-Main, 1966.

———. *Die Lesbarkeit der Welt.* Frankfurt-am-Main, 1981.

Bochner, Salomon. "The Emergence of Analysis in the Renaissance and After." *Rice University Studies* 64 (1978): 11–56.

Bock, Gisella. *Thomas Campanella, politisches Interesse und philosophische Spekulation.* Tübingen, 1974.

Bodin, Jean. *Methodus ad facilem historiarum cognitionem.* Strasbourg, 1907.

Boehm, A. *Le "vinculum substantiale" chez Leibniz: Ses origines historiques.* 2d ed. Paris, 1962.

Boehner, Philotheus, o.f.m. *Collected Articles on Ockham.* St. Bonaventure, N.Y., 1958.

———. *Ockham: Philosophical Writings.* Edinburgh, 1957.

Bohr, Niels. "Discussion with Einstein on Epistemological Problems in Atomic Physics." In *Albert Einstein: Philosopher-Scientist*, edited by Paul A. Schlipp, vol. 1. 3d ed. London, 1949.

Boler, John F. "Intuitive and Abstractive Cognition." *CHM*: 460–78.

Boll, Franz. "Die Lebensalter: Ein Beitrag zur antiken Ethnologie und zur Geschichte der Zahlen." *Neue Jahrbücher für das klassische Altertum*, ser. 2, 1 (1913): 89–145.

Bonansea, B. M., o.f.m. *Man and His Approach to God in John Duns Scotus.* Lanham, Md., 1983.

Bonaventura. *Opera omnia.* Edited by Quaracchi. 10 vols. Florence, 1882–1902.

Borkenau, Franz. *Der Übergang vom feudalen zum bürgerlichen Weltbild.* Paris, 1934.

Bos, H.J.M. "Arguments on Motivation in the Rise and Decline of a Mathematical Theory: The 'Construction of Equations,' 1637–ca. 1750." *Archive for History of the Exact Sciences* 30 (1984): 331–80.

Bossuet, Jacques-Bénigne. *Discours sur l'histoire universelle.* Edited by J. Truchet. Paris, 1966.

———. *Politique tirée des propres paroles de l'Écriture sainte.* Edited by J. le Brun. Geneva, 1967.

Bousset, Wilhelm. *Die Religion des Judentums in Späthellenistischer Zeit.* Edited by H. Gressmann. 3d ed. Tübingen, 1926.

Boyer, Carl B. "Galileo's Place in the History of Mathematics." In *Galileo, Man of Science,* edited by E. McMullin, p. 239. New York, 1967.

———. *The History of the Calculus and Its Conceptual Development.* New York, 1959.

Boyle, Robert. *Origins of Forms and Qualities According to the Corpuscular Philosophy* [1666]. In *The Works of the Honourable Robert Boyle,* Vol. 3. London, 1672.

Brandt, Frithiof. *Thomas Hobbes' Mechanical Conception of Nature.* Copenhagen and London, 1928.

Bréhier, Émile. "La création des vérités éternelles dans le système de Descartes." *Revue Philosophique de la France et de L'étranger* 113, 5–6, 7–8 (1937): 15–29. Translated by Willis Doney in *Descartes: A Collection of Critical Essays,* pp. 192–208. Garden City, N.Y., 1967.

Breidert, Wolfgang. *Das aristotelische Kontinuum in der Scholastik.* BGPhM, Neue Folge 1. 2d ed. Münster, 1979.

Brezzi, Paulo. "Ottone di Frisinga." *Bollettino dell'istituto storico italiano per il Medioevo e archivo Muratoriano* 54 (1939): 129ff.

Brown, Peter. *Augustine of Hippo: A Biography.* Berkeley and Los Angeles, 1969.

Brown, Stuart M. "The Taylor Thesis." In *Hobbes Studies,* edited by Keith C. Brown, pp. 31–34, 57–71. Oxford, 1965.

Buchdahl, Gerd. *Metaphysics and the Philosophy of Science: The Classical Origins: Descartes to Kant.* Cambridge, Mass., 1969.

Buchholz, Karl-Dietrich. *Isaac Newton als Theologe.* Witten, 1965.

Budde, Franz. *Historia ecclesiastica.* 3d ed. Jena, 1726.

Burckhardt, Jacob. *Die Kultur der Renaissance in Italien, Ein Versuch.* In *Werke,* Vol. 3. Darmstadt, 1962.

Buridan, Johannes. *Questiones super octo physicorum libros Aristotelis.* Paris, 1509.

———. *Questiones super libris quattuor de caelo et mundo.* Edited by Ernest A. Moody. Cambridge, Mass., 1942.

Burthogge, Richard. *An Essay upon Reason and the Nature of Spirits.* London, 1694.

Burtt, Edwin A. *The Metaphysical Foundations of Modern Science.* Garden City, N.Y., 1954.

Campanella, Tommaso. *Atheismus triumphatus.* Paris, 1636.

———. *Metaphysica.* Edited by G. di Napoli. Bologna, 1967.

———. *Rationalis philosophiae, part 5: Historiographia liber unus iuxta propria prin-*

cipia. In *Tutte le opere di Tommaso Campanella*, Vol. 1, edited by L. Firpo. Turin, 1954.

Caspar, Erich, ed. *Das Register Gregors VII*. 2 vols. MG Epistulae selectae in usum scholarum 2.1. Berlin, 1920.

Cassirer, Ernst. *Das Erkenntnisproblem in der Philosophie und Wissenschaft der neueren Zeit*. 3d ed. 4 vols. 1921; reprint Darmstadt, 1974.

——. *Leibniz' System in seinen wissenschaftlichen Grundlagen*. 1902; reprint Darmstadt, 1962.

——. *The Philosophy of Symbolic Forms*. Translated by R. Mannheim. 3 vols. New Haven and London, 1955.

——. *The Platonic Renaissance in England*. Translated by J. P. Pettegrove. Austin, 1953.

——. *Substance and Function*. Translated by William C. Swabey and Marie C. Swabey. Chicago, 1923.

Censorinus. *De die natali*. Edited by Friedrich Hulfsch. Leipzig, 1897.

Centiloquium theologicum. Edited by Philotheus Boehner. *Franciscan Studies* 17 (1942): 44ff.

Charles, R. H., ed. *The Apocrypha and Pseudepigrapha of the Old Testament*. 2 vols. 1913. Reprint. Oxford, 1963.

Chenu, Marie-Dominique. *Nature, Man, and Society in the Twelfth Century*. Selected, edited, and translated by Jerome Taylor and Lester K. Little. Chicago, 1968.

——. "La Théologie de la loi ancienne selon S. Thomas." *Revue Thomiste* 61 (1961): 485ff.

Cicero. *De natura deorum*. Edited by O. Plassberg. Stuttgart, 1959.

——. *De re publica*. Edited by K. Ziegler. Leipzig, 1960.

Clagett, Marshall. *The Science of Mechanics in the Middle Ages*. Madison, Wis., 1961.

Clarendon, Edward, Earl of. *A Brief Survey of the Dangerous and Pernicious Errors to Church and State in Mr. Hobbes' book entitled Leviathan*. Oxford, 1676.

Clauberg, Johann. *Differentia inter Cartesianam et in scholis vulgo usitatam philosophiam*. In *Opera omnia philosophica*. 2 vols. Amsterdam, 1691.

Clavelin, Maurice. *The Natural Philosophy of Galileo: Essay on the Origins and Formation of Classical Mechanics*. Translated by A. J. Pomerans. Cambridge, Mass., 1974.

Cnutonis regis gesta . . . auctore monacho sancti Bertini. Edited by Georg Heinrich Pertz. MG Script. in usu schol. Hanover, 1865.

Cochrane, Ch. N. *Christianity and Classical Culture: A Study of Thought and Action from Augustus to Augustine*. 2d ed. New York, 1957.

Cohen, Hermann. *Kommentar zu Immanuel Kants Kritik der Reinen Vernunft*. Leipzig, 1907.

Cohen, I. Bernard. *Introduction to Newton's Principia*. Cambridge, Mass., 1971.

——. "Newton's Use of 'Force,' or Cajori versus Newton." *Isis* 58 (1967): 226–30.

——. "Quantum in se est." *Notes and Records of the Royal Society* 19 (1964): 131–55.

Cohen, Jeremy. *The Friars and the Jews: The Evolution of Medieval Anti-Judaism.* Ithaca, N.Y. and London, 1982.

Condillac, Étienne Bonnot de. *Traité des systèmes.* In *Oeuvres complètes.* Paris, 1788.

Connell, Desmond. *The Vision in God Malebranche's Scholastic Sources.* Louvain and Paris, 1967.

Conway, Anne. *The Principles of the Most Ancient and Modern Philosophy.* Edited by P. Loptson. The Hague, 1982.

Conzelmann, H. *Die Mitte der Zeit: Studien zur Theologie des Lucas.* Tübingen, 1954.

Copenhauer, Brian P. "Jewish Theologies of Space in the Scientific Revolution: Henry More, Joseph Raphson, Isaac Newton, and Their Predecessors." *Annals of Science* 37 (1980): 489–548.

Copernicus, Nicolaus. *De revolutionibus orbium coelestium.* Thorn, 1873.

Cornford, Francis M. *From Religion to Philosophy: A Study in the Origins of Western Speculation.* New York, 1957.

——. *Plato's Cosmology: The Timaeus of Plato.* 1938; reprint New York, n.d.

Courtenay, William J. "The Critique of Natural Causality in the Mutakallimun and Nominalism." *Harvard Theological Review* 66.1 (1973): 77–94.

——. "The Dialectic of Divine Omnipotence." In *Covenant and Causality in Medieval Thought.* London, 1984.

——. "John of Mirecourt and Gregory of Rimini on Whether God Can Undo the Past." *Recherches de Théologie ancienne et médiévale* 39 (1972): 224–56; 40 (1973): 147–73.

——. "Necessity and Freedom in Anselm's Conception of God."*Analecta Anselmiana* 4.2 (1975): 39–64.

Couturat, Louis. *La Logique de Leibniz d'après des documents inédits.* Paris, 1901.

Cragg, G. R. *From Puritanism to the Age of Reason: A Study of Changes in Religious Thought within the Church of England 1660–1700.* Cambridge, 1966.

Cranz, F. Edward. "The Development of Augustine's Ideas on Society before the Donatist Controversy." *Harvard Theological Review* 58 (1954): 255ff.

——. "Kingdom and Polity in Eusebius of Caesarea." *Harvard Theological Review* 45 (1952): 51ff.

Crescas, Hisdai. *Or Adonai.* Vienna, 1859.

Crombie, A. C. *Robert Grosseteste and the Origins of Experimental Science 1100–1700.* 2d ed. Oxford, 1962.

Crosby, H. Lamar. *Thomas Bradwardine, His Tractatus de Proportionibus: Its Significance for the Development of Mathematics.* Madison, Wis., 1955.

Cross, Frank. *The Ancient Library of Qumran.* New York, 1961.

Cudworth, Ralph. *The True Intellectual System of the Universe.* London, 1678.

Cumont, Franz. *Die orientalischen Religionen im römischen Heidentum.* Translated and edited by August Burckhardt-Brandenburg. Darmstadt, 1959.

Curley, E. M. "Descartes on the Creation of Eternal Truths." *The Philosophical Review* 93.4 (1984): 569–97.

——. "The Roots of Contingency." In *Leibniz: A Collection of Critical Essays,* edited by H. G. Frankfurt, pp. 69–97. Garden City, N.Y., 1972.

Curley, E. M. *Spinoza's Metaphysics: An Essay in Interpretation.* Cambridge, Mass., 1969.

Curtius, Ernst Robert. *Europäische Literatur und lateinisches Mittelalter.* 3d ed. Bern, 1961.

Cusanus, Nicolaus. *Opuscula varia theologica et metaphysica.* Edited by Paul Wilpert. In *Nikolaus von Kues: Werke (Neuausgabe des Strassburger Drucks von 1488).* 2 vols. Berlin, 1967.

Damiani, Petrus. *De divina omnipotentia in reparatione corruptae, et factis infectis reddendis.* In *Lettre sur la tout-puissance divine,* edited by A. Cantin. Paris, 1972.

Dan, Joseph. *Hasipur ha'ivri biyme habenayyim* [The Hebrew Story in the Middle Ages]. Jerusalem, 1974.

Daniélou, Jean. "The New Testament and the Theology of History." *Studia Evangelica* 1 (1959): 25–34.

Daniels, A. *Quellenbeiträge und Untersuchungen zur Geschichte des Gottesbeweises im Mittelalter.* BGPhM 8.1–2. Münster, 1909.

Dante Alighieri. *Opere.* Edited by E. Moore and P. Toynbee. Oxford, 1924.

Davidson, Herbert. "Arguments from the Concept of Particularization." *Philosophy East and West* 18 (1968): 299ff.

Day, Sebastian, J., O.F.M. *Intuitive Cognition: A Key to the Significance of the Later Scholastics.* St. Bonaventure, N. Y., 1947.

DeJung, Ch. *Wahrheit und Häresie: Untersuchungen zur Geschichtsphilosophie bei Sebastian Franck.* Zurich, 1980.

Deku, H. "Possibile logicum." *Philosophisches Jahrbuch der Görres-Gesellschaft* 64 (1956): 1–21.

Dempf, Alois. *Sacrum Imperium.* 2d ed. Munich, 1949.

Denifle, Heinrich and Émile Chatelain. *Chartularium Universitatis Parisiensis.* 2 vols. 1889–1897; reprint Brussels, 1964.

Derrida, Jacques. *Of Grammatology.* Translated by G. Ch. Spivak. Baltimore and London, 1976.

Descartes, René. *Oeuvres de Descartes.* Edited by Charles Adam and Paul Tannery. 12 vols. 1897–1913; reprint Paris, 1973.

——. *Philosophical Works of Descartes.* Edited by E. S. Haldane and G.R.T. Ross. 2 vols. New York, 1934.

De Silvestris, Franciscus (Ferarensis). *Commentaria ad Summam contra gentiles.* In Thomas Aquinas, *Summa contra gentiles, Opera Omnia* (Leonine edition), vols. 13–15.

Dick, Steven J. *Plurality of Worlds: The Origins of the Extraterrestrial Life Debate from Democritus to Kant.* Cambridge, 1982.

Dick, W. *Die Gesta Romanorum nach der Innsbrucker Handschrift vom Jahre 1342.* Erlangen and Leipzig, 1890.

Diels, Hermann and Walter Kranz. *Fragmente der Vorsokratiker.* 3 vols. 6th ed. Berlin, 1952.

Diestel, Ludwig. *Geschichte des Alten Testaments in der christlichen Kirche.* Jena, 1869.

Dijksterhuis, E. J. *The Mechanization of the World-Picture.* Oxford, 1961.

Dilthey, Wilhelm. *Weltanschauung und Analyse des Menschen seit Renaissance und Reformation.* In *Gesammelte Schriften,* Vol. 2. Stuttgart, 1960.

Dixon, R. B. *Oceanic Mythology.* Boston, 1916.

Doren, A. "Campanella als Chiliast und Utopist." In *Kultur und Universalgeschichte* (Festschrift für W. Goetz), pp. 242–59. Leipzig and Berlin, 1927.

Douglas, Mary. *Purity and Danger: An Analysis of Concepts of Pollution and Tabu.* London, 1966.

Drake, Stillman. *Galileo at Work: His Scientific Biography.* Chicago, 1978.

——. *Galileo Studies.* Ann Arbor, Mich., 1970.

Dreiling, P. R., O.F.M. *Der Konzeptualismus in der Universalienlehre des Franziskanerbischofs Petrus Aureoli.* BGPhM 11.6. Münster, 1913.

Dronke, Peter. *Fabula: Explorations into the Uses of Myth in Medieval Platonism.* Leiden and Cologne, 1974.

Ducasse, Curt J. "William Whewell's Philosophy of Scientific Discovery." In *Theories of Scientific Method, The Renaissance through the Nineteenth Century,* edited by Edward H. Madden, pp. 183–217. Washington, D.C., 1966.

Duchaltelez, H. "La 'condescendance' divine et l'histoire du salut." *Nouvelle Revue Théologique* 95 (1973): 593–621.

Dühring, Ingemar. *Aristoteles, Darlegung und Interpretation seines Denkens.* Heidelberg, 1966.

Duhem, P.M.M., *Études sur Léonard de Vinci.* 3 vols. Paris, 1906–1913.

——. *Le Système du monde.* 10 vols. Paris, 1913–1959.

——. *To Save the Phenomena: An Essay on the Idea of Physical Theory from Plato to Galileo.* Translated by E. Donald and C. Maschler. Chicago, 1969.

Eibl, Hans. *Augustin und die Patristik.* Munich, 1923.

Einstein, Albert and Leopold Infeld. *The Evolution of Physics.* New York, 1966.

Eissfeldt, Otto. *The Old Testament, an Introduction.* Translated by P. R. Ackroyd. New York and Evanston, 1965.

Elders, L. *Aristotle's Cosmology: A Commentary on De Caelo.* Assen, 1966.

Eliade, Mircea. *Cosmos and History: The Myth of Eternal Return.* Translated by W. R. Trask. New York, 1959.

Elkana, Yehuda. "Science as a Cultural System: An Anthropological Approach." In *Scientific Culture in the Contemporary World,* edited by V. Mathieu and P. Rossi, pp. 269–89. Milan, 1979.

Elliger, Kurt. *Studien zum Habakuk—Kommentar vom Toten Meer.* Tübingen, 1953.

Enders, A. *Petrus Damiani und die weltliche Wissenschaft.* BGPhM 8.3. Münster, 1910.

Entsiklopedia talmudit le'inyene halacha [Talmudic Encyclopedia]. Edited by Meir Berlin and Shlomo Josef Zevin. Jerusalem, 1948ff.

Ettinger, Shmuel. "Jews and Judaism as Seen by the English Deists of the Eighteenth Century." *Zion* 29 (1964): 182ff.

Euchner, Walter. *Egoismus und Gemeinwohl: Studien zur Geschichte der bürgerlichen Philosophie.* Frankfurt, 1973.

Euclid. *The Thirteen Books of Euclid's Elements.* Translated by Thomas L. Heath. 2d ed. 1926; reprint New York, 1956.

Eusebius of Caesarea. *Commentaria in Psalmos*. Migne, *PG* 23: 66ff.

——. *Kirchengeschichte* (kleine Ausgabe) [*Historia ecclesiastica*]. Edited by Eduard Schwartz. 5th ed. Berlin, 1952.

Faj, Attila. "Vico as a Philosopher of Metabasis." In *Giambattista Vico's Science of Humanity*, edited by Giorgio Tagliacozzo and Donald Phillip Verene, pp. 87–109. Baltimore, 1976.

Farrington, B. *Greek Science*. 2d ed. London, 1953.

Faur, José. "Mekor Chiyyuban shel hammitsvot le'daat harambam." *Tarbiz* 38 (1969): 43–53.

Favaro, Antonio. *Galileo Galilei, Pensieri, motti e sentenze*. Florence, 1935.

Feldon-Eliav, Miriam. *Realistic Utopias*. Oxford, 1983.

Festugière, André Marie Jean. *Epicurus and His Gods*. Translated by C. W. Chilton. Oxford, 1952.

Feyerabend, Paul. "Consolations for the Specialist." In *Criticism and the Growth of Knowledge*, edited by Imre Lakatos and Alan Musgrave, pp. 197–230. Cambridge, 1970.

Fisher, N. W. and Sabetai Unguru. "Experimental Science and Mathematics in Roger Bacon's Thought." *Traditio* 27 (1971): 353–78.

Fletcher, Angus. *Allegory: The Theory of Symbolic Mode*. Ithaca, N.Y., 1964.

Fleury, l'Abbé. *Les Moeurs des Israelites*. 2d ed. Paris, 1712.

Flusser, David. "The Dead-Sea Sect and Pre-Pauline Christianity." *Scripta Hierosolymitana* 4 (1968): 220ff.

Fölsing, Albrecht. *Galileo Galilei: Prozess ohne Ende, Eine Biographie*. Munich and Zurich, 1983.

Foucault, Michel. *Archéologie de savoir*. Paris, 1969.

——. *Les Mots et les choses*. Paris, 1966.

Franciscus de Mayronis. *In quatuor libros sententiarum*. Venice, 1520.

Franck, Sebastian. *Paradoxa*. Edited by S. Wollgast. Berlin, 1966.

Frank, Manfred. *Was ist Neustrukturalismus?* Frankfurt-am-Main, 1984.

Frankfurt, Harry G. "Descartes on His Existence." *Philosophical Review* 75 (1966): 329–56.

Franklin, Julian H. *Jean Bodin and the Sixteenth-Century Revolution in the Methodology of Law and History*. New York, 1963.

Frechulf of Lisieux. *Chronicon*. Migne, *PL* 106: 917ff.

Freudenthal, Gideon. *Atom und Individuum in Zeitalter Newtons: Zur Genese der mechanistischen Natur- und Sozialphilosophie*. Frankfurt-am-Main, 1982.

Friedrich, Hugo. *Montaigne*. 2d ed. Bern and Munich, 1967.

Fritz, Kurt von. *Die griechische Geschichtsschreibung*. 2 vols. Berlin, 1967.

——. *Schriften zur griechischen Logik*. 2 vols. Stuttgart and Bad Cannstatt, 1978.

——. "Zenon von Sidon." Pauli-Wissowa, *RE* XA (2d ser., 19): 122–27.

Fromm, K. "Cicero's geschichtlicher Sinn." Ph.D. dissertation, University of Freiburg, 1954.

Frutolf of Michelsberg. *Chronicon*. Edited by G. Waitz. MG Script. 5: 33ff.

Fuchs, Harald. *Der geistige Widerstand gegen Rom in der Antiken Welt*. 2d ed. Berlin, 1964.

Funkenstein, Amos. "Changes in the Patterns of Christian Anti-Jewish Polemics in the 12th Century." *Zion* 23 (1968): 126–45.

———. "Descartes, Eternal Truths, and the Divine Omnipotence." *Studies in History and Philosophy of Science* 6.3 (1975): 185–99.

———. "The Dialectical Preparation for Scientific Revolutions." In *The Copernican Achievement*, edited by Robert Westman, pp. 163–203. Berkeley and Los Angeles, 1975.

———. "The Genesis of Rosenzweig's 'Stern der Erlösung': 'Urformell' and 'Urzelle.' " *Jahrbuch des Instituts für deutsche Geschichte* 4 (1983): 17–29.

———. "Gesetz und Geschichte: Zur historisierenden Hermeneutik bei Moses Maimonides und Thomas von Aquin." *Viator* 1 (1970): 147–78.

———. *Heilsplan und natürliche Entwicklung: Formen der Gegenwarts-bestimmung im Geschichtsdenken des Mittelalters*. Munich, 1965.

———. "Maimonides: Political Theory and Realistic Messianism." *Miscellanea Medievalia* 11 (1971): 81–103.

———. "Nachmanides' Typological Reading of History." *Zion* 45 (1960): 35–59. Translated in *Studies in Jewish Mysticism*, edited by J. Dan and F. Talmage, *AJS Review* (1982): 129–50.

———. "Patterns of Christian-Jewish Polemics in the Middle Ages." *Viator* 2 (1971): 373–82.

———. "Periodization and Self-Understanding in the Middle Ages and Early Modern Times." *Medievalia et Humanistica* 5 (1974): 3–23.

———. "The Persecution of Absolutes: On the Kantian and Neokantian Theories of Science." *The Kaleidosope of Science. The Israel Colloquium for the History and Philosophy of Science* 1 (1986): 39–63.

———. "Some Remarks on the Concept of Impetus and the Determination of Simple Motion." *Viator* 2 (1971): 329–48.

Funkenstein, Amos, and Jürgen Miethke. "Hugo von St. Viktor." In *Neue Deutsche Biographie* 10, pp. 19–22. Berlin, 1974.

Gabbey, Alan. "Force and Inertia in the Seventeenth Century: Descartes and Newton." In *Descartes: Philosophy, Mathematics, and Physics*, edited by Stephen Gaukroger, pp. 230–320. Brighton and Sussex, 1980.

Galileo Galilei. *Le Opere di Galileo Galilei, Edizione Nazionale*, edited by A. Favaro. 20 vols. Florence, 1891–1909.

Garnerius of Rochefort (?). *Contra Amaurianos*. Edited by Clemens Baeumker. BGPhM 25, 5-6. Münster, 1926.

Gassendi, Pierre. *De vita et moribus Epicure*. Lyons, 1647.

———. *Exercitationes paradoxicae adversus Aristoteleos*. Edited by Bernard Rochot. Paris, 1959.

———. *Syntagmatis philosophia*. Lyons, 1685.

Gay, Peter. *The Enlightenment: An Interpretation*, 1: *The Rise of Modern Paganism*. New York, 1968.

Geertz, Clifford. *Local Knowledge: Further Essays in Interpretive Anthropology*. New York, 1983.

Gelber, Hester G. *Exploring the Boundaries of Reason: Three Questions on the Na-*

ture of God by Robert Holcot OP. Studies and Texts of the Pontifical Institute of Medieval Studies 62. Toronto, 1983.

———. "Logic and Trinity: A Clash of Values in Scholastic Thought 1330–1335." Ph.D. dissertation, University of Wisconsin, Madison, 1974.

Gelber-Talmon, Yonina. "The Concept of Time in Primitive Mythus." *Iyyun, Philosophical Quarterly* 2 (1951): 201ff. (in Hebrew), 260 (English summary).

Gesta abbatum Trudonensium. Edited by Köpke. MG Script. in usu schol. 10.

Gesta Romanorum. Edited by H. Osterley. Berlin, 1872. Translated by Ch. Swan and W. Hooper, *Gesta Romanorum or Entertaining Moral Stories*. New York, 1959.

Ghellinck, J. de. *Le Mouvement théologique du 12ᵉ siècle*. 12th ed. Brussels and Paris, 1948.

Gierke, Otto von. *Das deutsche Genossenschaftsrecht*, III: *Die Staats- und Korporationslehre des Altertums und des Mittelalters und ihre Aufnahme in Deutschland*. 1881; reprint Darmstadt, 1954.

———. *Johannes Althusius und die Entwicklung der naturrectlichen Staatstheorien*. Breslau, 1913.

Gilson, Étienne. *Études sur le rôle de la pensée médiévale dans la formation du système Cartésien*. Paris, 1930.

———. *Index Scolastico-Cartésien*. 2d ed. Paris, 1979.

———. *Jean Duns Scot, Introduction à ses positions fondamentales*. Paris, 1952. Translated by Werner Dettloff, *Johannes Duns Scotus: Einführung in die Grundgedanken seiner Lehre*. Düsseldorf, 1959.

Ginzburg, Carlo. *The Cheese and the Worms: The Cosmos of a Sixteenth-Century Miller*. Baltimore, 1980.

Giraldus Cambrensis. *Topographia Hibernica*. In *Opera*, Vol. 5, edited by J. F. Dimock. London, 1867.

Glaber, Rudolfus. *Historiarum sui temporis libri quinque*. In *Raoul Glaber, les cinq livres de ses Histoires*, edited by M. Prou. Paris, 1896.

———. *Vita sancti Guillelmi abbatis Divionensis*. Migne, *PL* 142.

Glanvill, Joseph. *Plus Ultra or the Progress and Advancement of Knowledge since the Days of Aristotle*. London, 1668.

———. *Saducismus Triumphatus: or, Full and Plain Evidence Concerning Witches and Apparitions*. 2d ed. London, 1682.

Glossa ordinaria. Migne, *PL* 113.

Goddu, André. *The Physics of William of Ockham*. Leiden and Cologne, 1984.

Gössman, Elizabeth. *Metaphysik und Weilsgeschichte: Eine theologische Untersuchung der Summa Halensis (Alexander von Hales)*. Munich, 1964.

Goez, Werner. *Translatio imperii: Ein Beitrag zur Geschichte des Geschichtsdenken und der politischen Theorien im Mittelalter und in der frühen Neuzeit*. Tübingen, 1958.

Goldmann, Lucien. *Le dieu caché: Études sur la vision tragique dans les Pensées de Pascal et dans le théâtre de Racine*. Paris, 1959.

Goldsmith, M. M. *Hobbes' Science of Politics*. New York, 1966.

Goldzieher, Ignaz. *Vorlesungen über den Islam*. Heidelberg, 1910.

Goodman, Nelson. *Fact, Fiction, and Forecast*. 2d ed. Indianapolis, 1965.

Goretti, Maria. "Vico et le hétérogenèse de fins." *Les Études philosophiques* 3.4 (1968): 351–59.

Gossman, Lionel. *Medievalism and the Ideologies of the Enlightenment: The World and Work of La Curne de Sainte-Palaye.* Baltimore, 1968.

Gottleib, Abraham. *Mehkarim besifrut hakabala.* Tel-Aviv, 1976.

Grabmann, Martin. *Die Geschichte der scholastischen Methode.* 2 vols. 1911; reprint Basel and Stuttgart, 1961.

Graetz, Heinrich Z. *Die Konstruktion der jüdischen Geschichte.* Berlin, 1936.

Graiff, Cornelio Andrea, o.s.b., ed. *Siger de Brabant: Questions sur le Métaphysique.* Louvain, 1948.

Grant, Edward. "The Condemnation of 1277: God's Absolute Power and Physical Thought in the Middle Ages." *Viator* 10 (1979): 211–44.

——. "Medieval Explanations and Interpretations of the Dictum that Nature Abhors the Vacuum." *Traditio* 29 (1973): 327–55.

——. "Motion in the Void and the Principle of Inertia in the Middle Ages." *Isis* 55 (1964): 265–92.

——. *Much Ado about Nothing: Theories of Space and Vacuum from the Middle Ages to the Scientific Revolution.* Cambridge, 1981.

——. *A Sourcebook in Medieval Science.* Cambridge, Mass., 1974.

Grant, R. M. *Miracle and Natural Law in Greco-Roman and Early Christian Thought.* Amsterdam, 1952.

Gregori, Tullio. *Scetticismo e Empirismo: Studio su Gassendi.* Bari, 1961.

Gregory of Rimini. *Lectura super primum et secundum sententiarum.* Edited by A. D. Trapp, o.s.a. and V. Marcolino. 6 vols. Berlin and New York, 1979–1984.

Gregory the Great. *Homiliae in Ezechielem.* Migne, *PL* 76.

Griewank, R. *Der neuzeitliche Revolutionsbegriff, Entstehung und Entwicklung.* 2d ed. Frankfurt-am-Main, 1969.

Grondziel, Heinrich. "Die Entwicklung der Unterscheidung zwischen der potentia Dei absoluta und der potentia Dei ordinata von Augustin bis Alexander von Hales." D. Theol. dissertation, University of Breslau, 1926.

Grotius, Hugo. *De iure belli ac pacis.* Lausanne, 1751.

Gruenwald, Ithamar. *Apocalyptic and Merkavah Mysticism.* Leiden and Cologne, 1980.

Grundmann, Herbert. *Ketzergeschichte des Mittelalters.* In *Die Kirche in ihrer Geschichte,* edited by K. D. Schmitt and E. Wolf. Göttingen, 1963.

——. "Oportet et haereses esse: Das Problem der Ketzerei im Spiegel der mittelalterlichen Bibelexegese." *Archiv für Kulturgeschichte* 45 (1963): 129–64.

——. *Studien zu Joachim von Floris.* 1927; reprint Leipzig, 1966.

Guelincx, Arnold. *Annotationes ad Ethicam.* In *Opera Philosophica,* Vol. 3, edited by J.P.N. Land. The Hague, 1891–1893.

——. Γνῶθι σεαυτόν sive *Ethica.* Amsterdam, 1665.

Guerlac, H. "Copernicus and Aristotle's Cosmos." *Journal of the History of Ideas* 29 (1968): 109–13.

Gueroult, Martial. *Spinoza.* 2 vols. Paris and Hildesheim, 1968, 1974.

Gurewitsch, Aaron J. *Kategorii srednevekovoie kulture.* Moscow, 1972. Translated by Gabriele Lossak, *Das Weltbild des mittelalterlichen Menschen.* Dresden, 1978.

Gusdorf, G. *Dieu, la nature, l'homme au siècle des lumières.* Paris, 1972.

Guthrie, W.K.C. *A History of Greek Philosophy.* 5 vols. Cambridge, 1969.

Guttmann, Jakob. "Der Einfluss der Maimonideischen Philosophie auf das christliche Abendland." In *Moses ben Maimon, sein Leben, seine Werke und sein Einfluss,* edited by J. Guttmann, pp. 144–54. Leipzig, 1908.

Guttmann, Julius. "John Spencers Erklärung der biblischen Gesetze in ihrer Beziehung zu Maimonides." In *Festskrift i anleding af Professor David Simonsens 70-årige fødelsdag,* pp. 258–76. Copenhagen, 1923.

——. "Das Problem der Kontingenz in der Philosophie des Maimonides." *MGWJ* 83 (1939): 406ff.

Haber, Frank C. *The Age of the World: Moses to Darwin.* Baltimore, 1959.

Hägglund, B. *Theologie und Philosophie bei Luther und in der ockhamistischen Tradition.* Lund, 1935.

Häussler, A. "Vom Ursprung und Wandel des Lebensaltervergleichs." *Hermes* 92 (1964): 313ff.

Hahn, David E. *The Origins of Stoic Cosmology.* Athens, Ohio, 1976.

Hahn, T. *Tychonius-Studien: Ein Beitrag zur Kirchen- und Dogmengeschichte des vierten Jahrhunderts.* Leipzig, 1900.

Hallet, H. F. *Benedict de Spinoza.* London, 1957.

Harnack, Adolf von. *Marcion, Das Evangelium vom fremden Gott.* 2d ed. 1924; reprint Berlin, 1960.

——. "Porphyrius gegen die Christen." In *SB der königlichen Akademie der Wissenschaften, Phil- hist. Klasse I.* Berlin, 1916.

Harries, K. "The Infinite Sphere: Comments on the History of a Metaphor." *Journal of the History of Philosophy* 13 (1976): 5–15.

Hartke, Werner. *Römische Kinderkaiser: Eine Strukturanalyse römischen Denkens und Daseins.* Berlin, 1951.

Hassinger, Erich. *Empirisch-rationaler Historismus: Seine Ausbildung in der Literatur Westeuropas von Guiccardini bis Saint-Evremond.* Bern and Munich, 1978.

Hausamman, S. "Realpräsens in Luthers Abendmahllehre." In *Studien zur Geschichte und Theologie der Reformation, Festschrift für Ernst Bizer,* pp. 157–73. Neukirchen and Vluyn, 1969.

Hazard, Paul. *La Crise de la conscience européenne.* Paris, 1935.

Heath, Thomas L. *A History of Greek Mathematics.* 2 vols. Oxford, 1921.

——. *Mathematics in Aristotle.* Oxford, 1949.

Hegel, Georg Wilhelm Friedrich. *Philosophie der Geschichte.* Edited by F. Brunstädt. Reclam, 1961.

——. *Vorlesungen über die Geschichte der Philosophie.* In *Werke,* edited by Eva Moldenhauer and Karl Marcus Michel, vols. 18–20. Frankfurt-am-Main, 1971.

——. *Vorlesungen über die Philosophie der Geschichte.* Edited by H. Glockner. Stuttgart, 1928.

Hegler, Alfred. *Geist und Schrift bei Sebastian Franck: Eine Studie zur Geschichte des Spiritualismus in der Reformationszeit.* Freiburg im Breisgau, 1892.

Heidegger, Martin. *Sein und Zeit.* 9th ed. Tübingen, 1960.

Heinemann, Isaak. "Antisemitismus," Pauly-Wissowa, *RE*, Suppl. 5.3–43.

——. *Darche ha'agada.* 3d ed. Jerusalem, 1970.

——. *Ta'ame hamitsvot be-safrut Yisrael.* Jerusalem, 1959.

Heller-Willensky, Sara. "Isaac Ibn Latif." In *Jewish Medieval and Renaissance Studies,* edited by Alexander Altmann. Cambridge, Mass., 1967.

Henrich, Dieter. *Der ontologische Gottesbeweis: Sein Problem und seine Geschichte in der Neuzeit.* Tübingen, 1960.

Henry of Ghent. *Quodlibeta.* Venice, 1608.

Hervaeus Natalis. *In quatuor libros sententiarum commentaria.* Paris, 1647.

Heyd, Michael. *Between Orthodoxy and the Enlightenment: Jean-Robert Chouet and the Introduction of Cartesian Science in the Academy of Geneva.* Archives internationales d'histoire des idées 96. The Hague and Jerusalem, 1982.

Hieronymus. *Chronicon Eusebii.* Edited by R. Helms. GCS 24 (1913).

——. *Commentarium in Danielem.* CCSL 15A.

——. *Epistulae.* CSEL 56.

——. *Vita Malchi.* Migne, *PL* 23: 5ff.

Hildebertus Cenomanensis. *Tractatus theologicus.* Migne, *PL* 171: 1067ff.

Hintikka, Jaakko. "Cogito Ergo Sum as an Inference and a Performance." *Philosophical Review* 72 (1963): 487–96.

——. "Cogito Ergo Sum: Inference or Performance." *Philosophical Review* 71 (1962): 3–32.

——. "Kantian Intuitions." *Inquiry* 15 (1972): 341–45.

——. *Knowledge and the Known: Historical Perspectives in Epistemology.* Dordrecht, 1974.

——. "Leibniz on Plenitude, Relations, and the 'Reign of Law.' " In *Leibniz, A Collection of Critical Essays,* edited by H. G. Frankfurt, pp. 155–90. Garden City, N.Y., 1972.

——. *Time and Necessity: Studies in Aristotle's Theory of Modality.* Oxford, 1973.

Hipler, Franz. *Die Christliche Geschichtsauffassung.* Cologne, 1884.

Hobbes, Thomas. *The English Works of Thomas Hobbes.* Edited by Sir William Molesworth. 11 vols. London, 1839–1845.

——. *Opera Philosophica quae Latine scripsit omnia.* Edited by Sir William Molesworth. 5 vols. London, 1839–1845.

——. *Leviathan.* Edited by Michael Oakeshott. Oxford, 1955.

——. *Leviathan.* Edited by C. B. Macpherson. Baltimore, 1968.

Hochstetter, Erich. *Studien zur Metaphysik und Erkenntnislehre Wilhelms von Ockham.* Berlin, 1927.

Höffding, Harald. *History of Modern Philosophy.* New York, 1900.

Hönigswald, Richard. *Hobbes und die Staatsphilosophie.* Munich, 1924.

Holcot, Robert. *In quatuor libros Sententiarum.* Lyons, 1518.

Holton, Gerald. *Thematic Origins of Scientific Thought: Kepler to Einstein.* Cambridge, Mass., 1973.

Horning, G. *Wörterbuch der Philosophie.* Edited by Gerhard Ritter and Karl Gründer. Darmstadt, 1971.

Hugh of St. Victor. *De sacramentis Christianae fidis.* Migne, *PL* 176: 183ff.

———. *De scripturis.* Migne, *PL* 175: 9ff.

———. *De vanitate mundi.* Migne, *PL* 176: 703ff.

———. *Eruditionis didascalicae libri septem.* Migne, *PL* 176: 739ff.

Huizinga, Johann. *Zur Geschichte der Begriffes Mittelalter.* Stuttgart, 1954.

Husserl, Edmund. *Cartesianische Meditationen und Pariser Vorträge.* Edited by S. Strasser. In *Husserliana: In Gesammelte Werke,* vol. 1. Haag, 1963.

Huygens, Christiaan. *Oeuvres complètes.* 22 vols. La Haye, 1888–1950.

Ibn Khaldûn. *The Muqaddimah.* Edited by F. Rosenthal. 3 vols. Princeton, 1958.

Irenaeus of Lyons. *Sancti Irenaei episcopi Lugudunensis libri quinque adversus haereses.* Edited by W. W. Harvey. 2 vols., Cambridge, 1857.

Iserloh, Erwin, *Gnade und Eucharistie in der philosophischen Theologie des Wilhelms von Ockham, ihre Bedeutung für die Ursachen der Reformation.* Wiesbaden, 1956.

———. "Um die Echtheit des *Centiloquiums.*" *Gregorianum* 30 (1949): 78–103, 309–46.

Ishiguro, Hidé. *Leibniz's Philosophy of Logic and Language.* Ithaca, N.Y., 1972.

Isidore of Seville. *De ordo creaturarum.* Migne, *PL* 83.

———. *Etymologiarum sive originum libri XX.* Edited by W. M. Lindsay. 1911; reprint Oxford, 1957.

Jaeger, Werner. *Paideia: Die Formung des griechischen menschen.* 4th ed. (vol. 1), 5th ed. (vols. 2–3), in one volume. Berlin, 1973.

———. *Die Theologie der frühen griechischen Denker.* Stuttgart, 1953.

Jaki, Stanley L. *The Road of Science and the Ways of God.* Chicago, 1978.

James, William. *Pragmatism.* New York, 1955.

———. *The Varieties of Religious Experience.* London, 1902.

Jammer, Max. *Concepts of Space: A History of Theories of Space in Physics.* Cambridge, Mass., 1954.

Jardin, Nicholas. "The Significance of the Copernican Orbs." *Journal for the History of Astronomy* 13 (1982): 168–94.

Joachim of Fiore. *Enchiridion in Apocalypsim.* In *Joachim von Floris und die joachimitische Literatur,* edited by J. C. Huck. Freiburg, 1938.

Johannes de Bassolis. *In secundum sententiarum questiones.* Paris, 1516.

Johannes de Ripa. "Jean de Ripa I sent. dist. XXXVII: De modo inexistendi divine essentie in omnibus creaturis," edited by A. Combes and F. Ruello, *Traditio* 37 (1967): 161–267.

———. *Quaestio de gradu supremo.* Edited by A. Combes and P. Vignaux. Textes philosophiques du moyen âge 12, pp. 143–222. Paris, 1964.

Johannes Philoponos. *In Artist. Physicorum libros quinque posteriores commentaria.* Edited by Hieronymus Vitelli. CAG. Berlin, 1888.

John of Salisbury. *Historia pontificalis.* Edited by Marjorie Chibnall. London, 1956.

———. *Polycraticus.* Edited by C.C.I. Webb. 2 vols. Oxford, 1909.

Jordanes. *Romana et Getica.* Edited by Theodor Mommsen. MG AA (1882).

Josephus, Flavius. *De Iudaeorum vetustate sive contra Apionem.* In *Opera*, vol. 5, edited by B. Niese. 1889; reprint Berlin, 1955.

Junilius Africanus. *Instituta regularia divinae legis.* Migne, *PL* 68.

Kabitz, W. "Leibniz und Berkeley." *SB der Preuss. Akademie der Wissenschaften philosophisch-historische Klasse* 24 (1932): 636ff.

Kafka, Gustav and Hans Eibl. *Der Ausklang der antiken Philosophie und das Erwachen einer neuen Zeit.* Geschichte der Philosophie in Einzeldarstellungen 9. Munich, 1928.

Kamlah, Wilhelm. *Apokalypse und Geschichtstheologie. Die mittelalterliche Auslegung der Apokalypse vor Joachim von Fiore.* Historische Studien 285, edited by W. Andreas et al. Berlin, 1935.

——. *Christentum und Geschichtlichkeit: Untersuchungen zur Entsthehung des Christentums und zu Augustins "Burgerschaff Gottes."* 2d ed. Stuttgart, 1951.

Kant, Immanuel. *Idee zu einer allgemeinen Geschichte in weltbürglicher Absicht.* In *Werke*, edited by Wilhelm Weichschedel, vol. 11. Frankfurt-am-Main, 1964.

——. *Kritik der reinen Vernunft.* In *Werke*, edited by Weichschedel, vols. 3–4. Frankfurt-am-Main, 1964.

——. *Kritik der Urteilskraft.* In *Werke*, edited by Weichschedel, vol. 10.

Kantorowicz, Ernst H. *The King's Two Bodies: A Study in Medieval Political Theology.* Princeton, 1957.

Kapp, Ernst. *Greek Foundations of Traditional Logic.* New York, 1942.

Katz, Jacob. *Ben yehudim legoyyim.* Jerusalem, 1960.

Kelley, Donald R. *Foundations of Modern Historical Scholarship: Language, Law, and History in the French Renaissance.* New York, 1970.

——. "Klio and the Lawyers." *Medievalia et Humanistica* 5 (1975): 24–49.

Kenney, Anthony. *Descartes: A Study of His Philosophy.* New York, 1968.

Keohane, Nannerl O. *Philosophy and the State in France: The Renaissance to the Enlightenment.* Princeton, 1980.

Kepler, Johannes. *Harmonices Mundi Libri V.* In *Gesammelte Werke*, vol. 6. Edited by Max Caspar. Munich, 1940.

Kern, Fritz. *Humana Civilitas.* Leipzig, 1913.

Keuck, Karl. "Historia: Geschichte des Wortes und seine Bedeutung." Ph.D. dissertation, University of Münster, 1934.

Kirk, G. S. and J. E. Raven. *The Presocratic Philosophers.* Cambridge, 1957.

Kisch, Guido. *Erasmus und die Jurisprudenz seiner Zeit.* Basel, 1960.

Klein, Jakob. *Greek Mathematical Thought and the Origins of Algebra.* Cambridge, Mass., 1968.

Klempt, A. *Die Säkularisierung der universalhistorischen Auffassung.* Göttingen, 1960.

Kneale, William and Martha Kneale. *The Development of Logic.* Oxford, 1962.

Knuuttila, Simo. "Modal Logic." *CHM*: 342–57.

Koestler, Arthur. *The Sleepwalkers: A History of Man's Changing Vision of the Universe.* New York, 1963.

Kolakowski, Leszek. *Chrétiens sars église: La conscience religieuse et le lieu confessionel au xii* siècle.* Paris, 1969.

Koselleck, Reinhart. *Vergangene Zukunft: Zur Semantik geschichtlicher Zeiten.* Frankfurt-am-Main, 1979.

Koslow, A. "The Law of Inertia: Some Remarks on Its Structure and Significance." In *Philosophy, Science, and Method: Essays in Honor of Ernest Nagel,* edited by S. Morgenbesser et al., pp. 552–54. New York, 1969.

Koyré, Alexander. *The Astronomical Revolution: Copernicus-Kepler-Borelli.* Translated by R. E. W. Moddison. Paris and London, 1973.

——. *Descartes und die Scholastik.* 1923; reprint Darmstadt, 1971.

——. *From the Closed World to the Infinite Universe.* Baltimore, 1957.

——. "Galileo and Plato." *Journal of the History of Ideas* 4 (1943): 400ff.

——. *Metaphysics and Measurement: Essays in the Scientific Revolution.* Cambridge, Mass., 1968.

——. "Le vide et l'espace infini au xive siècle." *AHDL* 24 (1949): 45–91.

Koyré, Alexander, and I. Bernard Cohen. "Newton and the Leibniz-Clarke Correspondence with Notes on Newton, Conti, and Des Miazeaux." *Archives internationales d'histoire des sciences* 15 (1962): 69ff.

Kraus, Samuel. *Das Leben Jesu nach jüdischen Quellen.* Berlin, 1902.

Kristeller, Paul Oskar. *Eight Philosophers of the Italian Renaissance.* Stanford, 1964.

Kümmel, Werner Georg. *Einleitung in das Neue Testament.* 17th ed. Heidelberg, 1973.

Kuhn, Thomas. "A Function for Thought Experiments." In *The Essential Tension: Selected Studies in Scientific Tradition and Change.* Chicago, 1977.

Kurdzialek, M. "David von Dinant und die Anfänge der aristotelischen Naturphilosophie." In *La Filosofia della natura nel medioevo, Atti del terzo Congresso intern. di filos. med.,* pp. 407–16. Milan, 1966.

Lachterman, David R. "The Physics of Spinoza's Ethics." In *Spinoza: New Perspectives,* edited by R. W. Shahan and J. I. Biro, pp. 71–112. Norman, Okla., 1978.

Lactantius. *De divinis institutionibus.* CSEL 19.

Ladner, Gerhart B. *The Idea of Reform: Its Impact on Christian Thought and Action in the Age of the Fathers.* Cambridge, Mass., 1959.

Lampert of Hersfeld. *Annales.* Edited by Oswald Holder-Egger. MG Script. in usu Schol. Hanover, 1894.

Landau, A. *Die dem Raume entnommenen Synonyma für Gott in der hebräischen Literatur.* Zurich, 1888.

Landes, David S. *Revolution in Time: Clocks and the Making of the Modern World.* Cambridge, Mass., 1983.

Landsberg, Fritz. *Das Bild der alten Geschichte in mittelalterlichen Chroniken.* Berlin, 1934.

Langston, D. C. "Scotus and Ockham on the Univocal Concept of Being." *Franciscan Studies* 39 (1979): 105–29.

Lapidge, Michael. "Stoic Cosmology." In *The Stoics,* edited by John M. Rist, pp. 161–85. Berkeley and Los Angeles, 1978.

Lappe, J. *Nicolaus von Autrecourt, sein Leben, seine Philosophie, seine Schriften,* BGPhM 6.2. Münster, 1908.

Lasker, Daniel J. *Jewish Philosophical Polemics against Christianity in the Middle Ages.* New York, 1967.

Lassaux, Ernst von. *Philosophie der Geschichte.* Edited by E. Thurner. Munich, 1952.

Lasswitz, Kurd. *Geschichte der Atomistik vom Mittelalter bis Newton.* 2 vols. 1890; reprint Darmstadt, 1963.

Latta, Robert. *Leibniz: The Monadology and Other Philosophical Writings.* Oxford, 1898.

Lauterbach, Zwi. "The Saducees and Pharisees." In *Rabbinic Essays,* pp. 31ff. Cincinnati, 1951.

Lazarus-Yafe, Hava. "The Religious Problematics of Pilgrimage in Islam." *Proceedings of the Israel Academy of Sciences* 5.11 (1976): 222–43.

Leach, Edmund. "Anthropological Aspects of Language: Animal Categories and Verbal Abuse." In *Mythology,* edited by P. Maranda, pp. 39–67. Harmondsworth and Baltimore, 1972.

———. "Vico and Lévi-Strauss on the Origins of Humanity." In *Giambattista Vico: An International Symposium,* edited by Giorgio Tagliacozzo, pp. 309–18. Baltimore, 1976.

Leeuw, G. van der. *L'Homme primitif et la religion.* Paris, 1940.

Leff, Gordon. *William of Ockham: The Metamorphosis of Scholastic Discourse.* Manchester, 1975.

Leibniz, Gottfried Wilhelm. *The Leibniz-Arnauld Correspondence.* Edited by G.H.R. Parkinson. Edited and translated by H. T. Mason. Manchester, 1967.

———. *Die mathematische Schriften von G. W. Leibniz.* Edited by C. J. Gerhardt. 7 vols. Halle, 1849–1863.

———. *Opuscules et fragments inédits de Leibniz.* Edited by L. Couturat. Paris, 1903.

———. *Die philosophischen Schriften von Gottfried Wilhelm Leibniz.* Edited by C. J. Gerhardt. 7 vols. 1875–1890; reprint Hildesheim, 1965.

———. *Sämtliche Schriften und Briefe.* Darmstadt, 1923ff.

———. *Textes inédits d'après les manuscrits de la Bibliothèque Provinciale de Hanovre.* Edited by G. Grua. Paris, 1948.

Leszl, Walter. *Logic and Metaphysics in Aristotle.* Padua, 1970.

Lévi-Strauss, Claude. *La pensée sauvage.* Paris, 1962.

Levinger, Jacob. "Al tora shebe'al pe behaguto shel harambam." *Tarbiz* 37 (1968): 282–93.

———. *Darche hamachshava hahilchatit shel harambam* [Maimonides' Techniques of Codification]. Jerusalem, 1965.

Levy, Jochanan H. *Olamot nifgashim* [Studies in Jewish Hellenism]. Jerusalem, 1960.

Lewis, David K. *Counterfactuals.* Cambridge, Mass., 1973.

Lichtenstein, A. *Henry More: The Radical Theology of a Cambridge Platonist.* Cambridge, Mass., 1962.

Lieberman, Saul. *Yevanim ve yavnut be'erets yisrael* [Greeks and Hellenism in Jewish Palestine]. Jerusalem, 1962.

Livesey, Steven J. "Metabasis: The Interrelationship of Sciences in Antiquity and the Middle Ages." Ph.D. dissertation, UCLA, 1982.

Livius, Titus. *Ab urbe condita*. Edited by Robert S. Conway and Charles F. Walters. Oxford, 1958.

Lloyd, G.E.R. *Magic, Reason, and Experience: Studies in the Origins and Development of Greek Science*. Cambridge, 1979.

Lods, A. *Jean Astruck et la critique biblique au xviiie siècle*. Strasbourg and Paris, 1924.

Loewenstamm, S. A. *Masoret Jetsi'at mitsrayim behishtalsheluta*. 2d ed. Jerusalem, 1972.

Löwith, Karl. *Meaning in History*. Chicago, 1949.

——. "Vicos Grundsatz: *Verum et factum convertuntur*." In *Aufsätze und Vorträge 1830–1970*, pp. 157–88. Stuttgart, 1971.

Lombardus, Petrus. *Sententiae in IV libris distinctae*. Edited by Quaracchi. 2d. ed. Florence, 1916.

Lonergan, Bernard J. *Verbum: Word and Idea in Aquinas*. Notre Dame, Ind., 1967.

Loofs, F. *Theophilus von Antiochien "Adversus Marcionem" und die anderen theologischen Quellen bei Irenaeus*. Leipzig, 1930.

Lovejoy, Arthur O. *The Great Chain of Being: A Study of the History of an Idea*. New York, 1960.

Lovejoy, Arthur O. and G. Boas. *Primitivism and Related Ideas in Antiquity*. Baltimore, 1935.

Lubac, Henri de. *Exégèse médiévale: Les quartres sens de l'écriture*. 4 vols. Lyons, 1964.

Luce, A. A. *Berkeley and Malebranche*. Oxford, 1934.

Lucretius. *De rerum natura*. Edited by Cyrill Bailey. 2d ed. Oxford, 1922.

Luther, Martin. *Werke, Kritische Gesamtausgabe*. 54 vols. Weimar, 1883ff.

Lyttkens, Hampus. *The Analogy between God and the World: An Investigation of Its Background and Interpretation of Its Use by Thomas of Aquino*. Uppsala, 1953.

McCracken, Ch. J. *Malebranche and British Philosophy*. Oxford, 1983.

MacGinty, Francis P., O.S.B. "The Treastise De Mirabilibus Sacrae Scripturae." Ph.D. dissertation, National University of Ireland, 1971.

McGuire, J. E. "Neoplatonism and Active Principles: Newton and the *Corpus Hermeticum*." In *Hermeticism and the Scientific Revolution*, ed. J. E. McGuire and Robert Westman, pp. 95–150. Los Angeles, 1977.

——. "The Origins of Newton's Doctrine of Essential Qualities." *Centaurus* 12 (1968): 233–60.

Mach, Ernst. *Die Mechanik in ihrer Entwicklung*. 9th ed. 1933; reprint Darmstadt, 1976.

Macpherson, C. B. *The Political Theory of Possessive Individualism: Hobbes to Locke*. Oxford, 1962.

McVaugh, M. "Arnald of Villanova and Bradwardine's Law." *Isis* 58 (1966): 56–64.

——. "The Medieval Theory of Compound Medicines." Ph.D. dissertation, Princeton University, 1965.

Mahoney, Michael S. *The Mathematical Career of Pierre de Fermat (1601–1665)*. Princeton, 1973.

Maier, Anneliese. *Ausgehendes Mittelalter: Gesammelte Aufsätze zur Geistesgeschichte des 14. Jahrhunderts*. 3 vols. Rome, 1964, 1967, 1977.

——. *Kants Qualitätskategorien*. Kant-Studien 65. Berlin, 1930.

——. *Studien zur Naturphilosophie der Spätscholastik, I: Die Vorläufer Galileis im 14. Jahrhundert. II: Zwei Grundprobleme der scholastischen Naturphilosophie. III: An der Grenze von Scholastik und Naturwissenschaft. IV: Mataphysische Hintergründe der scholastischen Naturphilosophie. V: Zwischen Philosophie und Mechanik*. Storia e letteratura 22, 37, 41, 52, 69. 2d ed. Rome, 1951–1966.

Maier, F. G. *Augustin und das antike Rom*. Tübinger Beiträge zur Altertumswissenschaft 39. Stuttgart, 1955.

Maierù, Alfonso. "Logica Aristotelica e Teologia Trinitaria: Enrico Toffing da Oyta." In *Studi sul XIV secolo in Memoria di Anneliese Maier*, edited by A. Maierù et al., pp. 481-512. Rome, 1981.

Maimonides, M. (Moshe ben Maimon). *The Guide of the Perplexed*. Translated by S. Pines. Chicago, 1963.

——. *Iggrot harambam*. Edited by Josef Kafiḥ. Jerusalem, 1972.

——. *Liber More nebuchim, Doctor perplexorum*. Translated by Johannes Buxtorf. Basel, 1629.

——. *Mishne Tora*. Jerusalem, 1961.

——. *More ha nebuchim [Dalalat el Hairin]*. Edited by Josef Kafiḥ. 3 vols. Jerusalem, 1972.

——. *Perush ha Mishnayot*. Edited by Josef Kafiḥ. Jerusalem, 1963.

——. *Teshuvot ha Rambam (Responsa)*. Edited by Jehoshua Blau. 2 vols. Jerusalem, 1960.

Malcolm, Norman. "Anselm's Ontologial Arguments." *The Philosophical Review* 69 (1960): 41–62.

Malebranche, Nichole. *Nicholas Malebranche: The Search After Truth*. Translated by Th. M. Lennon. Columbus, Ohio, 1980.

——. *Oeuvres complètes de Malebranche*. Edited by A. Robinet et al. 21 vols. Paris, 1962–1964.

Mali, Josef. "Harehabilitatsia shel hamythos: Giambattista Vico vehamada hachadash shel hatarbut." Ph.D. dissertation, Tel-Aviv University, 1985.

Mannheim, Karl. *Ideologie und Utopie*. 4th ed. Frankfurt-am-Main, 1965.

Mansion, Suzanne, *Le Jugement d'existence chez Aristote*. Louvain and Paris, 1946.

Manuel, Frank E. and Fritzie P. Manuel. *Utopian Thought in the Western World*. Cambridge, Mass., 1979.

Marmorstein, A. *The Old Rabbinic Doctrine of God*. London, 1927.

Marrou, Henri Irénée. "Das Janusantlitz der historischen Zeit bei Augustin." In *Zum Augustin—Gespräch der Gegenwart*, edited by Carl Andresen. Wege der Forschung 5. Darmstadt, 1962.

Marsilius of Padua. *Defensor pacis*. Edited by R. Scholz. MG Fontes iuris Germanici antiqui in usu scholarum. Hanover, 1932–1933.

Martin, David. *A General Theory of Secularization*. Oxford, 1978.

Martin, Gottfried. *Immanuel Kant: Ontologie und Wissenschaftstheorie*. Berlin, 1969.

———. *Leibniz, Logic, and Metaphysics*. Translated by K. J. Northcott and P. G. Lucas. Manchester, 1964.

———. *Wilhelm von Ockham, Untersuchungen zur Ontologie der Ordnungen*. Berlin, 1947.

Martini, Raymundus. *Pugio fidei adversus Mauros et Judaeos*. Leipzig, 1687.

Marx, Karl. *Zur Judenfrage*. In *Die Frühschriften*, edited by Seigfried Landshut. Stuttgart, 1964.

Massaux, Édouard. *Influence de l'Évangile de Saint Matthieu dans la littérature chrétienne avant Saint Irenée*. Louvain, 1950.

Mates, Benson. *Stoic Logics*. 1953; reprint Berkeley and Los Angeles, 1973.

Mattheus ab Aquasparta. *Quaestiones Disputatae*. Edited by Quaracchi. Florence, 1903.

Maupertuis, Pierre Louis Moreau de. *Essai de Cosmologie*. Paris, 1951.

Mazat, H. L. "Die Gedankenwelt des jungen Leibniz." In *Beiträge zur Leibniz-Forschung*, edited by G. Schishkopf, pp. 37–67. Reutlingen, 1947.

Mazzarino, S. *Das Ende der antiken Welt*. Translated by Fritz Jaffé. Munich, 1961.

Megilot midbar Yehuda. Edited by S. Heberman. Tel-Aviv, 1959.

Meinecke, Friedrich. *Die Entstehung des Historismus*. In *Werke*, Vol. 2, edited by E. Hofer. Munich, 1955.

———. *Die Idee der Staatsraison in der neueren Geschichte*. 2d ed. Munich, 1960.

———. "Klassizismus, Romantizismus und historisches Denken im 18. Jh." In *Werke*, Vol. 4, edited by E. Kessel. Munich, 1959.

Melancthon, Philipp. *Chronicon Carionis*. In *Opera omnia*, Vol. 12, edited by C. G. Bretschneider. Halle/S., 1844.

Melville, G. *System und Diachronie: Untersuchungen zur theoretischen Grundlegung geschichtsschreiberischer Praxis im Mittelalter*. Historisches Jahrbuch 95. Berlin, 1975.

Mendelssohn, Moses. *Jerusalem oder über religiöse Macht und Judentum* (1783). *Gesammelte Schriften*. Edited by G. B. Mendelssohn. 3 vols. Leipzig, 1843–1845.

Menger, A. "A Counterpart of Ockham's Razor in Pure and Applied Mathematics: Ontological Uses." *Synthese* 12 (1960).

Merchavia, Ch. *Hatalmud bir'i hanatsrut: Hayaḥas Lesifrut Yisrael shele'ahar hamikra ba'olam hanotsri biyme habenayyim* [The Church versus Talmudic and Midrashic Literature]. Jerusalem, 1970.

Merton, Robert K. "Karl Mannheim and the Sociology of Knowledge." In *Social Theory and Social Structure*. 3d ed. New York and London, 1968.

Midrash bereshit rabba. Edited by Jehuda Theodor and Ḥanoch Albeck. Jerusalem, 1965.

Midrash vayyikra rabba, sifre bamidbar udebarim, Ms. Vat. Ebr. 32. Jerusalem, 1972.

Miethke, Jürgen. *Ockhams Weg zur Sozialphilosophie.* Berlin, 1969.

Miller, L. G. "Descartes, Mathematics, and God." *Philosophical Review* 66 (1957): 451–65.

Milton, John R. "The Origins and Development of the Concept of the 'Laws of Nature.' " *Archive of European Sociology* 22 (1981): 173–95.

Mommsen, Theodor E. "St. Augustine and the Christian Idea of Progress." *Journal of the History of Ideas* 12 (1951): 346ff.

Montaigne, Michel de. *Essais.* Edited by M. Rat. Paris, 1962. English translation by D. M. Frame. Stanford, 1958.

Montesquieu, Charles de Secondat, Baron de. *De l'esprit des lois.* In *Oeuvres,* edited by R. Callois. Paris. 1951.

Moody, Ernest A. *The Logic of William of Ockham.* New York, 1935.

——. *Studies in Medieval Philosophy, Science, and Logic: Collected Papers 1933–1969.* Berkeley and Los Angeles, 1975.

Moraux, J. "Saint Augustin et Malebranche." In *La Philosophie et ses problèmes: Ouvrage en homage à Mgr. Jolivet,* pp. 109–36. Lyons, 1960.

More, Henry. *An Antidote against Atheisme; or, an Appeal to the Natural Faculties of the Minde of Man, Whether There Be Not a God.* London, 1653.

——. *The Easie, True, and Genuine Notion . . . of a Spirit* [1700]. In *Philosophical Writings of Henry More,* edited by F. I. McKinnon. Oxford, 1925.

——. *Enchiridion metaphysicum.* London, 1671.

——. *The Immortality of the Soul, so farre forth as it is Demonstrable from the Knowledge of Nature and the Light of Reason.* London, 1669.

Morrow, G. R. *Proclus: A Commentary on the First Book of Euclid's Elements.* Princeton, 1970.

Moser, Simon. *Grundbegriffe der Naturphilosophie bei Wilhelm von Ockham.* Innsbruck, 1932.

Mueller, I. "Aristotle on Geometrical Objects." *Archiv für Geschichte der Philosophie* 52 (1970): 156–71.

Murdoch, John E. "*Mathesis in philosophiam scholasticam introducta*: The Rise and Development of the Application of Mathematics in Fourteenth-Century Philosophy and Theology." In *Arts libéraux et philosophie au moyen âge,* Actes du quatrième congrès de philosophie médiévale pp. 215–54. Montreal, 1967 and Pano, 1969.

Murdoch, John E. and Edith Sylla. "The Science of Motion." In *Science in the Middle Ages,* edited by D. C. Lindberg, pp. 249–51. Chicago, 1978.

Murray, G. *Five Stages of Greek Religion.* 3d ed. Garden City, N.Y., 1955.

Nachmanides (Moshe ben Nachman). *Perush haramban al hatora.* Edited by Ch. D. Chawel. 3 vols. Jerusalem, 1959.

Nacht-Eladi, S. "Aristotle's Doctrine of the *Differentia Specifica* and Maimon's Laws of Determinability." *Scripta Hierosolymitana* 6, *Studies in Philosophy,* edited by S. H. Bergman, pp. 222–48. Jerusalem, 1960.

Nagel, Ernest and James R. Newman. *Gödel's Proof.* New York, 1958.

Naphtali, Jehudith. "Ha'yachas sheben avoda le'erech bate'oriot hakalkaliot

388   BIBLIOGRAPHY

shel ha scholastika bameot ha-13 veha-14." [The Correlation between Labor and Value in the Scholastic Economic Theories of the 13th and 14th
Centuries]. Ph.D. dissertation, Tel-Aviv University, 1982.

Neugebaur, Otto. *A History of Ancient Mathematical Astronomy*. 3 vols. Berlin,
1975.

Newton, Sir Isaac. *Opera*. Edited by S. Horsley. 5 vols. London, 1779–1785.

——. *Opticks; or, a Treatise of the Reflections, Refractions, Inflections, and Colours
of Light* [1730]. Edited by D.H.D. Roller. New York, 1952.

——. *Philosophiae naturalis principia mathematica*. London, 1687.

——. *Philosophiae naturalis principia mathematica*. 3d ed. [1726]. Edited by
A. Koyré and I. B. Cohen. 2 vols. Cambridge, 1972.

——. *Sir Isaac Newton's Mathematical Principles of Natural Philosophy and His System of the World*. Translated by A. Motte [1729], revised by F. Cajori. 2
vols. Berkeley and Los Angeles, 1971.

——. *Unpublished Scientific Papers of Sir Isaac Newton*. Edited by A. R. Hall and
M. B. Hall. Cambridge, 1964.

Nicolini, Fausto. *Commento storico alla seconda Scienza nuova*. Rome, 1949.

Nock, Arthur D., ed. *Sallustius Concerning the Gods and the Universe*. Cambridge, 1926.

Norden, Eduard. *Die Geburt des Kindes: Geschichte einer religiösen Idee*. 1924; reprint Darmstadt, 1958.

Normore, Calvin. "Future Contingents." *CHM*: 358–81.

North, C. R. "Pentateuchal Criticism." In *The Old Testament and Modern
Study*, edited by H. H. Rowley, pp. 81ff. Oxford, 1951.

Nowak, L. "Laws of Science, Theories, Measurements (Comments on Ernest
Nagel's *The Structure of Science*)." *Philosophy of Science* 39.4 (1972): 533–48.

Oakley, Francis. "Medieval Theories of Natural Law: William of Ockham
and the Significance of the Voluntarist Tradition." *Natural Law Forum* 6
(1961): 65–83.

——. *Omnipotence, Covenant, and Order: An Excursion in the History of Ideas from
Abelard to Leibniz*. Ithaca, N.Y., and London, 1984.

Oberman, Heiko A. "*Facientibus Quod in se est Deus non Denegat Gratiam*: Robert Holcot O.P. and the Beginnings of Luther's Theology." *Harvard Theological Review* 55 (1962): 317–42.

——. *The Harvest of Medieval Theology: Gabriel Biel and Late Medieval Nominalism*. Cambridge, Mass., 1963.

Odo of Cluny. *Collationum libri tres*. Migne, *PL* 133.

Oetinger, Friedrich Christian. *Theologia ex idea vitae deducta*. Frankfurt and
Leipzig, 1765.

Oresme, Nicole. *Le Livre du ciel et du monde*. Edited by A. D. Menut and A. J.
Denomy. Madison, Wis., 1968.

——. *Nicole Oresme and the Medieval Geometry of Qualities and Motions*. Edited
by Marshall Clagett. Madison, Wis., 1968.

Origenes. *Contra Celsum*. Edited by P. Kötschau. GCS 30. Leipzig, 1899. Translated by Henry Chadwick. Cambridge, 1965.

———. *De principiis*. Edited by Herwig Görgemanns and Heinrich Karpp. Darmstadt, 1976.

Orosius, Paulus. *Historiarum adversus paganos libri VII*. Edited by C. Zangemeister. Leipzig, 1889.

Ortega y Gasset, José. *Der Prinzipienbegriff bei Leibniz und die Entwicklung der Deduktionstheorie [La Idea de Principio en Leibniz y la Evolución de la Teoría Deductiva]*. Translated by Evald Kirschnes. Munich, 1966.

Osler, Margaret J. "Providence and Divine Will in Gassendi's Views on Scientific Knowledge." *Journal of the History of Ideas* 44 (1983): 549–60.

Otto of Freising. *Chronicon sive historia de duabus civitatibus*. Edited by A. Hofmeister, MG Script. in usu schol. (1912).

Otto, St. *"Natura" und "dispositio": Untersuchungen zum Naturbegriff und zur Denkforms Tertullians*. Munich, 1960.

Owen, G.E.L. "Logic and Metaphysics in Some Earlier Works of Aristotle." In *Aristotle and Plato in the Mid-Fourth Century*, edited by I. Dühring and G.E.L. Owen, pp. 162–90. Göteburg, 1960.

———. "Tithenai ta phenomena." In *Aristote et les problèmes de la méthode, Symposium Aristotelicum*, pp. 83–103. Louvain, 1961.

Owens, Joseph. "Analogy as a Thomistic Approach to Being." *Medieval Studies* 24 (1962): 303–22.

———. *The Doctrine of Being in Aristotelian Metaphysics*. Toronto, 1951.

Ozment, Steven. *The Age of Reformation, 1250–1550*. New Haven and London, 1980.

Pacchi, A. *Convenzione e ipotesi nella formazione della filosofia naturale di Thomas Hobbes*. Florence, 1965.

Pannenberg, Wolfgang. *Die Prädestinationslehre des Duns Scotus im Zusammenhang der scholastischen Lehrentwicklung*. Göttingen, 1954.

Panofsky, Erwin. *Meaning in the Visual Arts*. New York, 1953.

Pape, Ingetrud. *Tradition und Transformation der Modalität, 1: Möglichkeit-Unmöglichkeit*. Hamburg, 1966.

Pappus. *Pappi Alexandri Collectionis quae supersunt*. Edited by F. Hultsch. Berlin, 1876.

Parker, Samuel. *A Free and Impartial Censure of the Platonick Philosophie*. Oxford, 1666.

Parsons, Charles. *Mathematics in Philosophy: Selected Essays*. Ithaca, N.Y., 1983.

Parsons, Terence. *Nonexistent Objects*. New Haven, 1980.

Pascal, Blaise. *Oeuvres complètes*. Edited by Jacques Chevalier. Bibliothèque de la Pléiade. Argenteuil, 1954.

Patzig, Günther. *Die Aristotelische Syllogistik*. Töttingen, 1969.

———. "Theologie und Ontologie in der 'Metaphysik' des Aristoteles." *Kants-Studien* 53 (1960–1961): 185–205.

Pelikan, Jaroslav. *The Christian Tradition: A History of the Development of Doctrine, III: The Growth of Medieval Theology (600–1300)*. Chicago and London, 1978.

Periera, Benedictus. *De communibus omnium rerum naturalium principiis et affectionibus libri quindecim*. Cologne, 1609.

Peterson, Erik. "Der Monotheismus als politisches Problem: Ein Beitrag zur Geschichte der politischen Theologie im Imperium Romanum." In *Theologische Traktate*, pp. 86ff. Munich, 1951.

Pettazzoni, Raffaele. *L'essere supremo nelle religioni primitivi: L'omniscienza di Dio*. Turin, 1955.

Peuckert, Will-Erich. *Die grosse Wende: Das apokalyptische Saeculum und Luther*. 1948; reprint Darmstadt, 1966.

Pfeiffer, R. *Geschichte der klassischen Philologie von den Anfängen bis zum Ende des Hellenismus*. Translated by M. Arnold. Hamburg, 1970.

Philo of Alexandria. *De confusione linguarum*. Edited by G. H. Whitaker and R. Marcus. Loeb Classical Library. Cambridge, Mass., and London, 1935.

———. *De opificio mundi*. Edited by L. Cohen and P. Wendland. Breslau, 1900.

———. *De somniis*. Edited by G. H. Whitaker and R. Marcus. Loeb Classical Library. Cambridge, Mass., and London, 1935.

———. *Legum allegoriae*. Edited by G. H. Whitaker and R. Marcus. Loeb Classical Library. Cambridge and London, 1962.

Philoponos. *In Aristotelis Physicorum libros . . . commentaria*. Edited by H. Vitelli. CAG 16–17. Berlin, 1887–1888.

Pinborg, Jan. *Logik und Semantik im Mittelalter: Ein Überblick*. Stuttgart, 1972.

Pirke de Rabbi Eliezer. Warsaw, 1882.

Plato. *Timaeus*. Edited by R. G. Bury. Loeb Classical Library. Cambridge, Mass., and London, 1951.

Pocock, J.G.A. *The Ancient Constitution and Feudal Law*. Cambridge, 1957.

Pohlenz, Max. *Die Stoa, Geschichte einer geistigen Bewegung*. 2 vols. 2d ed. Göttingen, 1959.

Polin, R. *Politique et philosophie chez Thomas Hobbes*. Paris, 1953.

Popkin, Richard H. *The History of Scepticism from Erasmus to Spinoza*. 2d ed. Berkeley and Los Angeles, 1979.

Popper, Karl. *The Logic of Scientific Discovery*. 2d ed. New York, 1968.

Potter, G. H. *Zwingli*. Cambridge, 1976.

Prantl, Carl. *Geschichte der Logik im Abendlande*. 4 vols. in 3. 1867; reprint Graz, 1955.

Pritchard, J. B., ed. *Ancient Near Eastern Texts Relating to the Old Testament*. 2d ed. Princeton, 1955.

Proclus Diadochus. *Procli Diadochi in primum Euclidis Elementorum librum commentarii*. Edited by G. Friedlein. Leipzig, 1874.

Prümm, K., s.j. "Göttliche Planung und menschliche Entwicklung nach Irenaeus Adversus Haereses." *Scholastik* 13 (1938): 206ff.

(Pseudo)Aristotle. *On the Cosmos*. Edited by D. J. Furley. Loeb Classical Library. Cambridge, Mass., 1945.

Quine, Willard V. O. "Two Dogmas of Empiricism." In *From a Logical Point of View*. Cambridge, Mass., 1953.

———. *Words and Objects*. Cambridge, Mass., 1960.

Rad, Gerhard von. *Deuteronomy: A Commentary*. Philadelphia, 1966.

Ramban. *See* Nachmanides.

Randall, John H., Jr. *The Career of Philosophy*. 2 vols. New York, 1962.

———. *The School of Padua and the Emergence of Modern Science*. New York, 1965.

Ravitsky, Ariezer. "Kefi koach ha'adam—yemot hamashiach bemishnat harambam." In *Meshichiyut ve'eschatologia*. Jerusalem, 1984.

Regino of Prüm. *Chronica*. Edited by F. Knopf. MG Script. in usu schol. (1890).

Reich, Klaus. "Der historische Ursprung des Naturgesetzbegriff." In *Festschrift für Ernst Kapp zum 70. Geburtstag*, edited by Hans Diller and Hartmut Erbse, pp. 121ff. Hamburg, 1958.

Reill, Peter H. *The German Enlightenment and the Rise of Historicism*. Berkeley and Los Angeles, 1975.

———. "History and Hermeneutics in the *Aufklärung*: The Thought of Johann Christoph Gatterer." *Journal of Modern History* 45 (1973): 24–51.

Reinhardt, Karl. *Poseidonius*. Munich, 1921.

———. "Poseidonius über Ursprung und Entfaltung." In *Orient und Occident*, edited by V. G. Bergesträsser et al. Regenbogen 6. Heidelberg, 1928.

Reiter, Josef. *System und Praxis: Zur kritischen Analyse der Denkformen neuzeitlicher Metaphysik im Werk von Malebranche*. Freiburg and Munich, 1972.

Rembaum, Joel. "The New Testament in Medieval Jewish Anti-Christian Polemics." Ph.D. dissertation, UCLA, 1975.

Renan, Ernest. *Averroès et l'Averroisme*. 3d ed. Paris, 1866.

Rescher, Nicholas. "Counterfactual Hypotheses, Laws, and Dispositions." *Noûs* 5.2 (1971): 157–78.

———. *Hypothetical Reasoning*. Amsterdam, 1964.

———. *The Philosophy of Leibniz*. Englewood Cliffs, N.J., 1967.

———. *Studies in the History of Arabic Logic*. Pittsburgh, 1963.

Reuchlin, Johannes. *Augenspiegel*. Tübingen, 1511.

Rist, J. M. "Forms of Individuals in Plotinus." *Classical Quarterly*, N.S. 13 (1963): 223–31.

———. *Plotinus: The Road to Reality*. Cambridge, 1967.

———. *Stoic Philosophy*. Cambridge, 1969.

Rivaud, Albert. "La Physique de Spinoza." *Chronicon Spinozanum* 4 (1924–1926): 24–57.

Rochot, Bernard. *Les Travaux de Gassendi sur Épicure et sur l'Atomisme*. Paris, 1944.

Rössler, D. *Gesetz und Geschichte: Untersuchungen zur jüdischen Apokalyptik und der pharisäischen Orthodoxie*. Neukirchen, 1960.

Rohr, G. *Platons Stellung zur Geschichte*. Glückstadt, 1931.

Rosenroth, Knorr von. *Kabbala denudata*. Sulzbach, 1677.

Rosenstock-Huessy, Eugen. *Die Europäischen Revolutionen*. Jena, 1931.

Rosenzweig, Franz. *Kleinere Schriften*. Berlin, 1937.

Ross, Sir David. *Aristotle*. 5th ed. London, 1966.

Rossi, Paolo. *Francis Bacon: From Magic to Science*. Chicago, 1968.

Rousset, P. "Raoul Glaber." *Revue d'histoire de l'église de France* 36 (1950): 5ff.

Russell, Bertrand. *A Critical Exposition of the Philosophy of Leibniz*. 2d ed. London, 1937.

Russell, D. R. *The Method and Message of Jewish Apocalyptic, 200 BC–AD 100.* London, 1964.

Sacksteder, William. "Spinoza on Part and Whole: The Worm's Eye View." In *Spinoza: New Perspectives,* edited by Robert W. Shahan and J. I. Biro, pp. 139–59. Norman, Okla., 1978.

Sackur, E. "Studien über Rudolphus Glaber." *Neues Archiv* 14 (1889): 377ff.

Sambursky, Shmuel. *Chukot shamayim vaarets* [Laws of Heaven and Earth]. Jerusalem, 1954.

———. *Das physikalische Weltbild der Antike.* Zurich and Stuttgart, 1965.

———. "Three Aspects of the Historical Significance of Galileo." *Proceedings of the Israel Academy of Sciences and Humanities* 2 (1964): 1–11.

Sarasohn, Lisa. "The Influence of Epicurean Philosophy on Seventeenth-Century Political Thought: The Moral Philosophy of Pierre Gassendi." Ph.D. dissertation, UCLA, 1979.

Scharl, E. *Recapitulatio mundi: Der Rekapitulationsbegriff des Hl. Irenaeus und seine Anwendung auf die Körperwelt.* Freiburg im Breisgau, 1941.

Scheffler, Israel. *The Anatomy of Inquiry.* Indianapolis and New York, 1963.

Schepers, Heinrich. "Zum Problem der Kontingenz bei Leibniz." In *Collegium Philosophicum,* pp. 326–50. Basel and Stuttgart, 1965.

Schmidt, R. "Aetates mundi, die Weltalter als Gliederungsprinzip der Geschichte." *Zeitschrift für Kirchengeschichte* (1955–1956): 299ff.

Schmitt-Biggermann, Wilhelm. *Maschine und Teufel: Jean Pauls Jugendsatiren nach ihrer Modellgeschichte.* Freiburg im Breisgau and Munich, 1975.

———. *Topica universalis: Eine Modellgeschichte humanistischer und barocker Wissenschaft.* Hamburg, 1983.

Schneider, A. *Der Gedanke der Erkenntnis des Gleichen durch Gleiches in antiker und patristischer Zeit.* In BGPhM Suppl. 2 (1923): 65–76.

———. *Geschichte und Geschichtsphilosophie bei Hugo von St. Viktor.* Münster, 1933.

Schofield, Robert E. *Mechanism and Materialism: British Natural Philosophy in the Age of Reason.* Princeton, 1970.

Scholder, Klaus. *Ursprünge und Probleme der Bibelkritik im 17. Jahrhundert: Ein Beitrag zur Entstehung der historisch-kritischen Theologie.* Forschungen zur Geschichte und Lehre des Protestantismus 10, no. 33. Munich, 1966.

Scholem, Gerschom. *Das Buch Sa'ar hasamayim oder Pforte des Himmels.* Frankfurt-am-Main, 1974.

———. "Zum Verständnis der messianischen Idee im Judentum." In *Judaica,* vol. 1, pp. 7ff. Frankfurt-am-Main, 1963.

Scholz, H. *Glaube und Unglaube in der Weltgeschichte: Ein Kommentar zu Augustins De civitate Dei.* Leipzig, 1911.

———. *Mathesis universalis: Abhandlungen zur Philosophie als strenge Wissenschaft.* 2d ed. Basel and Stuttgart, 1969.

Schramm, Mathias. "Roger Bacons Begriff vom Naturgesetz." In *Die Renaissance der Wissenschaften im 12. Jahrhundert,* edited by P. Weimar, pp. 197–209. Zurich, 1981.

Schulz, F. "Bracton on Kingship." *English Historical Review* 60 (1945): 136–76.

Schulz, Marie. *Die Lehre von der historischen Methode bei den Geschichtsschreibern des Mittelalters.* Berlin and Leipzig, 1909.

Schwarz, Eduard. *Charakterköpfe aus der Antike.* Edited by J. Stroux. 2d ed. Leipzig, 1943.

———. *Ethik der Griechen.* Edited by V. W. Richter. Stuttgart, 1951.

Scotus, Johannes Duns. *Opera Omnia.* Edited by P. C. Balić et al. 17 vols. (continued). Vatican City, 1950ff.

———. *Opera omnia.* Edited by L. Wadding. 12 vols. 1639; reprint Paris, 1891–1895.

Seder Eliyahu Rabba. Edited by M. Friedmann. 2d ed. Jerusalem, 1960.

Seeberg, Erich. *Gottfried Arnold: Die Wissenschaft und die Mystik seiner Zeit.* 1923; reprint Darmstadt, 1964.

Seeberg, Reinhold. *Lehrbuch der Dogmengeschichte.* 5 vols. 1953–1954; reprint Darmstadt, 1959.

Settle, Thomas B. "Galileo's Use of Experiment as a Tool of Investigation." In *Galileo, Man of Science,* edited by Ernan McMullin, pp. 315–37. New York, 1967.

Sextus Empiricus. *Adversus mathematicos.* Edited by H. Mitschmann. 2 vols. Leipzig, 1914.

Shapiro, Herman. "Motion, Time, and Place according to William of Ockham." *Franciscan Studies* 16 (1956): 213–303, 319–72.

Siegwart, Christian. *Spinozas neuentdeckter Tractat von Gott, dem Menschen und dessen Glückseligkeit.* Gotha, 1866.

Sigebert of Gembloux. *Epistula adversus Paschalem papam.* In MG LdL 2.

Silver, Daniel J. *Maimonides' Criticism and the Maimonidean Controversy, 1180–1240.* Leiden, 1965.

Simon, G. "Untersuchungen zur Topik der Widmungsbriefe mittelalterlicher Geschichtsschreibung bis zum Ende des 12. Jahrhunderts." *Archiv für Diplomatik* 4–6 (1958–1960): 52ff.

Simplicius. *In Aristotelis quatuor libros de caelo commentaria.* Edited by J. L. Heiberg. CAG 7. Berlin, 1894.

———. *In Aristotelis physicorum libros . . . commentaria.* Edited by H. Diels. CAG 10. Berlin, 1882, 1895.

Skinner, Quentin. *The Foundations of Modern Political Thought.* 2 vols. Cambridge, 1978.

Slomkowski, A. *L'État primitif de l'homme dans la tradition de l'Église avant St. Augustine.* Paris, 1928.

Smalley, Beryl. *The Study of the Bible in the Middle Ages.* Notre Dame, Ind., 1964.

———. "William of Auvergne, John of La Rochelle and St. Thomas Aquinas on the Old Law." In *St. Thomas Aquinas 1274–1974: Commemorative Studies* 2, pp. 11–73. Toronto, 1974.

Smith, Adam. *The Wealth of Nations.* Edited by E. Cannan. New York, 1937.

Solla Price, Derek J. de. "Clockwork before the Clock." *Horological Journal* 97 (1955): 810ff.

Solmsen, Friedrich. *Aristotle's System of the Physical World: A Comparison with His Predecessors*. Ithaca, N.Y., 1960.

——. *Science since Babylon*. New Haven, 1975.

Specht, Rainer. *Commercium mentis et corporis: Über Kausalvorstellungen im Cartesianismus*. Stuttgart and Bad Cannstatt, 1966.

Spencer, John. *De legibus Hebraeorum ritualibus et earum rationibus libri tres*. Cambridge, 1685.

Spengler, O. *Untergang des Abendlandes*. 2 vols. Munich 1923.

Spink, J. S. *French Free-Thought from Gassendi to Voltaire*. 1960; reprint New York, 1969.

Spinoza, B. *Opera quotquod reperta sunt*. Edited by J. van Vloten and J.P.N. Land. 3d ed. The Hague, 1914.

——. *The Collected Works of Spinoza*. Vol. 1. Edited and translated by Edwin Curley. Princeton, 1985.

——. *The Political Works: The Tractatus Theologico-Politicus in Part and the Tractatus Politicus in Full*. Edited and translated by A. G. Wernham. Oxford, 1958.

Spitzer, Leo. *Classical and Christian Ideas of World Harmony: Prolegomena to an Interpretation of the Word "Stimmung."* Baltimore, 1963.

Spörl, Johannes. *Grundformen mittelalterlicher Geschichtsschreibung*. Munich, 1935.

Spranger, Eduard. "Die Kulturzyklentheorie und das Problem des Kulturverfalls." *SB der Preuss. Akademie der Wissenschaften, philosophisch-historische Klasse* 35 (1926).

Stark, Werner. "Giambattista Vico's Sociology of Knowledge." In *Giambattista Vico: An International Symposium*, edited by Giorgio Tagliacozzo, pp. 297–307. Baltimore, 1969.

——. "Max Weber and the Heterogeneity of Purposes." *Social Research* 34.2 (1967): 249–64.

Stegmüller, F. "Die Zwei Apologien des Jean de Mirecourt." *Recherches de théologie ancienne et médiévale* 3 (1933): 40–78, 192–204.

Steinthal, H. *Geschichte der Sprachwissenschaft bei den Griechen und Römern*. 2d ed. 1891; reprint Hildesheim, 1961.

Stenzel, Julius. *Zahl und Gestalt bei Platon und Aristoteles*. 3d ed. Darmstadt, 1959.

Stern, Menahem. *Greek and Latin Authors on Jews and Judaism*, 1: *From Herodotus to Plutarch*. Jerusalem, 1976.

Stock, Brian. *Myth and Science in the Twelfth Century: A Study of Bernard Silvester*. Princeton, 1972.

Strack, Herman L. and P. Billerbeck. *Kommentar zum Neuen Testament aus Talmud und Midrasch*. Vols. 3–4. 2d ed. Munich, 1954–1956.

Straub, J. "Christliche Geschichtsapologetik in der Krisis des römischen Reiches." *Historia* 1 (1950): 52ff.

Strauss, Leo. *Natural Right and History*. Chicago, 1953.

——. *Spinoza's Criticism of Religion*. New York, 1965.

Stump, Eleonore. "Theology and Physics in *De sacramento altaris*." In *Infin-

ity and Continuity in Ancient and Medieval Thought, edited by N. Kretz-mann, pp. 207–30. Ithaca, N.Y., and London, 1982.

Stump, Eleonore and Paul Vincent Spade. "Obligations." *CHM*: 315–57.

Suarez, Franciscus. *Disputationes metaphysicae*. In *Opera omnia*, vol. 16, edited by C. Berton. Paris, 1866.

Swain, J. W. "The Theory of Four Monarchies: Opposition History under the Roman Empire." *Classical Philology* 35 (1940): 1ff.

Sylla, Edith. "Medieval Concepts of the Latitude of Forms: The Oxford Cal-culators." *AHDL* 40 (1973): 223–83.

——. "Medieval Quantifications of Qualities: The 'Merton School.' " *Archive for the History of Exact Sciences* 8 (1971): 9–39

Szabó, Arpád. *The Beginning of Greek Mathematics*. Dordrecht and Boston, 1978.

Tachau, Katherine H. "The Problem of *Species in Medio* at Oxford in the Gen-eration after Ockham." *Medieval Studies* 44 (1982): 394–443.

——. "Vision and Certitude in the Age of Ockham." Ph.D. dissertation, Uni-versity of Wisconsin, Madison, 1981.

Tacitus, Cornelius. *Historiarum libri*. Edited by C. D. Fisher. Oxford, 1911.

Tajo of Saragossa. *Sententiarum libri quinque*. Migne, *PL* 80: 727ff.

Taylor, A. E. "The Ethical Doctrine of Hobbes." *Philosophy* 13 (1938): 406–24.

Teitelbaum, Mordechai. *Harav miljadi umifleget habad*. 2 vols. Warsaw, 1913.

Telesio, Bernardino. *De rerum natura juxta propria principia*. 1st ed. 1565, 2d ed. 1570, 3d ed. 1586. Edited by V. Spampanato. 3 vols. Modena, 1910–1923.

Tertullian. *Adversus Judaeus*. CCSL 2: 1337ff.

——. *Adversus Marcionem*. CCSL 1: 437ff.

——. *Adversus Praxean*. CCSL 2: 1157ff.

——. *De anima*. CCSL 2: 779ff.

——. *De carne Christi*. CCSL 2: 871ff.

——. *De pallio*. CCSL 2: 731ff.

——. *De patientia*. CCSL 1: 297ff.

——. *De praescriptione haereticorum*. CCSL 1: 185ff.

——. *De testimonio animae*. CCSL 1: 173ff.

——. *De virginibus velandis*. CCSL 2: 1207ff.

Theodoret of Cyrrhus. *Graecorum affectionum curatio*. Migne, *PG* 83: 783ff.

——. *In Isaiam*. Migne, *PG* 81: 215ff.

——. *Questions in Octateuchum*. Migne, *PG* 80: 75ff.

Theophilus of Antioch. *Ad Autolicum*. Edited by J.C.T. Otto. *Corp. Apol.*, vol. 8. Jena, 1861.

Thomas, Ivor. *Selections Illustrating the History of Greek Mathematics*, 1: *From Thales to Euclid*. Loeb Classical Library. London, 1931.

Thomas Aquinas. *Opera Omnia*. 25 vols. Parma, 1852–1873.

——. *Opera Omnia* (Vives edition). 34 vols. Paris, 1872–1880.

——. *Opera Omnia: Iussu impensaque Leonis XIII, PM edita* (Leonine edition). 15 vols. Rome, 1882ff.

——. *De ente et essentia*. In *Le "De ente et essentia" de S. Thomas d'Aquin*, edited by M. D. Roland-Gosselin. Paris, 1948.

Thomas Aquinas. *In libros Aristotelis de anima.* Edited by R. M. Spiazzi. Turin, 1955.

——. *In libros Aristotelis de caelo . . . expositio.* Edited by R. M. Spiazzi, *Opera Omnia* (Leonine), vol. 3. Turin, 1952.

——. *Quaestiones disputatae.* Edited by R. M. Spiazzi et al. 2 vols. 10th ed. Rome, 1964–1965.

——. *Scriptum super libros sententiarum.* Edited by P. Mandonnet et al. Paris, 1929–1956.

——. *Summa contra gentiles.* In *Opera Omnia* (Leonine), vol. 13.

——. *Summa theologiae.* Edited by P. Casamello, *Opera Omnia* (Leonine), vols. 4–12.

Thomas Bradwardine. *De causa Dei contra Pelagium.* Edited by H. Savil. 1618; reprint Frankfurt, 1964.

Thomas of Strassburg. *Commentaria in IV libros sententiarum.* Venice, 1564.

Tierney, Brian. "The Prince Is Not Bound by the Laws." *Comparative Studies in Society and History* 5 (1963): 388ff.

Tocqueville, Alexis de. *L'Ancien régime et la révolution.* Translated by S. Gilbert. Garden City, N.Y., 1955.

Todd, Robert B. *Alexander of Aphrodisias on Stoic Physics.* Leiden, 1976.

——. "Monism and Immanence: The Foundation of Stoic Physics." In *The Stoics,* edited by J. M. Rist, pp. 137–60. Berkeley and Los Angeles, 1978.

Toulmin, Stephen. "Criticism in the History of Science: Newton on Absolute Space, Time, and Motion." *The Philosophical Review* 68 (1959): 1–29, 203–27.

Troeltch, Ernst. *Der Historismus und seine Probleme.* Tübingen, 1922.

Twersky, Isadore. *Introduction to the Code of Maimonides (Mishne Tora).* Yale Judaica Series, 22, New Haven and London, 1980.

Überweg, Friedrich and Bernhard Geyer. *Grundriss der Geschichte der Philosophie,* vol. 3: *Die patristische und scholastische Philosophie.* Basel and Stuttgart, 1960.

Underhill, Evelyn. *Mysticism: A Study in the Nature and Development of Man's Spiritual Consciousness.* 12th ed. New York, 1955.

Urbach, Ephraim E. *Ḥazal: Pirke 'emunot vedeot* [The Sages: Doctrines and Beliefs]. Jerusalem, 1969.

——. "Matay paska hanevu'a?" ["When Did Prophecy Cease?"]. *Tarbiz* 17 (1947): 1–11.

Vahinger, Hans. *Die Philosophie des Als Ob.* 2d ed. Berlin, 1913.

Van der Pot, J.H.J. *Die Periodisierung der Geschiedenis: Ein Overzicht der Theorien.* The Hague, 1951.

Vaux, Ronald de. *Ancient Israel.* 2 vols. New York and Toronto, 1965.

Verene, Donald P. *Vico's Science of the Imagination.* Ithaca, N.Y., 1981.

Vico, Giambattista. *Opere.* Edited by F. Nicolini. 8 vols. Bari, 1911–1941.

——. *Autobiography.* Translated by Thomas Goddard Bergin and Max Harold Fisch. London, 1945.

——. *Vico: Selected Readings.* Edited and translated by Leon Pompa. Cambridge, 1982.

Vignaux, Paul. *Nominalisme au XIVᵉ siècle.* Montreal and Paris, 1948.

Violet, B. *Die Esraapokalypse*, I: *Die Überlieferung*. Leipzig, 1910; II: *Die kritische Ausgabe*. Leipzig, 1927.

Vives, Johannes Ludovicus. *De veritate fidei Christianae*. In *Opera Omnia*, vol. 8. Valencia, 1790.

Vlastos, Gregory. "On the Pre-History of Diodorus." *American Journal for Philosophy* 67 (1946).

Vogelsang, M. "Rudolfus Glaber, Studien zum Problem der Cluniazensischen Geschichtsschreibung." Ph.D. dissertation, Munich, 1952. A summary appeared in *Studien und Mitteilungen zur Geschichte des Benediktinerordens* 67 (1956): 25–38, 277–97.

Volz, P. *Die Eschatologie der jüdischen Gemeinde im neutestamentlichen Zeitalter*. Tübingen, 1934.

Von den Brincken, Anna-Dorothe. *Studien zur lateinischen Weltchronistik bis in das Zeitalter Ottos von Freising*. Düsseldorf, 1957.

Von der Luft, Eric. "Sources of Nietzsche's 'God is Dead!' and Its Meaning for Heidegger." *Journal of the History of Ideas* 45 (1984): 263–76.

Vuillemin, Jules. *Mathématiques et métaphysique chez Descartes*. Paris, 1960.

Wachtel, Alois. *Beiträge zur Geschichtstheologie des Aurelius Augustinus*. Bonner historische Forschungen, 17. Bonn, 1960.

Walahfrid Strabo. *De exordium et incrementio quorundam in observationibus ecclesiasticio rerum*. Edited by A. Boretius and K. Krause, MG Capit. 2: 473ff.

Wallace, William A., o.p. "The Enigma of Domingo de Soto: *Uniformiter Difformis* and Falling Bodies in Late Medieval Physics." *Isis* 59 (1968): 384–401.

——. *Galileo and His Sources: The Heritage of the Collegio Romano in Galileo's Science*. Princeton, 1984.

——. *Galileo's Early Notebooks: The Physical Questions*. Notre Dame, Ind., 1978.

Walther, Manfred. *Metaphysik als Anti-Theologie: Die Philosophie Spinozas im Zusammenhang der religionsphilosophischen Problematik*. Hamburg, 1971.

Walzer, R. *Galen on Jews and Christians*. Oxford, 1949.

Warnock, G. J. "Concepts and Schematism." *Analysis* 9 (1948–1949): 77–82.

Warrender, Howard. *The Political Philosophy of Hobbes: His Theory of Obligation*. London, 1957.

Wartofsky, Marx. "Action and Passion: Spinoza's Construction of a Scientific Psychology." In *Spinoza: A Collection of Critical Essays*, edited by M. Grene, pp. 329–53. Garden City, N.Y., 1973.

Waszink, J. H. and P. J. Jansen, eds. *Timaeus a Calcidio translatus commentarioque instructus*. In *Plato Latinus*, Vol. 4, edited by R. Klibanksy. London and Leiden, 1962.

Watkins, J.W.N. *Hobbes' System of Ideas*. London, 1965.

Weber, Max. *Gesammelte Aufsätze zur Religionssoziologie*. 3 vols. Tübingen, 1920.

Weigelt, Horst. *Sebastian Franck und die lutherische Reformation*. Gütersloh, 1972.

Weinberg, Julius. *Nicolaus of Autrecourt: A Study in 14th Century Thought*. Princeton, 1948.

Weisheipl, James A. "The Spector of *Motor Coniunctus* in Medieval Physics." In *Studi sul XIV secolo in Memoria di Anneliese Maier*, edited by Alfonso Maieru and Agostino Paravicini Bagliani, pp. 81–104. Rome, 1981.

Wellek, René and Austin Warren. *Theory of Literature*. New York, 1956.

Werner, Karl. *Giambattista Vico als Philosoph und gelehrter Forscher*. Vienna, 1879.

———. *Die Scholastik des späteren Mittelalters*. Vienna, 1887.

Werner, Martin. *Die Entstehung des christlichen Dogmas*. 2d ed. Bern and Tübingen, 1953.

Wernham, A. G. *Spinoza: The Political Works*. Oxford, 1958.

Westfall, Richard S. *Force in Newton's Physics: The Science of Dynamics in the 17th Century*. London, 1971.

———. *Never at Rest: A Biography of Isaac Newton*. Cambridge, 1980.

Westman, Robert. "The Astronomer's Role in the Sixteenth Century: A Preliminary Study." *History of Science* 40 (1980): 105–47.

———. "The Copernicans and the Churches." In *God and Nature: Historical Essays on the Encounter between Christianity and Science*, edited by D. C. Lindberg and Ronald L. Numbers, pp. 76–113. Berkeley and Los Angeles, 1986.

Weyl, Hermann. *Philosophy of Mathematics and Natural Science*. New York, 1963.

Whewell, William. *On the Philosophy of Discovery*. 3d ed. London, 1860.

White, Lynn, Jr. *Medieval Technology and Social Change*. Oxford, 1962.

Wieder, N. "The Law Interpreter of the Sect of the Dead Sea Scrolls—the Second Moses." *Journal of Jewish Studies* 4 (1953): 158ff.

Wieland, Wolfgang. *Die aristotelische Physik: Untersuchungen über die Grundlegung der Naturwissenschaften und die sprachlichen Bedingungen der Prinzipienforschung bei Aristoteles*. Göttingen, 1962.

Wilks, Michael J. *The Problem of Sovereignty in the Later Middle Ages*. Cambridge, 1964.

William of Auvergne. *Guilelmi Aluerni episcopi . . . Opera*. Paris, 1674.

William of Ockham. *De sacramento altaris*. Edited by N. Birch. Burlington, Vt., 1930.

———. *In Aristotelis libros physicorum*. Ms. Berlin Lat. 2°41.

———. *Octo quaestiones de potestate papae*. In *Opera politica*, Vol. 1, edited by H. S. Offler. Manchester, 1974.

———. *Opus nonaginta dierum*. In *Opera politica*, Vol. 1.

———. *Philosophical Writings*. Edited by P. Boehner, O.F.M. Edinburgh, 1957.

———. *Quodlibeta septem*. Strassburg, 1491.

———. *Scriptum in librum primum sententiarum ordinatio*. In *Opera philosophica et theologica*, edited by G. Gal et al. St. Bonaventure, N.Y., 1967ff.

———. *Summa totius logicae*. Venice, 1508.

———. *Summulae in libros physicorum*. In *Philosophia Naturalis Guilielmi Occham*. 1637. Reprint. London, 1963.

———. *Tractatus de praedestinatione et de praescientia Dei respectu futurorum contingentium*. In *Opera theologica*, Vol. 2, edited by P. Boehner and S. Brown. St. Bonaventure, N.Y., 1967.

William of Tyre. *Historia rerum in partibus transmarinis gestarum.* Recueils des historiens des croisades, Historiens Occidentals vol. 1. Paris, 1844.

Wilson, Curtis. *William Heytesbury: Medieval Logic and the Rise of Mathematical Physics.* Madison, Wis., 1960.

Wilson, M. Dauler. *Descartes.* London, 1978.

Wirszubski, Ch. *Libertas as a Political Idea at Rome during the Late Republic and Early Principate.* Cambridge, 1968.

Witelo. *Liber de intelligentiis.* In *Witelo: Ein Philosoph und Naturforscher des XIII. Jahrhunderts,* edited by C. Baeumker. BGPhM. Münster, 1908.

Wölfflin, Heinrich. *Renaissance und Barock.* 7th ed. Basel and Stuttgart, 1968.

Wolff, Christian. *Deutsche Metaphysik.* 5th ed. Halle, 1751.

Wolff, Michael. *Fallgesetz und Massbegriff: Zwei wissenschaftshistorische Untersuchungen zur Kosmologie des Johannes Philoponos.* Berlin, 1971.

——. *Geschichte der Impetustheorie: Untersuchungen zum Ursprung der klassischen Mechanik.* Frankfurt-am-Main, 1978.

Wolfson, Harry A. *Crescas' Critique of Aristotle: Problems of Aristotle's Physics in Jewish and Arab Philosophy.* Cambridge, Mass., 1929.

——. "Infinite and Privative Judgements in Aristotle, Averroës, and Kant." *Philosophy and Phaenomenological Research* 7 (1947): 173–87.

——. *Philo: Foundations of Religious Philosophy in Judaism and Christianity.* 2 vols. Cambridge, Mass., 1947.

——. *The Philosophy of Spinoza.* 2 vols. 1934; reprint Cambridge, Mass., 1958.

——. "The Problem of the Souls of the Spheres from the Byzantine Commentaries on Aristotle through the Arabs and St. Thomas to Kepler." *The Dumbarton Oaks Papers* (1961): 67–93.

Wolters, H., s.j. "Geschichtliche Bildung im Rahmen der Artes Liberales." In *Artes liberales: Studien und Texte zur Geschichte des Mittelalters,* edited by J. Koch. Leiden and Cologne, 1959.

Woolcombe, K. J. "Biblical Origins and Patristic Developments of Typology." In *Essays on Typology,* pp. 39–75. London, 1957.

Woolhouse, R. S. *Locke's Philosophy of Science and Knowledge.* New York, 1971.

Yadin, Yigael. "The Dead Sea Scrolls and the Epistle to the Hebrews." In *Aspects of the Dead Sea Scrolls,* Scripta Hierosolymitana 4, edited by C. Rabin and Y. Yadin, pp. 36–55. Jerusalem, 1958.

Zeller, Eduard. *Die Philosophie der Griechen.* 6 vols. 4th ed. 1909; reprint Darmstadt, 1963.

Zeuthen, Hieronymus G. "Die geometrische Konstruktion als Existenzbeweis." *Mathematische Annalen* 47 (1896): 222–28.

Zimmermann, Harald. *Ecclesia als Objekt der Historiographie.* SB der Oesterreichischen Akademie der Wissenschaften, philosophisch-historische Klasse 235/4. Vienna, 1960.

Zohar, Tikkune hazohar. Edited by Reuben Margaliot. 4 vols. Jerusalem, 1940–1953.

Zwingli, Ulrich. *Sämtliche Werke.* Edited by E. Egli and G. Finsler. 13 vols. Leipzig and Berlin, 1905ff.

INDEX

A

Abaelard, Petrus: exegesis, 216n; *processus religionis*, 226n

Abarbanel, Yitschak, 242, 261n; on kingship, 133n (periodization)

Abelson, J., 47n

Abraham bar Hiyya, 261n

Abraham ibn Ezra: exegesis, 217; influence on Spinoza, 219; minimalistic accommodation, 215–19

acceleration: impetus theory, 165–66; Newton, 92–93; Oresme, 173, 174n

accommodation, principle of (*adequatio*), 11–12;

—Abraham ibn Ezra, 215–19; and anthropomorphism in the Bible, 214; and the "cunning" of God and reason, 345; by Christian authors, 223–27; from legal to exegetical principle, 213–15; Galileo, 217 and n; "heresies are necessary," 275–76; in Greek philosophy, 251–53; Oresme, 213n, 217 and n; Tertullian, 43; Thomas, 213n

—and history, 213; and the Church fathers, 253–56; Irenaeus of Lyons, 235, 253, 254–55 and nn. 36–40; John Spencer, 241–43; of monotheism, 235–36; of salvation, 253–54; political theology, 256–61; Tertullian, 253, 255–56 and nn. 42–46; Vico, 209 and n, 243

—and "reasons for the commandments": Maimonides, 227–28, 231–34; Sa'adia, 228; the Sabeans, 231–32, 234

—of sacrifices: by Christians, 223–26; by Jews, 222–23, 231–34 (Maimonides)

—"Scripture speaks the language of man": Jewish tradition, 213–15; maximalistic and minimalistic interpretations, 215–19 (Abraham ibn Ezra)

—the principle secularized: Spinoza, 219–21

—*See also* exegesis; God's providence; history

action at a distance, 54 and n (Ockham)

Adam, as prefiguration of Christ, 49

'adam kadmon, see God's body

Adams, Marilyn M., 138n, 140nn

Adams, Robert M., 146n, 196n

Aegidius Colonna, 17

Aiton, E. J., 74n, 322n

Akiba, 124; exegesis, 214 and n

Alanus ab Insulis, 49n ("book of nature"), 226n, 239n, 249n

Alexander of Hales, 53n (analogy), 55n (God's presence); *potentia absoluta* and *ordinata*, 129 and n

Alfunsi, Petrus: accommodation, 227 and n, 234n

allegory (allegorization), *see* exegesis

Altmann, Alexander, 360n

Amalric of Bena, 46, 52 (pantheism)

Ambrosius of Milan, 269; "fortuitous original sin," 203; Moses as historian, 208n; periodization, 262 and n

anachronism, 210 and n

analogy: and equivocation, 52 (Boethius); of being, *see* Thomas Aquinas; of nature, *see* Newton

analytic and synthetic a priori, *see* Kant

analyticity, 182 and n (Quine)

Anastos, Milton V., 220n

Anaxagoras, 36n

Anderson, Carol Susan, 127n

Andresen, Carl, 125n

annihilation, principle of, 57n; and hypothetical reasoning, 172; Descartes, 185–86 and n. 19; Hobbes, 334–35 and n. 22. *See also* William of Ockham

Anscombe, G.E.M., 136n, 146n, 180n

Anselm of Canterbury: God's power, 127–28 and n. 17; ontological proof, 25–26 and n. 10. 181 and n

Anselm of Havelberg: accommodation, 225 and n, 236–37 and n. 52 (and history), 256n (innovation), 264–65 and nn. 71–73 (typology)

Library of Congress Cataloging-in-Publication Data

FUNKENSTEIN, AMOS.
THEOLOGY AND THE SCIENTIFIC IMAGINATION FROM THE
MIDDLE AGES TO THE SEVENTEENTH CENTURY.

BIBLIOGRAPHY: P.
INCLUDES INDEX.
1. GOD—ATTRIBUTES—HISTORY OF DOCTRINES.
2. RELIGION AND SCIENCE. 3. KNOWLEDGE, THEORY OF—
HISTORY. 4. PHILOSOPHY—HISTORY. I. TITLE.
BT130.F86 1986   261.5   85-43281
ISBN 0-691-08408-4 (ALK. PAPER)

DATE DUE

DEMCO 38-296

Printed in the United States
890200001B

9 780691 024257